9xT10

Modern Operator Theory and Applications

The Igor Borisovich Simonenko Anniversary Volume

Ya. M. Erusalimsky
I. Gohberg
S. M. Grudsky
V. Rabinovich
N. Vasilevski
Editors

Birkhäuser Verlag
Basel · Boston · Berlin

Editors:

Yakob M. Erusalimsky
Mechanical-Mathematics Department
Rostov State University
Zorge Str. 5
Rostov-on-Don 344104
Russia
e-mail: dnjme@ns.math.rsu.ru

Vladimir Rabinovich
Instituto Politecnico Nacional
ESIME Zacatenco
Avenida IPN
Mexico, D. F. 07738
Mexico
e-mail: vladimir.rabinovich@gmail.com

Israel Gohberg
School of Mathematical Sciences
Raymond and Beverly Sackler
Faculty of Exact Sciences
Tel Aviv University
Ramat Aviv 69978
Israel
e-mail: gohberg@post.tau.ac.il

Sergei M. Grudsky
Nikolai Vasilevski
Departamento de Matemáticas
CINVESTAV
Apartado Postal 14-740
07000 Mexico, D.F.
Mexico
e-mail: grudsky@math.cinvestav.mx
           nvasilev@math.cinvestav.mx

2000 Mathematics Subject Classification 45E, 45P05, 47G, 47L

Library of Congress Control Number: 2006937466

Bibliographic information published by Die Deutsche Bibliothek
Die Deutsche Bibliothek lists this publication in the Deutsche Nationalbibliografie; detailed
bibliographic data is available in the Internet at <http://dnb.ddb.de>.

ISBN 978-3-7643-7736-6 Birkhäuser Verlag, Basel – Boston – Berlin

© 2007 Birkhäuser Verlag, P.O. Box 133, CH-4010 Basel, Switzerland
Part of Springer Science+Business Media
Printed on acid-free paper produced from chlorine-free pulp. TCF ∞
Cover design: Heinz Hiltbrunner, Basel
Printed in Germany
ISBN-10: 3-7643-7736-4
ISBN-13: 978-3-7643-7736-6

e-ISBN-10: 3-7643-7737-2
e-ISBN-13: 978-3-7643-7737-3

9 8 7 6 5 4 3 2 1

www.birkhauser.ch

Contents

Igor Borisovich Simonenko

Operator Theory:
Advances and Applications, Vol. 170, 1–26

Introduction

Ya.M. Erusalimsky

Life and Work of Igor Borisovich Simonenko

In August of 2005, the eminent Russian mathematician Dr. Igor Borisovich Simonenko celebrated his 70th Birthday.

Igor Borisovich was born in Kiev (Ukraine, former USSR), where he spent his childhood. Along with the majority of his contemporaries, he experienced all the difficulties of wartime, evacuation and occupation, together with his mother in the steppes of Salsk. In 1943, upon returning to Lugansk with his mother, he began school, entering the third grade. In 1947 he left the primary school and entered a machine-building technical school. Having graduated from school in 1953, Igor first worked in a factory and then began to study at the Physics and Mathematics Department of the Rostov State University.

The greatest influence on the young mathematician I.B. Simonenko was rendered by his teacher and supervisor, the brilliant scientist Fyodor Dmitrievich Gakhov, who managed to create three scientific schools: in Kazan, Rostov-on-Don and Minsk.

In Igor Borisovich's student years, the Physics and Mathematics Department of Rostov State University was on the rise. This had much to do with the presence of the talented young experts in mechanics, the Moscow State University graduates I.I. Vorovich, N.N. Moiseev and L.A. Tolokonnikov (later academicians of the Russian Academy of Science) and the arrival at RSU in 1953 of professor F.D. Gakhov from Kazan.

An active influence on the scientific life of the department was rendered by the scientific seminar "Boundary value problems" (headed by F.D. Gakhov) and the seminar "Theory of nonlinear operators" (headed by I.I. Vorovich and M.G. Khaplanov). The latter seminar became the source of ideas and methods in functional analysis and the starting point of a wide range of application of these methods by the Rostov mathematicians.

In his 1961 Ph.D. thesis "Treatise in the theory of singular integral operators" I.B. Simonenko followed the classical methods of the school of his teacher. After defending this thesis, I.B. worked for several years at the RSU computer center. During this period, the results on the problems of electrostatics were obtained (jointly with V.P. Zakharyuta and V.I. Yudovich), including a calculation of the capacity of condensers of complex form and dielectric materials with complex structure.

In 1967, at the age of 32, six years after he defended his Ph.D. thesis, I.B. Simonenko defended his thesis for a degree of Doctor of Science. In this thesis, entitled "Operators of local type and some other problems of the theory of linear operators," he sharply turned towards the wide usage of the general methods of functional analysis. In 1971 professor I.B. Simonenko became the head of the Numerical Mathematics Chair. The following year this chair was split into two; I.B. became the head of one of them, the Chair of Algebra and Discrete Mathematics.

The Chair of Algebra and Discrete Mathematics can be rightfully called the Chair of I.B. Simonenko. Here he worked together with his colleagues, students, and the students of his students. Here he fully developed his teaching talent. He lectured on "Algebra and geometry," "Mathematical logic," "Discrete mathematics," and "Mathematical analysis".

The scientific seminar of the Chair of Algebra and Discrete Mathematics is widely known both in Russia and abroad. Besides I.B. and his students, such well-known mathematicians as

> S.G. Mikhlin, I.Tz. Gokhberg, N.Ya. Krupnik, B.A. Plamenevsky,
> P.E. Sobolevsky, A.I. Volpert, A.S. Markus, A.P. Soldatov,
> R.V. Duduchava, A.S. Dynin, B. Silbermann, A. Böttcher,
> M.V. Fedoryuk, G.S. Litvinchuk, I.M. Spitkovsky, N.L. Vasilevski,
> A.B. Antonevich, N.N. Vragov, Yu.I. Karlovich, S.G. Samko,
> N.K. Karapetiantz

and others gave talks here.

While a reputed scientist and the head of widely known scientific school, Igor Borisovich remains a modest and charming man. If asked to describe him in several words, I would leave only two – the Scientist and the Teacher.

V.S. Pilidi

Operators of Local Type and Singular Integral Operators

We recall the main definitions from the theory of Fredholm operators. Let $\mathfrak{X}$ be a Banach space. Denote by $\mathcal{B}(\mathfrak{X})$ $(\mathcal{K}(\mathfrak{X}))$ the set of all linear continuous (all compact) operators acting on the space $\mathfrak{X}$. An operator $A \in \mathcal{B}(\mathfrak{X})$ is called *Fredholm* (Φ-operator) if its kernel is finite dimensional and the range is closed and has finite codimension.[1]. The Fredholm property of A is equivalent to the existence of operators R_1, $R_2 \in \mathcal{B}(\mathfrak{X})$ such that the following equalities hold: $R_1 A = I + T_1$, $AR_2 = I + T_2$, where $T_1, T_2 \in \mathcal{K}(\mathfrak{X})$. The operators R_1 and R_2 are called left and right regularizors of A. The existence of regularizors is evidently equivalent to the

---

[1]The two terms mentioned have practically superseded the earlier term "Noether operator," which was used by I.B. Simonenko in his classic paper "The new general method..."

invertibility of the residue class $A + \mathcal{K}(\mathfrak{X})$ in the quotient algebra $\mathcal{B}(\mathfrak{X})/\mathcal{K}(\mathfrak{X})$ (the "Calkin algebra"). Let us call operators A and B *equivalent* if $B - A \in \mathcal{K}(\mathfrak{X})$. We note the following trivial fact: if two operators are equivalent, then the Fredholm property of one of them implies this property for the other.

Classical Gelfand theory is in some sense a local principle, giving in the case of commutative Banach algebra conditions for invertibility in some "local" terms.

I.B. Simonenko's local method is, in essence, an analogue of this theory[2]. Speaking in algebraic terms, this principle permits to obtain criteria of invertibility of elements of Calkin algebras in noncommutative case.

We explain the definition of an operator of local type, given below, with the following example. Consider the singular integral operator

$$(Sf)(x) = \frac{1}{\pi i} \int_0^1 \frac{f(y)}{y - x} \, dy$$

acting on the space $L_2(0, 1)$, where the integral is understood in the sense of principal value. Let P_F be the operator of multiplication by the characteristic function of the measurable set $F \subset [0, 1]$ acting on the same space. If F_1 and F_2 are closed *nonintersecting* subsets of the segment $[0, 1]$, then the integral operator $P_{F_1} S P_{F_2}$ has bounded kernel, and therefore is compact. Note that the compactness of this operator is related to the fact that the strong singularity of the kernel lies on the diagonal of its domain of definition.

Now let us pass to the general definition of operator of local type. Let X be a compact Hausdorff space. Suppose that a σ-finite nonnegative measure is defined on this space, such that all open subsets of X are measurable. An operator $A \in \mathcal{B}(L_p(X))$ $(1 \le p < \infty)$ is called an operator *of local type* if for any two closed disjoint subsets F_1, $F_2 \subset X$, the operator $P_{F_1} A P_{F_2}$ is compact. This definition is equivalent to the following: for any continuous function φ on X, the commutator $\varphi A - A \varphi I$ is compact.

In the sequel we will suppose that the space X and the number p are fixed and that all operators under consideration are operators of local type. The notation $\mathcal{K}(L_p(X))$ will be shortened to $\mathcal{K}$.

Operators A and B are called locally equivalent at the point $x \in X$ when $\inf_u |(B - A)P_u| = 0$, where $|\cdot|$ denotes the seminorm modulo the set of all compact operators, and the greatest lower bound is taken over the set of all neighborhoods of x in X (this notion will be expressed as $A \overset{x}{\sim} B$).

An operator A is called *locally Fredholm* at the point $x \in X$ if there exist operators R_1, R_2 such that $R_1 A \overset{x}{\sim} I$, $A R_2 \overset{x}{\sim} I$.

The main assertion of the local principle is as follows: *an operator is Fredholm if and only if it is locally Fredholm at every point of X.*

The following statement plays an essential role: *if two operators are locally equivalent at some point, then the local Fredholm property for one of them implies*

---

[2]The idea of this comparison is mentioned in the book R. Hagen, S. Roch, B. Silbermann, *C*-Algebras and Numerical Analysis*, 2001, p. 204.

the same property for the other. This property allows us to reduce the local analysis to simpler operators. For example, the operator of multiplication by continuous function φ is locally equivalent at a point x_0 to the scalar operator $\varphi(x_0)I$.

Let Λ be the Banach algebra of all operators of local type. Denote by $\mathcal{I}_x$ the set of all operators locally equivalent to the zero operator at the point $x \in X$. When $\mathcal{I}_x = \Lambda$, all operators are locally Fredholm at this point. Such points are excluded, and below we suppose for simplicity that the set X has no such points. Thus the local Fredholm property of the operator A is equivalent to invertibility of the residue class $A + \mathcal{I}_x \in \Lambda/\mathcal{I}_x$. Then the basic theorem of the local principle may be reformulated as follows: the residue class $A + \mathcal{K} \in \Lambda/\mathcal{K}$ is invertible if and only if all the classes $A + \mathcal{I}_x \in \Lambda/\mathcal{I}_x$ $(x \in X)$ are invertible.

In the case of commutative Banach algebras with unit, the quotient algebra modulo the maximal ideal is isomorphic to the field of complex numbers. This property allows one to construct classical Gelfand transformation. We recall that in general this transformation is not monomorphic, and its range does not coincide with set of all continuous functions on the space of maximal ideals. For operators of local type, naturally, there is no canonical realization for the quotient algebras $\Lambda/\mathcal{I}_x$, moreover these algebras can be quite different for different $x \in X$, thus in each concrete case some additional analysis must be carried out. At the same time an analog of the Gelfand transformation still remains valid, i.e. there is a way (canonical in certain situations) to describe elements of $\Lambda/\mathcal{K}$ in terms of continuous (in appropriate topologies) families of elements of $\{\Lambda/\mathcal{I}_x\}_{x \in X}$ (a theorem on the enveloping operator and its various refinements).

The local principle of I.B. Simonenko makes it possible to investigate numerous classes of operators of convolution type (including associated boundary value problems for functions of several complex variables), one-dimensional and multidimensional singular integral operators, and some classes of pseudodifferential operators. Together with its numerous modifications[3], this local method has led to the convergence criteria of various approximation methods for operators of convolution type and for singular integral operators. We emphasize that the method has rendered unparalleled influence on the qualitative theory of operator equations.

The detailed presentation of the local method with applications to Fredholm theory of singular integral equations is given in the book of I.B. Simonenko and Chin' Ngok Min', "Local Method in the Theory of One-Dimensional Singular Integral Equations with Piecewise Continuous Coefficients," published in 1986 by the publishing house of Rostov State University.

Among numerous results obtained by I.B. Simonenko in the theory of singular integral equations, we mention only the "factorizational" criterion of Fredholmness for singular integral operators with measurable coefficients. Recall that classical Fredholm theory for singular integral operators (and associated boundary value problems) in the class of Hölder functions is based on the procedure of factorization, i.e., representation of the function as the product of boundary values of

---

[3]We mention here only local principles of I. Gohberg – N. Krupnik and B. Silbermann.

two analytic functions and the integer degree of the independent variable. Simonenko proved that the existence of some form of factorization is equivalent to the Fredholm property of the corresponding singular integral operator.

This investigation was continued by V.S. Pilidi, V.S. Rabinovich, and S.M. Grudsky. Pilidi studied bisingular operators, defending his Ph.D. thesis in 1972, and defended his Doctoral thesis, "Bisingular operators and operators of related classes," at the Tbilissi mathematical Institute in 1990. Rabinovich studied boundary value problems for pseudodifferential operators of convolution type in conic areas, presenting his Ph.D. thesis in 1968 and the Doctoral thesis "Limiting operators method in the problems of solvability of the pseudodifferential equations and the equations of convolution type" at the Institute for Low Temperature Physics and Engineering (FTINT, Kharkov) in 1993. Grudsky continued the investigation of one-dimensional singular integral operators with coefficients having non-standard discontinuities, and after his 1981 Ph.D. thesis, defended the thesis for a Doctoral thesis "Singular integral operators with infinite index and their application in the problems of diffraction theory" at St. Petersburg University in 1995.

An important modification of the local method was created by A.V. Kozak (Ph.D. thesis, 1974.). His approach allows one to obtain convergence criteria for the approximation methods for wide classes of operators of convolution type. Kozak's scheme was the starting point of numerous publications in this direction. In particular, this theory was carried over to the case of pseudodifferential operators by R.Ya. Doctorsky (Ph.D. thesis, 1978). In connection with investigation of Fredholmness for new classes of operators there arose the problem of index calculation. This problem was been solved for continual and discrete operators of convolution type by I.B. Simonenko and V.N. Semenyuta (Ph.D. thesis, 1972) and V.M. Deundyak (Ph.D. thesis, 1976).

V.B. Levenshtam, S.M. Zenkovskaya

An Averaging Method and its Application to Hydrodynamics Problems

A series of articles on the application of the Van-Der-Poll – Krylov – Bogolubov averaging method for nonlinear equations of parabolic type and the solution of hydrodynamic stability theory problems is one of the bright pages of the scientific biography of Igor Borisovich. He published several articles on this subject in the main journals, and in 1989 the monograph "Averaging method in theory of nonlinear equations of parabolic type with applications to hydrodynamic stability theory" was printed by the RSU publishing house.

A distinctive feature of this subject is that I.B., being a rigorous classical mathematician, this time acted as a physicist first. He posed a problem: how do high-frequency vibrations influence the stability of liquid motion? A classical pendulum model with a vibrating suspension point serves as a good guiding line when solving stability problems with high-frequency vibration influence. It was shown in the work of N.N. Bogolubov and P.L. Kapitsa that it is possible to make the upper position of a physical pendulum, which is unstable without the presence of vibration, stable with high-frequency vertical vibrations. They interpreted this effect theoretically by applying the averaging method to the pendulum oscillation equations.

In continuum mechanics, this method was first applied in the work of Academician V.N. Chelomey (1956) in the study of the influence of longitudinal vibrations on the dynamic stability of elastic systems.

I.B. Simonenko showed outstanding physical intuition perceiving the similarity between a pendulum on a vibrating support and a liquid in convective motion, driven by temperature gradient and container vibration.

In the paper of I.B. Simonenko and S.M. Zenkovskaya, "On the influence of vibration on the onset of convection," *Izv. AN USSR MZG*, 5, 1966, 51–55, the following problem was considered. A container D with solid impermeable boundary ∂D, filled with a viscous incompressible fluid, is subjected as a whole to translational vertical harmonic vibrations governed by the law $a/\omega \cos \omega t$. It is assumed that the frequency of vibrations ω is large, and the velocity amplitude a is finite, so that the amplitude of vibrations a/ω is small. A conclusion was reached on the basis of mathematical research: high-frequency vertical oscillations prohibit the onset of convection. Moreover, it is possible to select an amplitude of velocity of vibration such that the state of relative equilibrium will be stable at any temperature gradient (absolute stabilization). This theoretical conclusions and others were confirmed later by experiments conducted at Perm University (G.F. Putin and collaborators).

Numerous physical effects have now been discovered by applying the averaging method to convection problems. The guiding idea behind this asymptotical method is that under given conditions, the movement constitutes a superposition of a smooth (slow) motion and fast but small amplitude vibrations. This method permits dividing the motion into slow and fast components, and then expressing the fast component via the slow one. The averaged equations are then obtained, which are self-contained, and include additional forces which appear as an outcome of the interaction of vibration fields.

Let us show briefly how this was done in the paper cited above. The convection equations, written in a moving coordinate system, have the form

$$\frac{\partial \mathbf{v}}{\partial t} + (\mathbf{v}, \nabla)\mathbf{v} = -\frac{1}{\rho_0}\nabla p + \nu \Delta \mathbf{v} - (g - \mathbf{w}_e)\beta T, \tag{1}$$

$$\operatorname{div} \mathbf{v} = 0, \quad \mathbf{w}_e = -a\omega \cos \omega t,$$

$$\frac{\partial T}{\partial t} + (\mathbf{v}, \nabla T) = \chi \Delta T. \tag{2}$$

On the boundary ∂D the following condition must be satisfied: $\mathbf{v} = 0$, $T = h(s)$, $s \in \partial D$.

Here $\mathbf{v}$ is the relative velocity, p is the pressure, T is the temperature, and ρ_0 is the density. The values ν, β, χ are the coefficients of kinematic viscosity, thermal expansion and thermal diffusivity; $\mathbf{g} = \mathbf{k}g$, where $\mathbf{k} = (0, 0, 1)$ is the vertical unit vector, g is the acceleration due to gravity; $\mathbf{w}_e$ is the acceleration of translation. Equation (1) contains a rapidly oscillating coefficient $\cos \omega t$ which impedes, for example, the numerical solution of the system (1), (2). The averaging method was applied to this system.

Let us write $\mathbf{v} = \mathbf{u} + \xi$, $T = \theta + \eta$, $p = q + \delta$, where $\mathbf{u}$, θ, q are the smooth components, ξ, η, δ are the fast components of the unknowns, having zero average. Substituting these expansions into equations (1) and (2) and keeping the high order vibrational terms in ω we obtain expressions for the fast components:

$$\xi = -a\beta \sin \omega t \mathbf{w}, \quad \eta = -\frac{a\beta}{\omega} \cos \omega t (\mathbf{w}, \nabla \theta). \tag{3}$$

Here $\mathbf{w} = \Pi(\mathbf{k}\theta)$, where Π is the Weil orthogonal projector in $L_2(D)$ on the space S_2 of solenoidal vectors with zero normal component on the boundary, so that $\operatorname{div} \mathbf{w} = 0$, $w_n\big|_{\partial D} = 0$.

As the result of averaging, the following problem is obtained:

$$\frac{\partial \mathbf{u}}{\partial t} + (\mathbf{u}, \nabla)\mathbf{u} = -\frac{1}{\rho_0} \nabla q + \nu \Delta \mathbf{u} - \mathbf{g}\beta\theta + F_v;$$

$$F_v = \frac{1}{2} a^2 \beta^2 [\mathbf{k}(\mathbf{w}, \nabla \theta) - (\mathbf{w}, \nabla)\mathbf{w}], \tag{4}$$

$$\frac{\partial \theta}{\partial t} + (\mathbf{u}, \nabla \theta) = \chi \Delta \theta. \tag{5}$$

The boundary conditions have the form $\mathbf{u} = 0$, $\theta = h$, $w_n = 0$ on ∂D. Hence as the result of averaging, a self-contained system has been obtained, with an additional force F_v having a vibrational origin – it varies with the amplitude of vibration. Equations (4), (5) have become classical, and appear as the Simonenko-Zenkovskaya equations in textbooks and many articles. If dimensionless parameters are introduced in the problem (4), (5), we find that along with the known characteristic parameters $P = \nu/\chi$, the Prandlt and Raleigh numbers $R = (T_1 - T_2)\beta g l^3/\chi\nu$, one new parameter has appeared: the vibrational Raleigh number $\mu = (T_1 - T_2)^2 \beta^2 a^2 l^2/\chi\nu$. As is evident from its form, it doesn't depend on gravity and can characterize convection under conditions of weightlessness ($g = 0$). Naturally, these results were also used to study and explain the results of convection experiments on spacecraft. Knowing the stationary solution of the problem (4), (5), it is possible to find the additional high frequency terms using formula (3) and thereby the main terms of the corresponding periodic solution of the problem (1), (2).

This work was the first to apply the averaging method in hydrodynamics. Other work has appeared after this paper, and a new branch of research has formed – vibrational convection. A new effect of vibration in the case of nonvertical oscillations has been discovered in the work of S.M. Zenkovskaya: convection can occur not only from heating from below, but also from heating from above.

After the formal application of the averaging method, mathematical questions arise. How close is the periodical solution obtained to the exact one? How is the stability of the stationary solution of the averaged problem correlated with the stability of the corresponding periodical solution of the original problem? How to obtain the next approximations of the averaging method? These mathematical problems and others have been solved in work by I.B. on the foundations of the averaging method for partial differential equations and mathematical hydrodynamics problems. Let us describe this matter in more detail.

At the end of the 1960's I.B. Simonenko began to work on the justification of the averaging method for the problem of convection. There were few papers on this theme for partial differential equations at the time. In particular, parabolic equations were being considered only for the case of second order, linear (R.Z. Khasminskii) or semi-linear (S.D. Eidelman). The problem of convection (1)–(2) contains, aside from the heat equation, the Navier-Stokes system (1), which describes the motion of a viscous incompressible fluid. The Navier-Stokes system (as is known from mathematical hydrodynamics) with the help of Weyl's projector Π mentioned above, reduces to a differential equation in the Banach space S_2, whose principal operator coefficient $A_0 = \Pi\Delta$ generates the analytic semigroup in S_2 (P.E. Sobolevskii, V.I. Yudovich). These equations are called the abstract parabolic equations. I.B. Simonenko first proceeded to justify the averaging method for abstract parabolic equations of the form

$$\frac{dx}{dt} = Ax + f(x, \omega t), \quad \omega \gg 1 \tag{6}$$

in a complex Banach space B. Here A is a linear, in general unbounded, operator on B, and $f(x, \tau)$ is a nonlinear mapping subordinate (in a certain sense) to the operator A and possessing the mean

$$F(x) = \lim_{N \to \infty} \frac{1}{N} \int_0^N f(x, \tau) \, d\tau.$$

In applications to parabolic problems, the operator A is defined by the corresponding elliptic differential expression together with the boundary conditions. The subordination of f to the operator A means, roughly speaking, that the highest order of the unknown function contained in f is lower than the order of the main differential expression. In the case of the abstract equation (6), subordination of fs to the operator A is formulated in terms of fractional powers of the operator $-A$, which is considered to be positive without loss of generality. The exact statement also refers to certain Banach spaces B^δ ($\delta \geqslant 0$) of vectors x belonging to the domain of definition of the operator $(-A)^\delta$, with norm $\|x\|_{B^\delta} = \|(-A)^\delta x\|_B$.

Let $T > 0$. On the region $t \in [0, T]$ we consider the Cauchy problem for equation (6) with the initial condition

$$x(0) = x_0. \tag{7}$$

Under condition noted above, together with some additional conditions, I.B. Simonenko justified the averaging principle for that problem in the following form.

Let the averaged problem

$$\frac{dy}{dt} = Ay + F(y),$$
$$y(0) = x_0$$

(8)

have the solution $\overset{\circ}{y}(t)$ on the region $t \in [0, T]$. Then for sufficiently large ω on the same region, the perturbed problem (6)–(7) is also uniquely solvable; moreover its solution $x_\omega(t)$ satisfies the following relation,

$$\lim_{\omega \to \infty} \max_{t \in [0, T]} \| x_\omega(t) - \overset{\circ}{y}(t) \|_{B^1} = 0.$$

For the case when the mapping $f(x, \tau)$ is ℓ-periodic in τ, I.B. justified the averaging method for $\ell\omega^{-1}$-periodic solutions of equations of the form (6) as well. In this context it is supposed that the averaged equation (8) has a non-degenerate stationary solution $\overset{\circ}{y}$; i.e., $F(\overset{\circ}{y}) = 0$ and the Fréchet differential $((DF)(\overset{\circ}{y})$ is reversible. It is proved that for sufficiently large ω the perturbed equation has a relatively unique (i.e., unique in some sphere) $\ell\omega^{-1}$-periodic solution $x_\omega(t)$, and

$$\lim_{\omega \to \infty} \max_{t \in R} \| x_\omega(t) - \overset{\circ}{y} \|_{B^1} = 0.$$

(For brevity we have mentioned here only some results of I.B., falling far short of the profundity and generality which he achieved.)

Let us outline I.B. Simonenko's justification scheme for the averaging method for the problem (6)–(7). Because the operator A generates the analytic semigroup e^{tA}, problem (6)–(7) may be reduced to the integral equation

$$x(t) = e^{tA} x_0 + \int_0^t e^{(t-\tau)A} f[x(\tau), \omega\tau] \, d\tau \equiv$$
$$\equiv e^{tA} x_0 + N_\omega(x, t).$$

Selecting the Banach space of vector functions defined on the region $t \in [0, T]$ (this selection is an important component of the proof), I.B. Simonenko considers operators $M(\cdot, \omega)$ in that Banach space,

$$[M(x, \omega)](t) \equiv e^{tA} x_0 + N_\omega(x, t), \quad \omega \leqslant \infty,$$

where $N_\infty(x, t) = \int_0^t e^{(t-\tau)A} F[x(\tau)] \, d\tau$. He proves that the mapping $M(\cdot, \cdot)$ satisfies

in some neighborhood of the point $(\overset{\circ}{y}(t), \infty)$ the theorem of implicit operators (x is given implicitly), from which the averaging principle for the problem (6)–(7) follows.

The abstract results by I.B. Simonenko were transferred by him to wide classes of parabolic problems and the Navier–Stokes system. He used results by S. Agmon on algebraic conditions for generation of analytic semigroups by elliptic boundary-value problems and the results mentioned above by P.E. Sobolevskii and V.I. Yudovich on the operator A_0. In doing so, I.B. Simonenko had to restate abstract requirements for nonlinear parts of equations, expressed in terms of fractional powers of positive operators, in terms common in mathematical analysis. Thus natural conditions appeared for nonlinearities to belong to a Hölder space. We mention that the use of fractional powers of unbounded positive operators in the theory of equations in Banach spaces played an important role in the work of M.A. Krasnoselskii and his students. Moreover they proved several embedding theorems, involving domains of definition of fractional powers of positive operators together with Hölder and Sobolev spaces of functions. Analyzing the proofs of some of those theorems, I.B. Simonenko stated a new and important interpolation theorem, which in particular simplifies the statement of the part of this embedding theory. With the help of this theorem, Simonenko obtained results on embeddings which provide a constructive description of the domains of definition of fractional powers of elliptic operators and the operator $-A_0$.

It is well known that Mark Alexandrovich Krasnoselskii highly regarded Igor Borisovich's results on the averaging method.

Continuing to develop the theory of averaging methods for abstract parabolic equations, I.B. Simonenko turned to the construction of the high order approximations of solutions of the perturbed problem (6). He noted that the solution of the averaged problem $\overset{\circ}{y}$ may not be satisfactory for the two following reasons.

First, the norm in which the solutions x_ω and $\overset{\circ}{y}$ are close may be insufficient for practical purposes. It may happen, for instance, that one is interested in a functional of x_ω which is not continuous in that norm.

Secondly, we may need the highest order approximation relative to ω^{-1} rather than the norm of difference $x_\omega - \overset{\circ}{y}$. I.B. Simonenko notes that the second problem in the classical theory of the averaging method for the ordinary differential equations is well known, and is called the problem of construction of high-order approximations. It is solved there via classic changes of variables by Krylov-Bodoljubov. The first problem is typical for equations in infinite-dimensional spaces. He was engaged in solving both problems globally for the problem (6)–(7). As a result, for nonlinear problems of the form (6)–(7) a recurrent sequence of linear abstract parabolic problems with a common linear operator was constructed, whose solutions enables one to approximate x_ω in each of a sequence of norms and within any positive power of ω^{-1}.

Let us turn to the problem of justification of the averaging method directly for the problem of convection (1)–(2) with boundary conditions of the above-mentioned form

$$u|_{\partial D} = 0, \quad \theta|_{\partial D} = h(s), \quad \partial D \in C_3. \tag{9}$$

I.B. Simonenko considered an initial boundary-value convection problem on the finite time region $t \in [0, T]$ as well as the problem of solutions $2\pi\omega^{-1}$-periodic in time t. Here we concentrate on the latter. We say that its solution is the triple of fields $(\mathbf{v}, T, p)$, or just $(\mathbf{v}, T)$. I.B. Simonenko stated the following result.

Let the averaged problem (4), (5), (9) have the stationary solution $\overset{\circ}{\mathbf{v}}$, $\overset{\circ}{T}$ and suppose that the spectrum Λ of the problem linearized for stationary disturbances on that solution does not contain zero.

Then the following statements hold.

1. There exist r_0 and ω_0 such that for $\omega > \omega_0$ the problem (1), (2), (9) has the unique $2\pi\omega^{-1}$-periodic solution $(\mathbf{v}_\omega, T_\omega)$ in the sphere

$$\sup_{t\in[0,\infty)} \sup \; \|\mathbf{v} - (\overset{\circ}{\mathbf{v}} - a\beta\sin\omega t\Pi_j\,\overset{\circ}{T})\|_{L_q(D)} + \sup_{t\in[0,\infty)} \|T - \overset{\circ}{T}\|_{W_q^1(D)} \leqslant r_0,$$

and

$$\lim_{\omega\to\infty} \sup_{t\in[0,\infty)} (\|\mathbf{v}_\omega - (\overset{\circ}{\mathbf{v}} - a\beta\sin\omega t\Pi k\,\overset{\circ}{T})\|_{L_q(D)} + \|T_\omega - \overset{\circ}{T}\|_{W_q^1(D)}) = 0, \quad q > 9.$$

2. If the spectrum Λ is contained in the left complex half-plane, then the solution $(\mathbf{v}_\omega, T_\omega)$ is exponentially stable in the norm $L_q(D) \times W_q^1(D)$.

3. If the spectrum Λ has at least one point in the right complex half-plane, than the solution $(\mathbf{v}_\omega, T_\omega)$ is unstable in the norm $L_q(D) \times W_q^1(D)$.

We sketch the proof of result 1 only. In the first step of the proof we apply the projector Π to equation (1), taking into account the equalities $\Pi\mathbf{u} = \mathbf{u}$, $\Pi\dfrac{d\mathbf{u}}{dt} = \dfrac{d\mathbf{u}}{dt}$, $\Pi\nabla p = 0$. Then the system (1) will turn into an abstract parabolic equation in the Banach space S_2. Therefore the convection system (1), (2) will also take the form of an abstract parabolic equation, differing from an equation of the form (6) by the presence of a large summand, proportional to $\omega \gg 1$. In the second step the large summand is eliminated by a change of variables. This is similar to the classic change of variables by Krylov-Bogoljubov, but here the boundary conditions must also be satisfied. This makes the larger parameter ω penetrate a time as well as a space part of arguments of some summands of the transformed system, so that system is not a particular case of equation (6). In the third step, we refer to deep analytic results of Simonenko, where in particular his interpolation theorem is used systematically. Due to these results, the summands in a certain sense are minor and become small when $\omega \to \infty$. Then one may follow the approach used in the problem of periodic solutions of equation (6). Thus the fourth step consists of applying the theorem of implicit operators to the properly constructed operator equation.

Igor Borisovich's scientific interests are versatile. Thus he turned from the averaging method to other problems some time ago. However, this line of research has been carried on by his students and followers. S.M. Zenkovskaya defended the Ph.D. thesis, "Some questions of stability of periodic solutions of Navier–Stokes equations" (1971). She is currently involved in solving problems of vibrational convection in collaboration with her students. V.B. Levenshtam defended his Ph.D. thesis "Some problems of the theory of averaging method on the entire time axis" (1977) and Doctoral thesis "Averaging method in the theory of nonlinear parabolic equations with applications to the problems of hydrodynamics"

(2000, Novosibirsk). Currently V.B. Levenshtam and students are working on the development of the averaging method for some new problems. V.I. Yudovich recently joined the work on the averaging method. He knows Igor Borisovic's work thoroughly, taking part in the discussion of mathematical problems as well as in the selection of the models when solving the problems of hydrodynamic theory of stability. Yudovich developed the general theory of the averaging method applied to mechanical systems with relations. As a result it turned out that the problem of the pendulum, Chelomey's problem of the beam, as well as convection, are all particular cases of that theory. He introduced a concept of vibrogeneous force and studied its invariant geometric nature.

In conclusion, some words should be said about Igor Borisovich as a scientific advisor. We are his students, and we remember what an immense amount of time he devoted to us. Now, when we have students ourselves, we can appreciate the energy and time he devotes to work with undergraduate and graduate students, and how thoroughly he carries out this work. It is not by chance that his students get high prizes at student conferences, and that the best students enter his department.

List of Ph.D., whose supervisor was I.B. Simonenko

1. Saak E.M. (1967)
2. Zenkovskaya S.M. (1971)
3. Rode L.O. (1972)
4. Semenyuta V.N. (1972)
5. Pilidi V.S. (1972)
6. Deundyak V.M. (1976)
7. Boroditsky M.P. (1976)
8. Koledov L.V. (1976)
9. Erusalimsky Ya.M. (1976)
10. Levenshtam V.B. (1977)
11. Edelshtein S.L. (1980)
12. Kulikov I.V. (1981)
13. Janover V.G. (1982)
14. Nikolenko P.V. (1983)
15. Soibelman Ja. S. (1983)
16. Gordeyev S.R. (1986)
17. Falkovich I.M (1988)
18. Stukopin V.A. (1988)
19. Abramyan M.E. (1992)
20. Mikhalkovich S.S. (1994)
21. Bogachyev T.V. (2000)
22. Maksimenko E.A. (2004)

List of Ph.D., whose co-supervisor was I.B. Simonenko

1. Doktorsky R.Ja. (1978)
2. Khevelev A.B. (1979)
3. Myasnikov A.G. (1980)
4. Grudsky S.M (1981)
5. Levendorsky S.Z. (1981)
6. Shteinberg B.Ja. (1982)
7. Naumov V.V. (1987)
8. Olifer A.V. (1989)

List of D.Sc., whose advisor was I.B. Simonenko

1. Pilidi V.S. (1990)
2. Rabinovich V.S. (1993)
3. Grudsky S.M (1995)
4. Levenshtam V.B. (2000)

Principal Publications of I.B. Simonenko

[1] The Riemann boundary value problem with a continuous coefficient, Dokl. Acad. Nauk SSSR, 124 (1959), no. 2, 278–281. (Russian)

[2] On some integral-differential convolution type equations, Izv. Vyssh. Uchebn. Zaved. Mat., (1959), no. 2, 213–226. (Russian)

[3] (with V.V.Ivanov) On the approximate searching of all solutions of a given linear equation in the Banach spaces, Dokl. Acad. Nauk SSSR, 126 (1959), no. 6, 1172–1175. (Russian)

[4] A boundedness of singular integrals in Orlich spaces, Dokl. Acad. Nauk SSSR, 130 (1960), no. 5, 984–987. (Russian)

[5] The Riemann boundary value problem with a measurable coefficient, Dokl. Acad. Nauk SSSR, 135 (1960), no. 3, 538–541. (Russian); English transl. in Soviet Math. Dokl., 1 (1960), 1295–1298.

[6] The Riemann boundary value problem for n pairs of functions with continuous coefficients, Izv. Vyssh. Uchebn. Zaved. Mat., (1961), no. 1, 140-145. (Russian)

[7] The Riemann boundary value problem for n pairs of functions with measurable coefficients and its application to the study of singular integrals on weighted Lp-spaces, Dokl. Acad. Nauk SSSR, 141 (1961), no. 1, 36–39. (Russian); English transl. in Soviet Math. Dokl. 2 (1961) 1391–1394.

[8] On some boundary value problems of analytic functions, in "The investigations on modern problems of the theory of functions of complex variable" Moscow: Fizmatgiz, 1961, 392–398. (Russian)

[9] The Riemann and Riemann-Gazemann boundary value problems with continuous coefficients, in "The investigation on modern problems of the theory of functions of complex variable" Moscow: Fizmatgiz, 1961, 380–388. (Russian)

[10] On systems of convolution type equations, Izv. Vyssh. Uchebn. Zaved. Mat., (1962), no. 6, 119–130. (Russian)

[11] Interpolation and extrapolation of linear operators in the Orlich spaces, Dokl. Acad. Nauk SSSR, 151, (1963), no. 6, 1288–1291. (Russian)

[12] The Riemann boundary value problem for n pairs of functions with measurable coefficients and its application to the study of singular integrals in weighted spaces, Izv. Acad. Nauk SSSR Mat., 68 (1964), no. 2, 277–306. (Russian)

[13] Interpolation and extrapolation of linear operators on the Orlich spaces, Mat. Sbornik, 63 (1964), no. 4, 536–553. (Russian)

[14] On the maximal boundary property of functions possessing the integral representations of definite form, Mat. Sbornik, 65 (1964), no. 3, 390–398. (Russian)

[15] On the maximal boundary property of functions possessing the integral representations of a definite form, Dokl. Acad. Nauk SSSR, 157, (1964), no. 6, 1301–1302. (Russian)

[16] A theorem from the theory of commutative normed rings, Nauchn. soobshch. za 1963, Rostov-na-Donu: IRU, 1964, p. 19–20. (Russian)

[17] A new general method for investigation of the linear operator equations of singular integral equation type, Dokl. Acad. Nauk SSSR Mat., 158 (1964), no. 4, 790–793. (Russian)

[18] Singular integral equations with continuous and piecewise continuous symbols, Dokl. Acad. Nauk SSSR Mat., 159, no. 2, (1964), 279–282. (Russian)

[19] (with A.I. Kravchenko, G.G. Bondarenko, N.F. Demin, G.P. Doroshenko, D.E. Sinelnikov) The investigation of horizontal static electric locomotive 2-2-2-track interaction, the work has been registered by the USSR Inventions and Discoveries State Committee at February, 20, 1964, no. 31118. (Russian)

[20] (with V.P. Zaharuta, L.S. Shatskih, V.I. Yudovich) The Green function for domain with dielectric layer, Izv. Vyssh. Uchebn. Zaved. Electromekhanika, (1964), no. 9, 1052–1056. (Russian)

[21] (with V.P. Zaharuta, V.I. Yudovich) Point charge method for capacity calculation, Izv. Vyssh. Uchebn. Zaved. Electromekhanika (1964), no. 11, 1305–1310. (Russian)

[22] (with V.P. Zaharuta, V.I. Yudovich) Calculation of capacities of three infinite strips, Izv. Vyssh. Uchebn. Zaved. Electromehanika, (1965), no. 1, 20–23. (Russian)

[23] (with V.P. Zaharuta, V.I. Yudovich) Approximate method of calculation of capacities of conductors' systems situated on the dielectric layer, Izv. Vyssh. Uchebn. Zaved. Electromekhanika, (1965), no. 3, 247–253. (Russian)

[24] (with V.P. Zaharuta, A.A. Chekulaeva, V.I. Yudovich) Capacity of the round disk on the dielectric layer, Izv. Vyssh. Uchebn. Zaved. Electromekhanika, (1965), no. 5, 487–494. (Russian)

[25] (with V.P. Zaharuta, E.S. Chubukova, V.I. Yudovich) Capacity of two rectangles, Izv. Vyssh. Uchebn. Zaved. Electromekhanika, (1965), no. 5, 487–494. (Russian)

[26] A new general method for investigation of linear operator equations of singular integral equation type. I, Izv. Acad. Nauk SSSR Mat. 29 (1965), no. 3, 567–586. (Russian)

[27] A new general method for investigation of linear operator equations of singular integral equation type. II, Izv. Acad. Nauk SSSR Mat. 29 (1965), no. 4, 757–782. (Russian)

[28] (with A.A. Chekulaeva) Computation of mutual capacities of the two and more than two disks on dielectric layer, Sbornik "Voprosi vichislitelnoy matematiki i vichislitelnoy tekhniki", Rostov-na-Donu: IRU, 1965, 35–40. (Russian)

[29] (with D.I. Zaks, N.P. Gaponenko) For the problem of calculation of heat regime of some solid circuits class, Spec. Sbornik, Izd-vo MViSSO RSFSR, 1966. (Russian)

[30] (with S.M. Zenkovskaya) On the influence of high frequency vibration on the origin of convection, Mekhanika zhydkosti i gaza, 1966, no. 5, 51–56. (Russian)

[31] (with V.V. Bublik, E.N. Bozdarenko) The method of calculation of electrostatic fields for the large thickness layer, Izv. Vyssh. Uchebn. Zaved. Electromekhanika, (1967), no. 5, 503–512. (Russian)

[32] (with V.N. Ped) The steady-state problem of heat conductivity in the layer with heat transfer conduction on the boundaries, Prikladnaya Matematica i Mekhanika, 31 (1967), no. 2, 320–327. (Russian)

[33] Convolution type operators in cones, Dokl. Acad. Nauk SSSR Mat., 176 (1967), no. 6, 1251–1254. (Russian); English transl. in Soviet Math. Dokl. 1967. V.8, N5. P. 1320–1323.

[34] Operators of convolution type in cones, Mat. Sbornik., 74 (1967), no. 2, 298–313. (Russian); English transl. in Math. USSR Sbornik. 1967. V.3, N2. P. 279–193.

[35] (with L.O. Rode) Multidimensional singular integrals in the classes of highest general modules of smoothness, Sib. Math. Zhurnal, 9 (1968) , no. 4, 928–936. (Russian).

[36] On the multidimensional discrete convolutions, Mat. Issled., Kishinev: Shtiintsa, 3 (1968), no. 1, 108–122. (Russian)

[37] Some general questions of the theory of the Riemann boundary value problem, Izv. Acad. Nauk SSSR Mat, 32 (1968), no. 5, 1138-1146. (Russian)

[38] (with V.N. Semenuta) On the indices of multidimensional discrete convolutions, Mat. Issled., Kishinev: Shtiintsa, 4:2(12), (1969), 88–94. (Russian)

[39] A justification of the averaging method for convection problem, Mat. Analiz i ego Prilozhenia, Rostov-on-Don: IRU, (1969), 185–189. (Russian)

[40] (with V.N. Semenuta) Calculation of index of multidimensional discrete convolutions, Mat. Issled., Kishinev: Shtiintsa, 4:4, (1969), 134–141. (Russian)

[41] For the heat computation of solid circuits, Radiotekhnika i electronika, 15 (1970), no. 6, 1232–1240. (Russian)

[42] The capacity of round conductive disk situated on the dielectric layer of small thickness, Mat. Analiz i ego Prilozheniya, Rostov-na-Donu: IRU, 2 (1970), 117–131. (Russian)

[43] A justification of the averaging method for the abstract parabolic equations, Mat. Sbornik, 81, no. 1, (1970), 53–61. (Russian); English transl. in Math. USSR Sbornik. 1970. V.10, N1. P. 51–59.

[44] A justification of the averaging method for the abstract parabolic equations, Dokl. Acad. Nauk SSSR, 191 (1970), no. 1, 33-34. (Russian); English transl. in Soviet Math. Dokl. 1970. V.11, N2. P. 323–325.

[45] On a question of index, Mat. Analiz i ego Prilozhenia, Rostov-on-Don: IRU, (1970), 99–105. (Russian)

[46] (with O.K. Kolibelnikova) On the capacity of conductive disk situated on the dielectric layer, Izv. Vyssh. Uchebn. Zaved. Electromekhanika, (1971), no. 1, 720–724. (Russian)

[47] For the question of solvability of bisingular and polysingular equations, Funk. Analiz i ego Prilozhenia, 5 (1971), no. 1, 93–94. (Russian)

[48] Boundary value problems of analytic functions of two variables and their associated integral equations, Dokl.Acad. Nauk SSSR, 199 (1971), no. 3, 551–552. (Russian); English transl. in Soviet Math. Dokl. 1971. V.12, N4, P. 1131–1133.

[49] Approximate-moment scales and interpolation theorems, Mat. Analiz i ego Prilozhenia, 3 (1971), 37–52. (Russian)

[50] (with A.A. Checulaeva) On the capacity of condenser consisting of infinite strips, Izv. Vyssh. Uchebn. Zaved. Electromekhanika, (1972), no. 4, 362–365. (Russian)

[51] A justification of the averaging method for a problem of convection in a field of rapidly oscillating forces and other parabolic equations, Mat. Sbornik, 87 (1972), no. 2, 236–253. (Russian); English transl. in Math. USSR. Sbornik. 1972. V. 16, N 2, p. 245–263.

[52] Introduction to topology, Rostov-on-Don: IRU, 1973, 100p. (Russian)

[53] Higher approximations of the averaging method for parabolic equations. Dokl.Acad. Nauk SSSR, 213 (1973), no. 6, 1255–1257. (Russian); English transl. in Soviet Math. Dokl. 1973. V. 14, N 6, p. 1884–1886.

[54] Higher approximations of the averaging method for abstract parabolic equations, Mat. sbornik, 92 (1973), no. 4, 541–549. (Russian); English transl. in Math. USSR. Sbornik. 1973. V. 21, N 4, P. 535–543.

[55] Problems of electrostatics in inhomogeneous medium. The case of thin dielectric with large dielectric constant. I, Differentsyalnie Uravnenia, 10 (1974), no. 2, 301–309. (Russian)

[56] Characteristic bisingular equations in the spaces of measurable functions, Izv. Vyssh. Uchebn. Zaved. Mat., (1974), no. 2, 115–119. (Russian)

[57] Problems of electrostatics in inhomogeneous medium. The case of thin dielectric with large dielectric constant. II, Differentsyalnie Uravnenia, 11 (1975), no. 10, 1870–1878. (Russian)

[58] On a limit problem of heat conductivity in inhomogeneous medium, Sib. Math. Zhurnal, 16 (1975), no. 6, 1291–1300. (Russian)

[59] (with M.P. Boroditskiy) On a differently dimensional variational problem, Funk. Analiz i ego Prilozhenia, 9 (1975), no. 4, 63–64. (Russian)

[60] (with M.P. Boroditskiy) Problems of electrostatics in inhomogeneous medium. The case of thin dielectric with large dielectric constant, Rost. Universitet, Rostov-na-Donu, 1975, 36p. Dep. In VINITI 03.11.75, N3167-75, RZh Fisika 1976, 2B108. (Russian)

[61] Some more on the Treftz method, Izv. SKNC VS, Seria Estestv. Nauk, 1976, no. 2, 10–12. (Russian)

[62] Some estimations for power quasipolynomials, Mat. Sbornik, 100 (1976), no. 1, 89–101. (Russian)

[63] An observation on random quasipolynomials, Mat. Analiz i ego Prilogenia, Rostov-na-Donu:IRU, 1978, 122–124. (Russian)

[64] On the convergence of Biberbach polynomials in the case of Lipschitz domains, Izvestia Acad. Nauk SSSR Mat., 42 (1978), no. 4, 870–878. (Russian)

[65] (with A.V. Kozak) On the projection methods of studying two-dimensional singular equations on the tore, Funk. Analiz i ego Prilogenia, 12 (1978), no. 1, 74–75. (Russian)

[66] On the application of square minimization principle to approximate searching the Christoffel-Schwartz constants, Izvestia SKNC VS, Seria Estestv. Nauk, (1978), no. 3, 7–10. (Russian)

[67] Some new results on a convergence of approximate methods for conformal mappings, Actual. Voprosy Mat. Analiza, Rostov-na-Donu: IRU, 1978, 150–160. (Russian)

[68] (with S.L. Edelshtein) Exponentially convergent method of searching proper numbers of Laplace operator in the case of plane polygonal domain, Rost. Univer., Rostov-na-Donu, 1978, 41p. Dep. In VINITI 20.07.78, N2438-78, RZh Matematika 1978, 11B1441. (Russian)

[69] (with V.B. Levenschtam) The highest approximations of the averaging method for parabolic equations, Trudi Vsesouznoy konf. po uravneniyam s chastnimi proizvodnimi, Moskva: Izd. MGU, 1978, 443–445. (Russian)

[70] Exponentially convergent linear method of searching the Christoffel-Schwartz constants and Markov type estimation for potential with quasipolynomial densities, Rost. Univer., Rostov-na-Donu, 1979, 24p. Dep. In VINITI 17.01.79, N209-79 RZh Matematika 1979, 4B1266. (Russian)

[71] Exponentially convergent linear method of searching the Christoffel-Schwartz constants and Markov type estimation for harmonic power quasipolynomials, Sib. Math. Zhurnal, 22 (1981), no. 3, 188–196. (Russian)

[72] On the one estimation of conformal mapping from the circle to the Lipschitz domain, Rost. Univer., Rostov-na-Donu, 1979, 19p. Dep. In VINITI 24.04.79, N1465-79 RZh Matematika 1980, 6B121. (Russian)

[73] (with A.V. Kozak) Projection methods of the studying multidimensional discrete equations of convolution type, Sib. Math. Zhurnal, 21 (1980), no. 2, 119–127. (Russian)

[74] (with A.V. Kozak) Invertibility of the convolution type operators in the large domains, Mat. Issled., Kishinev: "Shtiintsa", 1980, v. 54, 56–66. (Russian)

[75] On the factorization and the local factorization of measurable functions, Dokl. Acad. Nauk SSSR, 250 (1980), no. 5, 1063–1066. (Russian)

[76] On a relation between factorizability and local Noethericity property, Soobshch. Acad. Nauk Gruzin. SSR, 98 (1980), no. 2, 281–283. (Russian)

[77] Equivalence of factorizability and local factorizability of measurable functions, defined on the contour of R type, Rost. Univer., Rostov-na-Donu, 1987, 51p. Dep. In VINITI 02.06.80, N2193-80, RZh Matematika 1980, 11B818. (Russian)

[78] Equivalence of factorizability of measurable function and the local Noethericity property of the singular operator generated by this function, Rost. Univer., Rostov-na-Donu, 1987, 51p. Dep. In VINITI 02.06.80, N2194-80, RZh Matematika 1980, 11B819. (Russian)

[79] (with V.I. Azamatova, A.A. Kilbas, G.S. Litvinchuk, O.I. Marichev, M.D. Martinenko, V.S. Rogozhyn, S.G. Samko, A.S. Fedenko, Yu.I. Cherskiy) Fedor Dmitrievich Gakhov, Izv. Acad. Nauk BSSR, (1980), no. 4, 130–132. (Russian)

[80] (with N.P. Vekua, G.S. Litvinchuk, S.M. Nikolskiy, V.S. Rogogin, S.G. Samko, S.G. Khvedelidze, U.I. Cherskiy) Fedor Dmitrievich Gakhov (obituary), Uspehi Mat. Nauk, 36 (1981), no. 1, 193–194. (Russian)

[81] On the closure of the set $\{(\rho, \gamma) : |\,\rho|^{1+\gamma} \in W_\rho(\Gamma)\}$ and some another properties of weight functions for singular integral Cauchy in the case of R type contour, Rost. Universitet, Rostov-na-Donu, 1981, 27p. Dep. In VINITI 27.02.81, N956-81, RZh Matematika 1981, 6B221. (Russian)

[82] Some properties of singular integral with continuous density in the case of R type contour, Izv. SKNC VS, Seria Estestv. Nauk, 1982, no. 2, 17–19. (Russian)

[83] (with Chin Ngok Minh) On the index and the homotopy classification of families of singular operators with Carleman shift, Dokl. Acad. Nauk SSSR, 263 (1982), no. 5, 1070–1073. English transl. in Soviet Math. Dokl. 1982. V. 25, N 2, P. 503–506. (Russian)

[84] (with Chin Ngok Minh) On the index of families of singular operators with Carleman shift, Rost. Universitet, Rostov-na-Donu, 1982, 40p., Dep. In VINITI 20.05.82, N2558-82, RZh Matematika 1982, 9B693. (Russian)

[85] A defining role of the values of the natural additive mappings of the K^* – functor on the sphere, Rost. Universitet, Rostov-na-Donu, 1982, 16p., Dep. In VINITI 20.05.82, N2560-82, RZh Matematika 1982, 9A459. (Russian)

[86] Some more on the Muckenhoupt conditions, Rost. Universitet, Rostov-na-Donu, 1982, 17p. Dep. In VINITI 20.05.82, N2557-82, RZh Matematika 1982, 9B46. (Russian)

[87] The weight properties of exponent of multidimensional singular integral with continuous and bounded density, Rost. Universitet, Rostov-na-Donu, 1982, 35p. Dep. In VINITI 20.05.82, N2559-82, RZh Matematika 1982, 9B41. (Russian)

[88] Stability of the weight relative to singular integral Cauchy properties of the functions, Mat. Zametki, 33 (1983), no. 3, 409–416. (Russian)

[89] On the global and local factorability of the measurable matrix-function and Noethericity of the generated by this matrix-function singular operator, Izv. Vissh. Uchebn. Zaved. Mat., (1984), no. 4, 81–87. (Russian)

[90] An example of function that satisfies Muckenhoupt condition, but is not a weight function for the Cauchy singular integral in the case of contour with a turning point, Rost. Universitet, Rostov-na-Donu, 1983, 20p. Dep. In VINITI 18.05.83, N2659-83, RZh Matematika 1983, 9B780. (Russian)

[91] On the global and local factorability of the measurable matrix-function and Noethericity property of the generated by this matrix-function singular operator in the scales of L_p-spaces in the case of R type contour, Rost. Universitet, Rostov-na-Donu, 1983, 38p. Dep. In VINITI 18.05.83, N2660-83, RZh Matematika 1983, 9B802. (Russian)

[92] On the one effective method of calculating the Christoffel-Schwartz integral constants in the case of half-plane, perturbed by cross-cuts, Rost. Universitet, Rostov-na-Donu, 1983, 16p. Dep. In VINITI 18.05.83, N2658-83 RZh Matematika 1983, 9B133. (Russian)

[93] (with S.L. Edelshtein) Estimates for quasipolynomials and the convergence highest norms of approximation methods of solving elliptic problems in domains with corners, Probleme und Methoden der Mathematischen Physik. 8. Tagung. Karl-Marx-Stadt, 1983. B. 63, p. 46–52.

[94] (with S.L. Edelshtein) Estimates for quasipolynomials and the convergence higher norms of approximation methods of solving elliptic problems in domains with corners, Vortragauszüge 8. TMP. Technische Hochschule K-M-S, 1983, 53–54. (Russian)

[95] Convergence of the asymptotic expansions for some types of elliptic problems in the singular perturbed domains, Rost. Universitet, Rostov-na-Donu, 1983, 30p. Dep. In VINITI 05.12.83, N6544-83, RZh Matematika 1984, 3B440. (Russian)

[96] (with R.A. Simonenko) Conformal mapping of domain with singular perturbed corner portion of the boundary, Rost. Universitet, Rostov-na-Donu, 1983, 29p. Dep. In VINITI 05.12.83, N6543-83, RZh Matematika 1984, 3B177. (Russian)

[97] (with Chin Ngok Minh) A local method in the theory of one-dimensional singular integral equations with piecewise continuous coefficients, Rostov-na-Donu: IRU, 1986, 60p. (Russian)

[98] Weight properties of the exponent of the multidimensional singular integral, Funk. Analiz i ego Prilozhenia, 18 (1984), v. 3, 92–93. (Russian)

[99] (with R.A. Simonenko) Conformal mapping of the singular perturbed half-plane. Calculating of the Christoffel-Schwartz constants in the case of polygonal perturbances, Rost. Universitet, Rostov-na-Donu, 1984, 32p. Dep. In VINITI 13.12.84, N7979-84, RZh Matematika 1985, 4B187. (Russian)

[100] (with R.A. Simonenko) Conformal mapping of the domain with singular perturbed straight-line portion of the boundary, Rost. Universitet, Rostov-na-Donu, 1984, 27p. Dep. In VINITI 13.12.84, N7980-84, RZh Matematika 1985, 4B188. (Russian)

[101] (with Chin Ngok Minh) A local method in the theory of one-dimensional singular integral equations with piecewise continuous coefficients, Nauchnie Trudi Yubileynogo Seminara po Kraeviv Zadacham, Minsk: Izdatelstvo "Universitetskoe", 1985, 115-120. (Russian)

[102] (with V.M. Deundyak, Chin Ngok Minh) Index and homotopy classification of families of one-dimensional singular operators with piecewise continuous coefficients, Rost. Universitet, Rostov-na-Donu, 1985, 46p. Dep. In VINITI 12.05.85, N3204-85, RZh Matematika 1985, 9B834. (Russian)

[103] (with R.A. Simonenko) Conformal mapping of the singular perturbed half-plane, Mejvuzovskiy Sbornik Nauchnih Trudov, KGU-RGU, Elista, 1985, 149–159. (Russian)

[104] (with V.M. Deundyak, Chin Ngok Minh) Homotopy classification of families of one-dimensional singular operators with piecewise continuous coefficients, Dokl. Acad. Nauk SSSR, 289 (1986), no. 3, 521–524. (Russian); English transl. in Soviet Math. Dokl. 1987. V.34, N1, P. 100–103.

[105] (with R.A. Simonenko) Conformal mapping of the singular perturbed domain, Dokl. Acad. Nauk SSSR, 289 (1986), no. 2, 302-305. (Russian); English transl. in Soviet Math. Dokl. 1987. V.34, N1, P. 93–95.

[106] The averaging method in the theory of nonlinear equations of parabolic type with application to the problems of hydrodynamic stability, Rost. Universitet, Rostov-na-Donu, 1986, 184p., Dep. In VINITI 22.10.86, N7357-86, RZh Matematika 1987, 2B498. (Russian) a) The averaging method in the theory of nonlinear equations of parabolic type with application to the problems of hydrodynamic stability, Rostov-na-Donu: IRU, 1989, 112p. (Russian)

[107] (with S.M. Zenkovskaya) Application of averaging method for the solution of the problem of convection in the field of rapidly oscillating forces and for the solution of other parabolic equations, Proceedings of the 4 International Conference on Boundary and Interior Layers, Boole press (Ireland), 1986, 436–441.

[108] (with V.A. Stukopin) On the one universal algebra of singular integral operators on the segment, Izv. SKNC VS, Seria Estestv. Nauk, 1987, no. 3, 52–55. (Russian)

[109] (with I.S. Soybelman) On the methods of computation of capacity, Izv. Vissh. Uchebn. Zaved., Electromekhanika, 1987, no. 11, 33–39. (Russian)

[110] (with V.A. Stukopin) On a universal algebra of singular integral operators on the segment, Rost. Universitet, Rostov-na-Donu, 1987, 38p. Dep. In VINITI 23.07.87, N5327-V87, RZh Matematika 1987, 11B1071. (Russian)

[111] (with V.M. Deundyak, Chin Ngok Minh) Symbols and homotopy classification of families of one-dimensional singular operators with piecewise continuous coefficients, Izv. Vissh. Uchebn. Zaved., Mat., 1988, no. 12, 17–27. (Russian)

[112] (with V.M. Deundyak, Chin Ngok Minh) The calculation of index of the families of one-dimensional singular operators with piecewise continuous coefficients, "Dif., Integral. Uravneniya i Komplexnyi Analiz", Mejvuzovskiy Sbornik Nauchnih Trudov, KGU-RGU, Elista, 1988, 18–28. (Russian)

[113] On the one algorithm of searching linear functional by its value on the image. Estimation of error, Rost. Universitet, Rostov-na-Donu, 1988, 20 p. Dep. In VINITI 17.11.88, N8174-88, RZh Matematika 1989, 2G264. (Russian)

[114] On an algorithm of solving systems of linear equations. Estimation of error, Rost. Universitet, Rostov-na-Donu, 1988, 14p. Dep. In VINITI Dep. 16.11.88, N8147-V88, RZh Matematika 1989, 2G14. (Russian)

[115] (with M.E. Abramyan) Conditionality of the one special system of functions, Rost. Universitet, Rostov-na-Donu, 1989, 13p. Dep. In VINITI 28.10.88, N7732-V88, RZh Matematika 1989, 2B212. (Russian)

[116] (with A.V. Kozak, V.S. Pilidi) Asymptotic of the solutions of some integral equations in the large domains, Dokladi Rasshirennih Zasedanii Seminara Institute Pricladnoi Matematiki im. I.N. Vekua, Tbilisi: Izdatelstvo Tbilisskogo Universiteta, 3 (1988), no. 1, 100–103. (Russian)

[117] Propagation of sound in the few-mode hydroacoustic wave conductors, in "Akustika Okeanskoy Sredy. Acad. Nauk SSSR. Komitet po Problemam Mirovogo Okeana", Moskva: Nauka, 2 (1989), 39–46. (Russian)

[118] The acoustic field in wave conductor with diaphragm having a small hole, Rost. Universitet, Rostov-na-Donu, 1989, 39p. Dep. In VINITI 21.02.89, N1155-V89, RZh Fizika 1989, 6P22. (Russian)

[119] (with M.E. Abramyan) Conditionality of the special function systems. Two-sided estimates, Rost. Universitet, Rostov-na-Donu, 1989, 17p. Dep. In VINITI 15.06.89, N3990-V89, RZh Matematika 1989, 10B51. (Russian)

[120] (with Mikhalkovich) Asymptotics of the acoustic fields in wave conductor with absolutely rigid small inclusion, Rost. Universitet, Rostov-na-Donu, 1989, 79p., Dep. In VINITI 01.11.89, N6617-V89, RZh Fizika 1990, 2P161. (Russian)

[121] (with V.A. Babeshko, I.I. Vorovich, G.S. Litvinchuk, I.V. Ostrovskiy, V.S. Rogozhyn, V.S. Samko, B.V. Khvedelidze) Hikolay Vasilievich Govorov, UMN, 44 (1989), no. 5, 187–190. (Russian)

[122] (with M.E. Abramyan) Estimates of condition number of some function systems, Izv. SKNC VS, Seria Estestv. Nauk, 1990, no. 2, 63–66. (Russian)

[123] (with Mikhalkovich) Asymptotics of the acoustic fields in wave conductor with small inclusion of different types, Rost. Universitet, Rostov-na-Donu, 1989, 79p., Dep. In VINITI 19.07.90, N4071-V90, RZh Fizika 1990, 10P166 DEP. (Russian)

[124] Algorithms of solving the approximate problems with improperly stipulated Gram matrix, Rost. Universitet, Rostov-na-Donu, 1991, 28p., Dep. In VINITI 05.02.91, N591-V91, RZh Matematika 1991, 6G129. (Russian)

[125] The sections' method with a multiple domains' superposition in application to computation of conformal mapping of the square, Rost. Universitet, Rostov-na-Donu, 1991, 33p., Dep. In VINITI 05.02.91, N592-V91, RZh Matematika 1991, 6G208. (Russian)

[126] Harmonic Dirichlet problem in the domain with fine-grained structure of the boundary. The uniform convergence, Rost. Universitet, Rostov-na-Donu, 1991, 34p., Dep. In VINITI 05.02.91, N590-V91, RZh Matematika 1991, 6B360. (Russian)

[127] (with M.E. Abramyan) Fredholm property and solvability of problems on waveguide joint, Rost. Universitet, Rostov-na-Donu, 1991, 24p. Dep. In VINITI 23.05.91, N2120-V91, RZh Matematika 1991, 9B754. (Russian)

[128] (with I.M. Erusalimskiy) 35 Lectures on discrete mathematics. 1. Propositional algebra, Rostov-na-Donu: MGP "Gaudeamus–XX1", 1991, 51p. (Russian)

[129] (with I.M. Erusalimskiy) 35 Lectures on discrete mathematics. 2. Algebra of predicates and sets, Rostov-na-Donu: MGP "Gaudeamus–XXI", 1991, 44p. (Russian)

[130] (with I.M. Erusalimskiy) 35 Lectures on discrete mathematics. 3. The elements of combinatorial analysis, Rostov-na-Donu: MGP "Gaudeamus–XXI", 1991, 53 p. (Russian)

[131] (with I.D. Jerdenovskaya) The sections' method with a multiple superposition of decomposition domains in application to computation of conformal mapping of polygons, Rost. Universitet, Rostov-na-Donu, 1992, 22p., Dep. In VINITI 08.01.92, N71-V92, RZh Matematika 1992, 5B139. (Russian)

[132] (with I.M. Erusalimskiy) 35 Lectures on discrete mathematics. 4. Boolean functions, Rostov-na-Donu: MGP "Gaudeamus–XX1", 1992, 35 p. (Russian)

[133] (with I.M. Erusalimskiy) 35 Lectures on discrete mathematics. 5. The elements of the automata theory, Rostov-na-Donu: MGP "Gaudeamus–XXI", 1992, 33p. (Russian)

[134] (with I.M. Erusalimskiy) 35 Lectures on discrete mathematics. 5. The elements of the graph theory, Rostov-na-Donu: MGP "Gaudeamus–XXI", 1992, 48p. (Russian)

[135] (with I.D. Jerdenovskaya) The sections' method of conformal mapping of polyhedral surfaces, Rost. Universitet, Rostov-na-Donu, 1992, 13p., Dep. In VINITI 30.12.92, N3723-V92, RZh Matematika 1993, 5B94. (Russian)

[136] (with V.M. Deundyak, V.A. Stukopin) Homotopy classification of the families of Noether singular operators with piecewise continuous coefficients on a composite contour, Dokl. Rossiyskoy Acad. Nauk, 329 (1993), v. 5, 540–542.

[137] Ideals of a Banach Algebra of Singular Integral Operators with Piecewise Continuous Coefficients in the Space Lp on a Composite Contour. Funct. Analiz i ego Prilozheniia 1993. V. 4. P. 69–71. (Russian) English transl. in Plenum Publishing Corporation. 1994. P. 277–279.

[138] (with I.D. Jerdenovskaya) The sections' method with a multiple superposition of decomposition domains in application to computation of conformal mapping of polygons. Quantization of residual functional, Rost. Universitet, Rostov-na-Donu, 1993, 14p., Dep. In VINITI 03.02.93, N269-V93, RZh Matematika 1993, 6B142. (Russian)

[139] Estimation of error of arithmetical algorithms with application to least-squares method, Rostov-na-Donu: UPL, 1993, 58p. [Programma "Universiteti Rossii"]. (Russian)

[140] Elements of the sets theory and combinatorial analysis, Rostov-na-Donu: UPL, 1993, 36 p. [Programma "Universiteti Rossii"]. (Russian)

[141] (with V.M. Deundyak, V.A. Stukopin) Symbol calculation and the structure of ideals of algebra of singular operators on the beam system, Sbornik "Dif. and Integral. Uravneniia i Komplexnyi Analiz", Elista: Izdat. Kalmitskogo Gos. Universiteta, 1993, 36–47. (Russian)

[142] (with V.M. Deundyak, V.A. Stukopin). The Index of Families of Noether Singular and Bisingular Integral Operators with Piecewise Continuous Coefficients on Composite Contour. Dokl. Rossiiskoi Acad. Nauk. 1994. V. 338, N 2. p. 158–161. (Russian); English transl. in Russian Acad. Sci. Dokl. Math. 1995. Vol. 50, N 2. p. 210–214.

[143] (with V.M. Deundyak) Models of local theory and local method in the pairs of Lp-spaces, Rostov-na-Donu, 1994, 31p., Dep. In VINITI 08.11.94, N2525-B94, RZh Mat 1995, N3B558. (Russian)

[144] (with S.M. Zenkovskaya, V.B. Levenshtam) Averaging method for differential equations in Banach spaces with applications to convection problems. Proceeding of the

14th IMACS World Congress on Computation and Applied Mathematics. July 11–15, 1994, Georgia Institute of Technology, Atlanta, Georgia, USA 1994. Vol. 2. p. 922-925.

[145] (with I.D. Jerdenovskaya) On a direct method of computation of conformal mapping of polygons, Izv. Vyssh. Uchebn. Zaved. Mat., no. 2(393)(1995), 27–36. (Russian)

[146] (with V.P. Gusakov) On some modification of the method of separation of variables for Helmholtz equation with a wave number depending on the angular coordinate, Sbornik "Sovremennie Problemy Mekhaniki Sploshnoi Sredy", Rostov-na-Donu: MP "Kniga", 1995, 173–177. (Russian)

[147] (with V.M. Deundyak) Index of families of Noether convolution type operators in the pair of spaces, Rostov-na-Donu, 1995, 17p., Dep. In VINITI 21.04.95, N1138-V95, RZh Matematika 1995, N8B478. (Russian)

[148] (with V.P. Gusakov) On the special systems of functions connected with Helmholtz equation with wave number depending on the angular coordinate, Rostov-na-Donu, 1995, 20p., Dep. In VINITI 06.06.95, N1655-V95, RZh Matematika 1995, N9B370. (Russian)

[149] On the one method of computation of wave conductor with diffuser, Rostov-na-Donu, 1995, 24p., Dep. In VINITI 28.11.95, N3132-V95, RZh Fizika. 1996, N4P41. (Russian)

[150] Some remarks on the minimum problem for quadratic functional depending on parameter and close problems, Rostov-na-Donu, 1996, 24p., Dep. In VINITI 17.01.96, N197-V96, RZh Matematika 1996, N6B504. (Russian)

[151] Local type operators theory and its applications, Rostov-na-Donu, 1996, 74p., Dep. In VINITI 23.01.96, N275-V96, RZh Matematika 1996, N10B592. (Russian)

[152] (with V.M. Deundyak, V.A. Stukopin) The structure of set of ideals and classification of algebras of singular integral operators in the space L_p on a composite contour, Rostov-na-Donu, 1996, 50p., Dep. In VINITI 19.03.96, N828-V96, RZh Matematika 1996, N9B584. (Russian)

[153] (with V.M. Deundyak) Local method in the pairs of Lp-spaces and index, Dokladi Rus. Acad. Nauk, 349 (1996), no. 5, 592–595. (Russian)

[154] (with V.P. Gusakov) On some modification of the method of separation of variables for equation $\Delta U + k^2(\varphi)U = 0$. The correct and incorrect Dirichlet problems, Rostov-na-Donu, 1996, 23p., Dep. In VINITI 31.10.96, N3185-V96. RZh Matematika 1997, N5B291. (Russian)

[155] (with V.M. Deundyak, V.A. Stukopin) Local analysis of singular operators with piecewise-continuous coefficients on compound contour. I. Noethericity, Don. Gos. Universitet, Rostov-na-Donu, 1997, 60p., Dep. in VINITI 22.11.96, N3389-V96. RZh Matematika 1997, 5B689.

[156] On the one method of computation of fundamental frequencies of membrane having the shape of right triangle, Rostov-na-Donu, 1996, 31p., Dep. In VINITI 25.12.96, N3810-V96. RZh Matematika 1997, N6B363. (Russian)

[157] Traces of W_1^2-functions on Lipschitz lines, Rostov-na-Donu, 1997, 47p. Dep. In VINITI 24.03.97, N896-V97. RZh Matematika 1997, 11B65. (Russian)

[158] (with V.M. Deundyak, V.A. Stukopin) Local analysis of singular operators with piecewise-continuous coefficients on compound contour. II. Homotopy classification

and calculation of index of families, Don. Gos. Universitet, Rostov-na-Donu, 1997, 60p., Dep. in VINITI 25.07.97, N2495-V97. RZh Matematika 1998, 2B810. (Russian)

[159] (with U.V. Lazarev, A.V. Marchenko) Differential equation for determinant of integral Fredholm operator of second type and its applications, Rostov-na-Donu, 1997, 14p, Dep. in VINITI 31.10.97, N3229-V97. RZh Matematika 1998, 3B465. (Russian)

[160] Some remarks on normalizations of Sobolev spaces and possible definitions of variational and generalized problems, Rostov-na-Donu, 1997, 27p., Dep. In VINITI 31.10.97, N3230-V97. RZh Matematika 1998, 4B694. (Russian)

[161] (with U.V. Lazarev, A.V. Marchenko) Determinant of integral Fredholm equation of second type with continuous kernel, Rostov-na-Donu, 1997, 11p., Dep. In VINITI 19.11.97, N3402-V97. RZh Matematika 1998, 5B291. (Russian)

[162] The sections' method with superposition of domains in application to the problem of computation of fundamental frequencies of the membrane with a polygon boundary, Rostov-na-Donu, 1998, 42p., Dep. In VINITI 10.02.98, N365-V98. (Russian)

[163] An integral equation of the theory of interaction of wave conductor and diffuser, Izv. Vyssh. Uchebn. Zaved. Severo-Kavkazskyi Region, 2000, no. 3, 155–158, (Russian)

[164] Multidimensional variant of G. Szegő's theorem on the limit spectral density of truncated Toeplitz operators. The case of Wiener symbol, Rostov-na-Donu, 1998, 10p., Dep. In VINITI 21.08.98, N2635-V98. RZh Matematika 1999, 1B744. (Russian)

[165] Multidimensional variant of G. Szegő's theorem on the limit spectral density of truncated Toeplitz operators. The case of measurable bounded symbol, Rostov-na-Donu, 1998, 19p., Dep. In VINITI 30.10.98, N3165-V98. RZh Matematika 1999, 4B606. (Russian)

[166] Multidimensional variant of G. Szegő's theorem on the limit spectral density of truncated Toeplitz operators. The case of compound convolutions, Rostov-na-Donu, 1998, 11p., Dep. In VINITI 29.12.98, N3930-V98. RZh Matematika 1999, 6B674. (Russian)

[167] (with U.V. Lazarev, A.V. Marchenko) On the calculation of determinants of truncated one-dimensional continual convolutions with rational symbols, Izv. Vyssh. Uchebn. Zaved. Mat., (1999), no. 4(443), 77–78. (Russian).

[168] Some results on the stability of limit Szegő's set and its applications, 12p., Rostov-na-Donu, 1999, Dep. In VINITI 09.07.99, N2256-V99, 12p. RZh Matematika 2000, 00.01-13B, 641DEP. (Russian)

[169] Limit Szegő-type theorems for compound and generalized convolution type operators, 61p., Rostov-na-Donu, 1999, Dep. In VINITI 12.08.99, N2639-V99. 61p. RZh Matematika 2000, 00.02-13B, 619DEP. (Russian)

[170] On one effective method of computation of fundamental frequencies of membrane with polygon boundary, Dokl. Rus. Acad. Nauk, 367 (1999), no. 5, 603–605. (Russian)

[171] Limit Szegő-type theorems and asymptotic of determinants for the generalized discrete convolution type operators, 24p., Rostov-na-Donu, 1999, Dep. In VINITI 29.11.99, N3528-V99. 24p. RZh Matematika 2000, 00.05-13B, 729DEP. (Russian)

[172] Limit Szegő-type theorems for determinants of truncated generalized multidimensional discrete convolutions, Dokl. Rus. Acad. Nauk, 373 (2000), no. 5, 588–589. (Russian)

[173] (with O.N. Zabroda) Limit asymptotic spectral behavior of truncated generalized one-dimensional discrete convolution type operators. I. 25p., Rostov-na-Donu, 2001, Dep. in VINITI 19.01.01, N145-V2001. 25p. RZh Matematika 2001, 01.07-13B, 595DEP. (Russian)

[174] Szegő-type Limit Theorems for Multidimensional Discrete Convolution Operators with Continuous Symbols. Funk. Analiz i ego Prilozheniya. V. 35, no. 1, 2001, Pp.91-93. (Russian); English transl. in Functional Analysis and its Applications, Vol. 35, No. 1, p. 77–78, 2001

[175] (with V.A. Vasiliev) On the one Widom's theorem, Rostov-na-Donu, 2001, Dep. in VINITI 09.04.01, N906-V2001, 21p. RZh Matematica 2001, 01.11-13B, 75Dep. (Russian)

[176] Lections on mathematical analysis. Part 1. Differential and integral calculation for the functions of one variable, Rostov-na-Donu: Rost. Gos. Universitet, 2001, 132p. (Russian)

[177] Limit Szegő-type theorem for perturbed Toeplitz matrixes, Rostov-na-Donu, 2001. Dep. In VINITI 13.06.01, N1425-V2001. 30p. RZh Matematika 2002, 02.02-13, 291DEP. (Russian)

[178] (with O.N. Zabroda) Collective asymptotic spectral behavior of linear truncated generalized one-dimensional discrete convolution type operator. II. Rostov-na-Donu, 2001, Dep. in VINITI 25.12.01, N2677-V2001. 30p. RZh Matematika 2002, 02.07-13B, 648DEP. (Russian)

[179] (with I.A. Kazmin) Asymptotic decomposition of higher coefficients of characteristic polynomial of truncated discrete convolution, Rostov-na-Donu, 2001, Dep. in VINITI 21.12.01, N2645-V2001. 13p. (Russian)

[180] (with O.N. Zabroda) Collective asymptotic spectral behavior of linear truncated generalized one-dimensional discrete convolution type operator. III. Rostov-na-Donu, 2002, Dep. in VINITI 13.05.02, N824-2002. 27 p. (Russian)

[181] (with E.A. Maximenko) Asymptotic of multiple continual convolutions, truncated by expanding polyhedrons, Rostov-na-Donu, 2002, Dep. in VINITI 14.05.02, N846-V2002. 34 p. (Russian)

[182] (with O.N. Zabroda) Asymptotic invertibility of truncated generalized one-dimensional discrete convolutions and Szegő-type limit theorem, Rostov-na-Donu, 2002, Dep. in VINITI 02.07.02, N1230-V2002, 21 p. (Russian)

[183] (with I.A. Kazmin) Asymptotic of trace of matrix inverse to truncated Toeplitz matrix with rational symbol, Rostov-na-Donu, 2002, Dep. in VINITI 16.09.02, N1571-V2002. 10 p. (Russian)

[184] (with O.N. Zabroda) Asymptotic invertibility of truncated generalized one-dimensional discrete convolutions and Szegő-type limit theorem. II., Rostov-na-Donu, 2002, Dep. in VINITI 11.10.02, N1721-V2002, 24 p. (Russian)

[185] Elements of the theory of convex sets and asymptotic behavior of integer-valued volumes, Rostov-na-Donu, Dep. In VINITI 13.11.2002, N1964-V2002. 78 p. (Russian)

[186] (with E.A. Maximenko, O.N. Zabroda) Szegő-type limit theorems for Toeplitz matrixes and generalized convolutions, Izv. Vyssh. Uchebn. Zaved. Severo-Kavkazskiy Region. Estestvennie Nauki, 2002, Yubileyniy Vypusk, 100–103. (Russian)

[187] (with V.A. Vasiliev, E.A. Maximenko) On a Szegő-Widom's theorem, Dokl. Akad. Nauk, 393 (2003), no. 3, 307–308. (Russian)

[188] (with A.U. Novoseltsev) Relationship between asymptotic of higher proper values of truncated continual convolution and rate of maximization of symbol, Rostov-na-Donu, Dep. in VINITI 29.05.2003, N1050-V2003. 9 p. (Russian)

[189] (with O.N. Zabroda) Asymptotic invertibility of truncated generalized one-dimensional discrete convolutions and Szegő-type limit theorem. III., Rostov-na-Donu, Dep. in VINITI 08.08.2003, N1557-V2003. 24 p. (Russian)

[190] (with A.U. Novoseltsev) Relationship between asymptotic of extreme eigenvalue of truncated Toeplitz matrixes and rate of maximization of symbol, Rostov-na-Donu, Dep. in VINITI 29.05.2003, N1050-V2003. 9 p. (Russian)

[191] (with O.N. Zabroda) Asymptotic Invertibility and the Collective Asymptotic Spectral Behavior of Generalized One-Dimensional Discrete Convolutions, Functional Analysis and its Applications, Vol. 38, No. 1, pp. 65–66, 2001

[192] (with E.A. Maximenko) Asymptotic of iterated truncated discrete convolutions, Rostov-na-Donu, Dep. in VINITI 04.12.2003, N2101-V2003. 61p. (Russian)

[193] Problems of electrostatics and heat conductivity in inhomogeneous medium with thin inclusions possessing the large physical characteristics, Izv. Vyssh. Uchebn. Zaved. Severo-Kavkazskiy Region. Est. Nauki. Matematika i Mekhanika Sploshnoy sredy. Special. Vypusk. 2004, 207–209.

[194] (with A.U. Novoseltsev) Relationship between asymptotic of extreme eigenvalue of truncated Toeplitz matrixes and rate of maximization of symbol, Algebra i Analiz, 16 (2004), no. 4, 146–152.(Russian)

[195] Szegő-Type Limit Theorem for the Operators of Generalized Multidimensional Discrete Convolution. Matrix Case. Russian Journal of Mathematical Physics. V. 11, No. 4, 2004, pp. 498–500.

[196] Szegő-Type Limit Theorems for Generalized Discrete Convolution Operators. Matem. Zamet, V. 78, no. 2, P. 265–277. 2005. (Russian); English transl. in Mathematical Notes. Volume 78, Numbers 1-2. Date: July 2005, pp. 239–250

[197] (with V.M. Deundyak) On homotopy properties and indices of families of singular and bisingular operators with piecewise-continuous coefficients. Journal of Mathematical Sciences. V. 126, No. 6, 2005, pp. 1593–1599.

[198] (with A.U. Novoseltsev) Relationship between asymptotic of higher proper values of truncated continual convolution and rate of maximization of symbol, Izv. Vyssh. Uchebn. Zaved. Mat., (2005), no. 9 (520), 52–56. (Russian)

Operator Theory:
Advances and Applications, Vol. 170, 27–41

Coefficients Averaging for Functional Operators Generated by Irrational Rotation

A.B. Antonevich

To Professor I.B. Simonenko on the occasion of his 70th birthday

Abstract. The problem under consideration can be posed in the following manner: what is the procedure of changing the coefficients of a functional operator to the coefficients with a simpler behavior under which the principle properties of the initial operator are preserved?

In the paper we consider a number of variants of precise formulation of the problem for the model functional operators generated by an irrational rotation of the circle. In particular we obtain the description of all the possible changes of coefficients under which the spectrum of the operator preserves.

Mathematics Subject Classification (2000). Primary 47B37; Secondary 34C29.

Keywords. Weighted shift operator, irrational rotation, spectral radius, geometric mean, averaging theory.

1. Introduction

A number of mathematical and physical problems reduce to investigation of the spectral properties of bounded operators acting in different spaces $F(X)$ of functions on a set X and having the form

$$Au(x) = \sum_{k=1}^{m} a_k(x)\, u(\alpha_k(x)), \quad u \in F(X), \tag{1.1}$$

where $\alpha_k : X \to X$ are given mappings and a_k are given functions. Operators of the form (1.1) are called *functional operators, operators associated with dynamical system,* or *transfer operators.* Functional operators that have the only one summand, that is the operators of the form

$$Au(x) = a(x)\, u(\alpha(x)), \tag{1.2}$$

are called *weighted shift operators* or *composition operators with a weight*. Operators of the form

$$T_\alpha u(x) = u(\alpha(x)), \qquad (1.3)$$

are called *shift operators, composition operators* or *internal superposition operators*. The properties of functional operators have been investigated from the different points of view.

In a number of applications the coefficients a_k describe the properties of a media where the process of particles transformation takes place. If the media is nonhomogeneous then the coefficients may be functions with a rather complicated behavior and calculation exploiting these operators can be difficult. Therefore it is natural to try to reduce the investigation to the case of operators having coefficients with simpler behavior. In the differential equations theory there exist at least two variants of the corresponding reduction – reducing the equation to the canonical form and averaging theory.

Reducing the equation to the canonical form means the construction of a conjugate operator with simpler coefficients where the conjugation is implemented by means of variables transformation operators.

The subject of investigation in averaging theory are differential equations such that their right-hand side is a function with a complicated behavior and the problem is to construct an averaged equation with a simpler structure in such a way that the solutions to the initial equation are close to the solutions of the averaged one. This problem is the subject of numerous investigations among which a substantial place is occupied by the work by I.B. Simonenko (see, for example [7]) and his followers.

It is natural to consider the problem – what is the analogy of the averaging theory or the reduction to the canonical form for functional operators? The following essential difference should be mentioned at once. At the base of the averaging theory for differential equations lies a simple fact that the integration operator (the inverse to the differentiation operator) maps a rapidly oscillating function to a function with a small norm. Under the action of the inverse to the shift operator we do not have this effect but averaging effects take place as a result of multiple application of the shift operator and are described by the ergodic theorems.

Therefore it is natural to look for the reasonable setting (first of all) when we describe the properties depending on the asymptotic behavior of the powers of the operator.

The most important characteristic of the asymptotic behavior of the powers of an operator A is its spectral radius

$$r(A) = \lim_{n \to \infty} \|A^n\|^{1/n}.$$

Therefore among the properties, the preservation of which should be demanded under the averaging of coefficients, the first place can be occupied by the preservation of the spectral radius.

This is especially essential for modelling of the processes of particles transformation. In such models the function u defines the distribution of the particles in a space and the operator A describes the transformation of this distribution in a time unit. Thus depending on $r(A)$ we have two qualitatively different behaviors of the process: for $r(A) < 1$ the number of particles decreases rapidly (the process damps), for $r(A) > 1$ we have an exponential growth of the number of particles (explosion). The natural demand to the procedure of coefficients averaging is the preservation of qualitatively the same behavior which is ensured by preservation of the spectral radius.

In general the problem can be posed in the following manner: what is the procedure of changing the coefficients of a functional operator for coefficients with simpler behavior which preserves the principle properties of the initial operator? In the form formulated the problem is not precisely posed: it needs the explanation – what coefficients in a concrete situation can be considered as coefficient with simple behavior and preservation of what properties can be demanded. In the present article we consider for the model functional operators generated by an irrational rotation of the circle a number of variants of precise formulation of the problem. In particular we obtain the description of all the possible changes of coefficients under which the spectrum of the operator does not change. Here one can consider as the main result the uncovering of rather delicate questions arising in connection with the problem posed.

2. Weighted composition operators generated by irrational rotation

Let us consider the mentioned questions for the weighted shift operators generated by an irrational rotation of the circle which are the most popular functional operators. Let $S^1 = \mathbb{R}/\mathbb{Z}$, and $\alpha(x) = x + h$, and h be an irrational number. Let us consider in the space $L_2(S^1)$ the weighted shift operators of the form

$$Au(x) = a(x)\, u(x + h), \; x \in S^1. \tag{2.1}$$

In fact the questions arising in connection with the averaging problem for the operators of the form (2.1) have been considered earlier in different context. Thus in this section we present in principle the known facts interpreting them from the point of view of the averaging problem and the problem of reduction to the canonical form.

First let us demonstrate an essential difference between the objects under consideration and differential equations. Let us take the simplest example of averaging for differential equations. Let the sequence of differential equations of the form

$$\frac{du_n}{dx} = a(nx)\, u_n(x), \tag{2.2}$$

be given, where $a(x) = 1 + \frac{1}{2}\cos(2\pi x)$, and let us consider the Cauchy problem $u_n(0) = u_0$.

For large n the functions $a(nx)$ oscillate rapidly but at the same time the function $a(x)$ possesses the mean value

$$\widetilde{a} = \lim_{t \to +\infty} \frac{1}{t} \int_0^t a(x)dx = 1.$$

Therefore there is a natural construction of the equation with averaged coefficient

$$\frac{du_0}{dx} = \widetilde{a}u_0(x).$$

Averaging theorems tell us that the solutions $u_n(x)$ tend to the function $u_0(x)$ – the Cauchy problem solution for the equation with the averaged coefficient.

Now let us consider the sequence of weighted shift operators

$$A_n u(x) = a_n(x) \, u(\alpha(x)),$$

where the coefficients $a_n(x) = 1 + \frac{1}{2}\cos(2\pi nx)$ are the same as in differential equations (2.2). At first glance it is quite natural to change the coefficient (in the way analogous to the case of differential equation (2.2)) for its mean value, that is for the constant 1. But in this situation we obtain the construction which is evidently false: the spectral radius of the averaged operator constructed in this way (its spectral radius is equal to 1) differs essentially from the spectral radii of the operators A_n.

Indeed. For irrational h the spectral radius of the operators of form (2.1) with continuous (and also with Riemann integrable) coefficients [2,3] is

$$r(A) = \exp\left[\int_0^1 \ln|a(x)|dx\right]. \qquad (2.3)$$

This equality means that the spectral radius is not the arithmetical mean value but the geometrical mean value of the coefficient. In the example considered the geometrical mean values of the coefficients a_n do not depend on n and we have the strict inequality

$$\exp\left[\int_0^1 \ln|a_n(x)|dx\right] = \text{const} < 1.$$

Formula (2.3) implies that the coefficient averaging procedure for the operators of the form (2.1) guaranteeing the preservation of the spectral radius reduces to the change of the coefficient for the geometric mean of its modulus.

However here the most essential fact is that in reality we have *much stronger averaging effect*: the relation of the initial operator and the operator with the averaged coefficient does not reduce only to the coincidence of the spectral radii but is deeper.

Proposition 2.1. *Let $a \in C(S^1)$ be such that $\forall x \, |a(x)| > 0$, set*

$$\widetilde{a} = \exp\left[\int_0^1 \ln|a(x)|dx\right]$$

and define the averaged operator as the weighted shift operator with the constant coefficient: $\widetilde{A} = \widetilde{a}T_h$. *Then*

$$\sigma(\widetilde{A}) = \sigma(A).$$

The statement follows from the next description of the spectrum of the operator under consideration.

Proposition 2.2. *If the number h is irrational and the coefficient a is Riemann integrable then in the case when* $\operatorname{ess\,inf} |a(x)| > 0$ *we have*

$$\sigma(A) = \{\lambda : |\lambda| = \exp\left[\int_0^1 \ln|a(x)|dx\right]\}, \tag{2.4}$$

and in the case when $\operatorname{ess\,inf}|a(x)| = 0$ *we have*

$$\sigma(A) = \{\lambda : |\lambda| \leq \exp\left[\int_0^1 \ln|a(x)|dx\right]\}. \tag{2.5}$$

This description was obtained for continuous coefficients in [3], the proof can be carried over to the case of Riemann integrable functions.

Let us consider the question about the conjugation between operator (2.1) and an operator with a constant coefficient. Moreover the conjugation will be assumed to be defined by means of an invertible bounded operator S of multiplication by a function $s(x)$. It turns out that under the additional assumptions on the smoothness of the coefficient a the operators aT_h and $\widetilde{a}T_h$ are conjugated in the sense mentioned and thus the operator $\widetilde{a}T_h$ with the constant coefficient can be considered as the *canonical form* of the operator aT_h with a varying coefficient.

The condition that an operator S defines a conjugation between two weighted shift operators, that is the equality $Sa_1T_hS^{-1} = a_2T_h$, is equivalent to the equality

$$a_1(x)\frac{s(x)}{s(x+h)} = a_2(x). \tag{2.6}$$

Functions a_1 and a_2 will be called *homologically equivalent* if there exists a function s satisfying the equality (2.6). In particular, if the coefficient a is homologically equivalent to a constant then the operator aT_h is conjugate to an operator with a constant coefficient.

Now we shall discuss a different way of formulation of the homological equivalence of a positive function a to a constant C, that is

$$a(x) = C\frac{s(x)}{s(x+h)}. \tag{2.7}$$

A representation of the function a in the form (2.7) is called *factorization with the shift.*

Let $a \in C(S^1)$ and $a(x) > 0$. Let us introduce the new functions $\varphi(x) = \ln a(x)$ and $d(x) = \ln s(x)$. Logarithmic equality (2.7) implies the so-called *homological equation*

$$d(x) - d(x+h) = \varphi(x) - \ln C, \tag{2.8}$$

where function d and constant C are unknown. We remark that operator S is invertible if and only if the function d is bounded. Equation (2.8) arises in a number of different problems and is well investigated [1, 2, 4–6]. We shall denote by $C_0(S^1)$ the subspace of $C(S^1)$ consisting of the functions satisfying the condition

$$\int \varphi(x)\, dx = 0. \tag{2.9}$$

The necessary condition for solvability of equation (2.8) is the condition $\varphi - \ln C \in C_0(S^1)$ determining the constant C in a unique way. Naturally this constant coincides with the spectral radius of the initial operator.

If one expands formally a solution to the homological equation into the Fourier series then

$$d(x) = \sum_{k \neq 0} \frac{\varphi_k}{1 - \exp i2\pi kh} \exp i2\pi kx,$$

where φ_k are the Fourier coefficients of the function φ. This series contains small denominators and may diverge.

Bounded solutions of equation (2.8) do exist iff the function $\varphi - \ln C$ belongs to a certain (difficult to describe) vector subspace M_h of $C_0(S^1)$. The general picture of solvability of equation (2.8) is rather complicated and depends on the relation between the arithmetic nature of the irrational number h and the properties of the function φ. The main known results are the following (see [1,2,4,5]).

Proposition 2.3. *If the function φ is a trigonometric polynomial, then the solution to (2.8) does exist and is a trigonometric polynomial as well.*

For each irrational number h there exists a continuous function $\varphi \in C_0(S^1)$ such that equation (2.8) has no bounded solution.

If a function φ is triply differentiable then for almost all h there exists a continuous solution to the homological equation.

From Proposition 2.3 it follows that the vector subspace M_h is not closed and is dense in $C_0(S^1)$ and $C_0^3(S^1) \subset M_h$ for almost all h.

Proposition 2.4. *For each function φ which is not a trigonometric polynomial there exists an irrational number h such that homological equation (2.8) has no bounded solutions.*

Proposition 2.5. *There exists a continuous positive function a such that the operator (2.1) is not conjugate as an operator in Hilbert space to a multiple to a unitary operator, in particular, it is not conjugate to any operator of the form (2.1) whose coefficient has constant modulus.*

Proof. Without loss of generality the consideration can be reduced to the case $r(A) = 1$, that is to the situation when $\ln a \in C_0(S^1)$. We remark that if $r(A) = 1$ and $|a(x)| \equiv \text{const}$ then $|a(x)| \equiv 1$ and this operator is unitary.

Let us suppose that each weighted shift operator with a continuous positive coefficient and satisfying the condition $r(A) = 1$, is conjugate to a certain unitary operator.

Since all the powers of a unitary operator are uniformly bounded it follows that for each of the operators A under consideration all the powers A^n, $n \in \mathbb{Z}$, are uniformly bounded. This is equivalent to the condition that for each function $\varphi \in C_0(S^1)$ the sequence

$$\varphi_n(x) := \sum_0^{n-1} \varphi(x + kh)$$

is bounded in the space $C_0(S^1)$. Since

$$\varphi_n(x) = S_n\varphi,$$

where

$$S_n := \sum_0^{n-1} T_h^k,$$

it follows from the Banach-Steinhaus theorem that the sequence of operators S_n is bounded with respect to the norm in the space $C_0(S^1)$. But one can easily verify that $\|S_n\| = n$ thus arriving at a contradiction. $\qquad\square$

If the coefficient a is a complex-valued function then the foregoing reasoning should be modified. It can happen that for a smooth coefficient a a continuous branch of the logarithm does not exist and then the function φ is discontinuous; the homological equation for such φ has no bounded solution.

The obstacle to the existence of a continuous branch of the logarithm for a function a is its Cauchy index $\chi(a)$ (recall that $\chi(a)$ is the increment of the argument of a under the circuit of the circle divided by 2π). A continuous branch of the logarithm does exist iff $\chi(a) = 0$. But if $\chi(a) \neq 0$ then a different factorization with the shift is possible:

$$a(x) = Cz^m(x)\frac{s(x)}{s(x + h)},$$

where $m = \chi(a)$, $z(x) = \exp i2\pi x$.

Indeed, the function $\varphi(x) = \ln[a(x)z^{-m}]$ is continuous and for the function $d(x) = \ln s(x)$ we obtain the homological equation (2.8). But now the function $\varphi(x)$ is complex valued and the number C may be complex. Thus if the corresponding solution to the homological equation does exist we have

$$SaT_hS^{-1} = Cz^nT_h$$

and therefore the operator aT_h is conjugate to the operator Cz^mT_h, that can be considered in this case as the *canonical form* of the operator aT_h with a varying coefficient. In particular if $\chi(a) = 0$ then the operator is conjugate to a shift operator with a complex constant coefficient. For this operator there is a base consisting of eigenfunctions but the eigenvalues differ from those of the shift operator. This example shows that the change of the coefficient for the geometric mean of its modulus preserves the spectrum but brings us to a nonconjugate operator even in good conditions. Therefore a 'more correct' averaging procedure in the case

$m = 0$ is the change of the coefficient for its geometric mean value (but not for the geometric mean value of its modulus).

If $\chi(a) \neq 0$ then in the situation when the homological equation is resolvable the operator is conjugate to a shift operator with a complex coefficient of a special form $Cz^m T_h$. This operator is a multiple of a unitary operator but is not conjugate to a weighted shift operator with a constant coefficient. This follows, in particular, from the fact that for $m \neq 0$ an operator $Cz^m T_h$ has no eigenvalues while an operator with a constant coefficient possesses a base consisting of eigenvectors.

Thus even for the simplest functional operators, namely weighted shift operators generated by an irrational rotation different settings of the problem are possible and the averaging picture is quite complicated. Let us summarize the results of consideration under the condition $a \in C(S^1)$, $|a(x)| > 0$.

I. *If we demand only the preservation of the spectrum under the change of the coefficient for a simpler one then the coefficient can be changed for a positive constant – the geometric mean of its modulus.*

II. *If we pose a question about reduction to the canonical form, that is the construction of a conjugate operator with the simplest coefficient by means of a multiplication operator S, then for sufficiently smooth coefficients and for almost all h an operator of the form $Cz^m T_h$ can serve as the canonical form of an operator of the form (2.1) and this canonical form depends on the Cauchy index $\chi(a)$.*

III. *The set of all operators of the form (2.1) with arbitrary continuous coefficients decomposes by means of relation (2.6) into a wide family of equivalence classes, this family is parameterized by elements of an infinitely dimensional space $C(S^1)/M_h$.*

The question: whether two operators from different classes are nonconjugate as operators in a Hilbert space is still open. Positive answer is equivalent to the following proposition: if two operators of form (2.1) are conjugate as operators in a Hilbert space then these operators are conjugate by means of an operator of multiplication.

Note in conclusion that the factorization with a shift method was used by N.K. Karapetiants in [6], where a particular case of Proposition 2.2 was obtained for almost all h under the additional conditions of positivity and smoothness of the coefficient.

3. Two-term functional operators, generated by an irrational rotation

Let us consider in $L_2(S^1)$ the operators of the form

$$Au(x) = a_0(x)\, u(x) + a_1(x)\, u(x + h), \quad x \in S^1, \tag{3.1}$$

where $S^1 = \mathbb{R}/\mathbb{Z}$ and h is an irrational number.

First we obtain the general description of the spectra of the operators under consideration.

Theorem 3.1. *Let the coefficients a_0, a_1 of an operator A of the form (3.1) are Riemann integrable functions, in particular continuous or piecewise continuous and let B_j be the spectrum of the operator of multiplication by the function $a_j(x)$. Then*

$$\sigma(A) = \Sigma_1 \bigcup \Sigma_2 \bigcup \Sigma_3,$$

where

$$\Sigma_1 = \{\lambda : \int \ln|a_0(x) - \lambda|dx = \int \ln|a_1(x)|dx\},$$

$$\Sigma_2 = \{\lambda : \lambda \in B_0, \int \ln|a_0(x) - \lambda|dx > \int \ln|a_1(x)|dx\},$$

$$\Sigma_3 = \begin{cases} \{\lambda : \int \ln|a_0(x) - \lambda|dx < \int \ln|a_1(x)|dx\}, & \text{if } 0 \in B_1, \\ \emptyset, & \text{if } 0 \notin B_1. \end{cases}$$

The theorem follows immediately from the next proposition, which contains the invertibility conditions for two-term functional operators.

Proposition 3.2. *Under the assumptions on the coefficients mentioned above operator (3.1) is invertible in the space $L_p(S^1)$ iff one of the following two conditions is satisfied:*

i) $0 \notin B_0$, $\int \ln|a_0(x)|dx > \int \ln|a_1(x)|dx$;

ii) $0 \notin B_1$, $\int \ln|a_0(x)|dx < \int \ln|a_1(x)|dx$.

In the case of continuous coefficients this proposition is proved in [3, see also 2], and the proof can be carried over to the case of Riemann integrable coefficients.

Let us analyze what information on the form of the spectrum of the operator can be derived from the theorem presented and on what properties of the coefficients does the spectrum depend.

First of all let us note that for the description of the spectrum among the properties of the function a_1 one uses the geometric mean of this coefficient and the condition $0 \in B_1$. If $0 \notin B_1$, then the change of the coefficient a_1 for the constant C equal to the geometric mean of this coefficient does not change the spectrum of the operator. Let us consider now the case when $0 \notin B_1$.

Here the dependence of the spectrum on the coefficient a_0 is more complicated. To start with let us consider two simple examples. Let us take as a_0 the following piecewise constant function: $a_0(x) = b_1$, if $0 \leq x \leq 1/2$, and $a_0(x) = b_2$, if $1/2 < x \leq 1$.

Then the condition $\lambda \in \Sigma_1$, that is the condition

$$\int \ln|a_0(x) - \lambda|dx = \int \ln|a_1(x)|dx,$$

has the form

$$\frac{1}{2}\ln(|b_1 - \lambda||b_2 - \lambda|) = \ln C,$$

which is equivalent to

$$|b_1 - \lambda||b_2 - \lambda| = C^2. \tag{3.2}$$

Equation (3.2) defines the known curve called the *Cassini oval* or *Cassini curve*.

The subset Σ_2 of the spectrum of the operator A, in this case has the form

$$\{\lambda : \lambda \in B_0, \int \ln|a_0(x) - \lambda|dx > \int \ln|a_1(x)|dx\}$$

and therefore is empty. Indeed for the function under consideration $B_0 = \{b_1, b_2\}$ and for $\lambda \in B_0$ we have

$$\int \ln|a_0(x) - \lambda|dx = -\infty < \int \ln|a_1(x)|dx.$$

Thus in the example discussed the spectrum coincides with the Cassini curve.

As a rule the Cassini curve is given by the following verbal description: the *Cassini curve* is the set of points of the plane such that the product of the distances from a point to two given points b_1 and b_2 is a given constant C^2. For large C this curve is really an oval and differs slightly from the circle of the radius C and having the center at b_1. But when C decreases the qualitative restructuring of the curve takes place: for $C = (1/2)|b_1 - b_2|$ the curve turns into a figure eight curve, and for $C < (1/2)|b_1 - b_2|$ the curve decomposes into two components, where the first one is an oval containing in its inferior the point b_1, and the second one is an oval containing in its inferior the point b_2.

It is easy to check that the points b_1 and b_2 can by restored by means of an oval in a unique way. Can we restore the coefficient a_0 by means of this curve? Clearly, no. For example, for a different function $a_0(x) = b_2$ for $0 \leq x \leq 1/2$ and $a_0(x) = b_1$, $1/2 < x \leq 1$ we get the same spectrum of the operator. Also any function a_0 having the value b_1 for a certain union of segments of measure $1/2$ and taking the value b_2 on the complement of this union generates the operator with the same spectrum.

Let us take as a_0 a more complicated piecewise constant function:

$$a_0(x) = b_k, \quad \text{if} \quad \frac{k-1}{N} \leq x < \frac{k}{N}.$$

Then the condition

$$\int \ln|a_0(x) - \lambda|dx = \int \ln|a_1(x)|dx$$

can be written in the form

$$\frac{1}{N}\sum_1^N \ln|b_k - \lambda| = \ln C.$$

which is equivalent to

$$\left(\prod_1^N |b_k - \lambda|\right) = C^N. \tag{3.3}$$

and this condition defines the spectrum of the operator.

We shall call the curve given by equation (3.3) the *Cassini curve of order N*. It is convenient to give its verbal description in somewhat different way: this curve is the set of points of the plane such that the geometric mean of the distances from N given points b_k is a given constant C.

According to Theorem 3.1 in the general case the first component of the spectrum is a level line of the function

$$F(\lambda) = \int \ln |a_0(x) - \lambda| dx,$$

that is the curve given by the equation $F(\lambda) = \ln C$.

The examples considered show that the function a_0 is not defined in a unique way by the function $F(\lambda)$. So what is the invariant? What properties of a_0 define $F(\lambda)$? The answer can be obtained in the following way. Let us define on the plane a measure ρ generated by the function a_0: for each Borel set E we set

$$\rho(E) = \mu(a_0^{-1}(E)),$$

where μ is the Lebesgue measure on S^1.

Clearly the support of the measure constructed coincides with the set B_0, and the measure itself can be interpreted as a certain mass distribution on B_0. Since $\rho(B_0) = 1$, it follows that it is a probability measure. If a_0 is a real valued function then the support of the measure belongs to a certain segment of the real line and the measure ρ is given by the distribution function $g(t) = \mu(\{x : a_0(x) < t\})$. Moreover it is clear that different functions a_0 can generate the same measures ρ.

Further, implementing the change of variables $t = a_0(x)$, we have

$$F(\lambda) = \int_0^1 \ln |a_0(x) - \lambda| dx = \int_{B_0} \ln |t - \lambda| d\rho(t), \tag{3.4}$$

which implies that the function $F(\lambda)$ is determined precisely by the associated measure ρ.

Let us note that the function $exp[F(\lambda)]$ is the geometric mean with respect to the measure ρ of the distances from the point λ to the points of the set B_0 ($=$ the support of the measure ρ).

We shall say that Λ is a *generalized Cassini curve generated by a measure ρ* if Λ is the set of points λ such that the geometric mean (with respect to the measure ρ) of the distances from λ to the points of the set B_0 is a given constant.

Thus the component Σ_1 of the spectrum of A is a generalized Cassini curve generated by the measure ρ.

It is a note to the point that the coincidence of the associated measures for two given functions is equivalent to the fact that the operators of multiplication by these functions are conjugate. The conjugacy can be given by means of the shift operators generated by (discontinuous) transformation of the circle preserving the Lebesgue measure. Thus the conjugacy of the first coefficients of two operators of the form (3.1) (under the condition that the second coefficients are equal) brings us to the coincidence of the spectrum.

Depending on the form of the measure ρ there may arise qualitatively different situations. If as in examples considered there exist points of finite measure ρ, then $F(\lambda) = -\infty$ at these points. Thus for arbitrary small positive value of the geometric mean C of the function a_1 the level line $F(\lambda) = \ln C$ is nonempty and the spectrum differs from the set B_0.

In another extreme case when the measure ρ is absolutely continuous with respect to the Lebesgue measure on B_0 and the distribution function is continuous: the function $F(\lambda)$ has the minimum $m_0 > -\infty$. Therefore for $C < m_0$ the level line is empty. Thus in the cases mentioned the spectrum of the operator coincides with the set B_0 and does not depend on C. If $C > m_0$ then the level line is nonempty and the spectrum depends on C.

These remarks give us an opportunity to paraphrase Theorem 3.1 and describe the dependence of the spectrum on the coefficient in more detail.

Theorem 3.3. *Let us suppose that the coefficients of the operator A of the form (3.1) be piecewise continuous and $0 \notin B_1$. Denote $\ln C = \int \ln |a_1(x)| dx$,*

$$F(\lambda) = \int_{B_0} \ln |t - \lambda| d\rho(t),$$

where ρ is the measure generated by the coefficient a_0.

The spectrum of the operator A is the union of a generalized Cassini curve $\{\lambda : F(\lambda) = \ln C\}$ and the set $B_0 \bigcap \{\lambda : F(\lambda) > \ln C\}$, moreover one of the sets mentioned can be empty.

In particular for sufficiently large C the set $B_0 \bigcap \{\lambda : F(\lambda) > \ln C\}$ is empty and the spectrum is a smooth curve.

Let us consider the inverse problem: what information on the coefficients can be obtained by means of the spectrum of the operator? In particular, can we restore the measure ρ by means of the spectrum? The answer to this question gives us a possibility to describe the changes of the coefficients of a functional operator under which the spectrum is preserved. It appears that for sufficiently large C and a real valued function a_0 the answer to the last question is positive.

Theorem 3.4. *Let the coefficients of operator (3.1) are piecewise continuous, $0 \notin B_1$ and the coefficient a_0 takes only the real values. If the spectrum of the operator A is a smooth curve then there exists the only one measure ρ and the only one value C, which generate this spectrum.*

Proof. The expression for the function $F(\lambda)$ can be written in the form of convolution $F = \ln |x| * \rho$. It is known that the function

$$E_2(x_1, x_2) = -\frac{1}{2\pi} \ln |x|$$

is the fundamental solution to the Laplace operator on the plane, and the function $F(\lambda)$, being the convolution with the fundamental solution, is the solution to the Poisson equation

$$\Delta F = -2\pi \rho,$$

where the differentiation is understood in the sense of generalized functions [8]. Thus by means of the function F the measure ρ is restored in a unique way.

Note that in the case considered the result of convolution in the sense of generalized functions is a regular generalized function, therefore in what follows we consider $F(\lambda)$ as an ordinary locally integrable function. This function may have singularities only at the points of the support of the measure ρ, and out of the support of ρ it satisfies the Laplace equation and is a real analytic and is defined uniquely by its values in the upper and lower half-planes.

Let us prove that the function F is restored uniquely by means of the spectrum of the operator: by the only one level line of the function mentioned.

According to Theorem 3.1 the spectrum of the operator in the general case is the union of a generalized Cassini curve and the set $B_0 \cap \{\lambda : F(\lambda) \geq C\}$. If the latter set is nonempty then, since $B_0 \subset \mathbb{R}$, it follows that B_0 is a segment or a union of a number of segments and so the spectrum is not a smooth curve. The conditions of the theorem imply that the support of the measure ρ lies in the generalized Cassini curve and the spectrum is the generalized Cassini curve.

Let for two measures ρ_1 and ρ_2 the conditions of the theorem be satisfied, then the corresponding spectra are the generalized Cassini curves and the supports of the measures lie in the interiors of these curves. Let us suppose that the spectra of the corresponding operators coincide. The (common) spectrum can be given in two ways:

$$\sigma(A) = \{\lambda : F_1(\lambda) = \ln C_1\} = \{\lambda : F_2(\lambda) = \ln C_2\}.$$

Let D be the external domain with respect to this curve. Consider the difference $F_1 - F_2$ in this domain. The function $F_1 - F_2$ is continuous in the closure of D, satisfies the Laplace equation in the domain D and takes the constant value $\ln C_1 - \ln C_2$ on the boundary. Thus one can use the uniqueness of the solution to the external Dirichlet problem of the Laplace equation. Recall that the uniqueness takes place in the class of bounded at infinity functions [8].

Note that each of the functions $F(\lambda)$ under consideration has the asymptotics of the form [8]

$$F(\lambda) = \ln |\lambda| + O\left(\frac{1}{\lambda}\right). \tag{3.5}$$

Indeed. If $s \in \operatorname{supp}\rho$ we have that $|s| \leq d$ and for large values of $|\lambda|$ the following relations are satisfied

$$\left| \ln |\lambda - s| - \ln |\lambda| \right| = \ln \frac{|\lambda - s|}{|\lambda|} \leq \ln \left(1 + \frac{d}{|\lambda|}\right) \leq \frac{d}{|\lambda|},$$

which implies (3.5).

The asymptotics obtained implies that the difference $F_1 - F_2$ tends to 0 at infinity and in particular it is bounded. By the uniqueness of the external Dirichlet problem for harmonic functions we have that the solution (in the class of bounded functions) to the external Dirichlet problem with the constant boundary value is the only one constant:

$$F_1(\lambda) - F_2(\lambda) = \ln C_1 - \ln C_2 \quad \text{for } \lambda \in D.$$

But this function should tend to zero at infinity. Therefore $F_1(\lambda) - F_2(\lambda) \equiv 0$ in D. So by the real analyticity the functions F_1 and F_2 coincide in the domain of analyticity and in particular they coincide in the upper and lower half planes. This means that there exists the only one function of the form mentioned which takes a certain constant value on a given generalized Cassini curve. In particular this constant value can be restored in a unique way as well. $\square$

Remark 3.5. The statement of Theorem 3.4 can be carried over to the coefficients of a more general form but it needs a certain additional condition. The thing is that in the case of a complex valued function a_0 the support of the measure ρ can be a smooth curve and the form of the spectrum does not tell us whether the spectrum is a generalized Cassini curve or is the support of the measure ρ. Moreover the next example shows that even in the case when the spectrum is a generalized Cassini curve the measure ρ cannot be restored by means of the spectrum in a unique way.

Example. Let $a_0(x) = \exp[i2\pi x]$. Then the measure ρ is a normalized measure on the circle. The function $F(\lambda)$ is invariant with respect to the rotations around 0. Thus for large C the generalized Cassini curve ($=$ the spectrum) is a circle with the center at 0. But this circle is also the spectrum in the case of the coefficient $a_0(x) \equiv 0$ which corresponds to a different measure, namely the measure concentrated at the point 0.

If we return to the problem of the change of coefficients for simpler ones with the preservation of the spectrum for the operators of the form (3.1) then we have only a small arbitrariness – one can allow only transformations of the coefficient a_0 that preserve the associated measure. The best result that can be achieved by means of such transformations is a monotone function. This means that for the operators considered the averaging effect appears for the coefficients of the shift operators and there is practically no averaging for the coefficients of the identity operator.

The question on the conjugation of operators of the form (3.1) with coinciding spectra is more complicated and is still open.

References

[1] D.V. Anosov, *The additive functional homology equation that is connected with an ergodic rotation of the circle.* Izv.Akad. Nauk SSSR, Ser. Mat. **37** (1973), 1259–1274 (Russian).

[2] A.B. Antonevich, *Linear Functional Equations. Operator Approach.* Operator Theory: Advances and Applications **83**. Birkhäuser Verlag, Basel 1996. Russian original: University Press, Minsk, 1988.

[3] A.B. Antonevich and V.B. Ryvkin, *The normal solvability of the problem of the periodic solutions of a linear differential equations with deviating argument*, Differentsial'nye Uravneniya. **10** (1974), 1347–1353 (Russian).

[4] V.I. Arnol'd, *Supplementary Chapters of the Theory of Ordinary Differential Equations*. Nauka, Moscow, 1978 (Russian).

[5] A.Ya. Gordon, *On sufficient condition for the nonsolvability of the additive functional homological equation associated with an ergodic rotation of the circle*. Funktsional. Anal. i prilozhen. **9** (1975), no. 4, 71–72 (Russian).

[6] N.K. Karapetyants, *On a class of discrete convolution operators with oscillating coefficients*. DAN SSSR, **216** (1974), 28–31 (Russian).

[7] I.B. Simonenko, *The averaging method in the theory of nonlinear parabolic equations with application to the problems of hydrodynamic stability*. Rostov University Press, Rostov-na Dony, 1989 (Russian).

[8] V.S. Vladimirov, *The equations of Mathematical Physics*. Nauka, Moscow, 1967 (Russian).

A.B. Antonevich
Department of Mathematics and Mechanics
Belarussian State University
Av. Niezavisimosti 4
220050, Minsk, Belarus

and

Institute of Mathematics
University of Bialystok, Poland
e-mail: `antonevich@bsu.by`

Operator Theory:
Advances and Applications, Vol. 170, 43–51

On the Verification of Linear Equations and the Identification of the Toeplitz-plus-Hankel Structure

Albrecht Böttcher and David Wenzel

Dedicated to Igor Borisovich Simonenko on his 70th birthday

Abstract. Testing whether a given matrix is a Toeplitz-plus-Hankel matrix amounts to the verification of a system of linear equations for the matrix entries. If the matrix dimension is large, we are forced to work with the computer and hence cannot check whether something is exactly zero. We provide bounds such that if a test quantity is smaller than the bound, then the system of linear equations may be accepted to be valid and the probability for erroneously accepting the validity of the system is smaller than a prescribed value.

Mathematics Subject Classification (2000). Primary 47B35; Secondary 15A24, 65F35.

Keywords. Displacement matrix, Toeplitz-plus-Hankel, computer verification.

1. Introduction

A real Toeplitz-plus-Hankel matrix (T+H matrix for short) is a matrix of the form $(t_{i-j} + h_{i+j})_{i,j=1}^{n}$ with real numbers t_k and h_k. In contrast to the pure Toeplitz or pure Hankel structures, it is not immediately seen whether a given $n \times n$ matrix X is T+H. However, Heinig and Rost [4] observed that X is T+H if and only if the central $(n-2) \times (n-2)$ submatrix of $W_n X - X W_n$ is zero, where W_n is the $n \times n$ matrix with ones on the first superdiagonal and the first subdiagonal and with zeros elsewhere. Thus, letting $S_n := \mathrm{diag}\,(0, 1, \ldots, 1, 0)$, we obtain that X is T+H if and only if

$$D_n X := S_n(W_n X - X W_n)S_n = 0. \tag{1}$$

If n is large, we may be caused to check the equation $D_n X = 0$ using the computer. As a computer cannot test whether something is exactly zero, we must ask whether

$D_n X$ is small in some sense. But is it justified to assume that X is approximately a T+H matrix if $D_n X$ is small?

In [1] it was shown that the answer to this question is negative theoretically but in the affirmative practically and optimistically. To be more precise, denote by $M_n(\mathbf{R})$ the linear space of all real $n \times n$ matrices and think of D_n as a linear operator of $M_n(\mathbf{R})$ into itself. The T+H matrices are just the matrices in the null space $\operatorname{Ker} D_n$ of D_n. Let $|\cdot|$ be the ℓ^2 norm (= Frobenius norm = Hilbert-Schmidt norm) on $M_n(\mathbf{R})$ and put

$$\operatorname{dist}(X, \operatorname{Ker} D_n) = \min_{Y \in \operatorname{Ker} D_n} |X - Y|.$$

A result of [1] says that there are constants $0 < C_1 < C_2 < \infty$ such that*

$$C_1 \, n^2 \leq \sup_{X \notin \operatorname{Ker} D_n} \frac{\operatorname{dist}(X, \operatorname{Ker} D_n)}{|D_n X|} \leq C_2 \, n^2.$$

Thus, although $|D_n X|$ is small, the distance of X to the set of all T+H matrices may be large. This is what we mean by saying that the answer to the above question is theoretically negative. However, another result of [1] states that if X is randomly drawn from the unit sphere $\mathbf{S}^{n^2-1}$ of $M_n(\mathbf{R})$ with the uniform distribution, then

$$P\left(\frac{\operatorname{dist}(X, \operatorname{Ker} D_n)}{|D_n X|} > 10\right) < \frac{79}{n^2}$$

for $n \geq 10$, where $P(E)$ here and in the following denotes the probability of the event E. This makes precise our statement that practically and optimistically the answer to the question is yes.

In this paper we present an alternative probabilistic approach to the question raised above. As in [1] and [3], our starting point is an observation of [2], according to which the values of the random variable

$$\xi = \frac{|D_n X|^2}{|X|^2}$$

are sharply concentrated around its expected value $E\xi$. Incidentally, this is true for every linear operator D_n on $M_n(\mathbf{R})$, and not only for the D_n given by (1). Given any prescribed probability P^*, for instance, $P^* = 0.05$, we determine an $\varepsilon \in (0, 1)$ such that

$$P\left(\xi < (1 - \varepsilon)E\xi\right) \leq P^*.$$

If D_n is defined by (1), we have $E\xi = 4(n - 2)^2/n^2$. The conclusion is as follows. We check the inequality $|D_n X|^2/|X|^2 < 4(1 - \varepsilon)(n - 2)^2/n^2$ with the aid of the computer. If the inequality is satisfied, we accept X to be T+H. The probability for accepting X as T+H although it is not T+H is then at most P^*.

---

*The proof of this result makes use of quarter-plane Toeplitz operators and is thus related to an area that was pioneered by I.B. Simonenko in the 1960s.

2. General homogeneous linear equations

Fix a real matrix $A = (a_{ij})_{i,j=1}^N$ and put

$$|A|^2 = \sum_{i,j=1}^N a_{ij}^2, \quad \lceil A \rceil^4 = \sum_{j=1}^N \left(\sum_{i=1}^N a_{ij}^2 \right)^2.$$

The matrix $A^\top$ is the transpose of A, $A^\top = (a_{ji})_{i,j=1}^N$. We think of A as a linear operator on $\mathbf{R}^N$. The ℓ^2 norm on $\mathbf{R}^N$ will also be denoted by $|\cdot|$.

Let $x = (x_1, \ldots, x_N) \in \mathbf{R}^N$ be a random vector. We assume that $x_1, \ldots, x_N$ are identically distributed and that x_1 is symmetric about zero. This implies in particular that $E(x_j) = 0$ for all j and $E(x_j x_k) = 0$ for all $j \neq k$. We put

$$c_2 = E\left(\frac{x_j^2}{|x|^2}\right), \quad c_{22} = E\left(\frac{x_j^2 x_k^2}{|x|^4}\right), \quad c_4 = E\left(\frac{x_j^4}{|x|^4}\right),$$

where $j \neq k$ in the definition of c_{22}, and we suppose that c_2, c_{22}, c_4 are finite. We do not assume that the $x_1, \ldots, x_N$ are independent. Examples of admissible distributions are the uniform distributions on the unit sphere $\mathbf{S}^{N-1} := \{x \in \mathbf{R}^N : |x| = 1\}$ or the unit ball $\mathbf{B}_N := \{x \in \mathbf{R}^N : |x| \leq 1\}$ and the case where $x_1, \ldots, x_N$ are subject to the normal distribution $N(0, \sigma)$.

Theorem 2.1. *For the random variable $\xi := |Ax|^2/|x|^2$ we have*

$$E\xi = |A|^2 c_2, \quad E(\xi^2) = \lceil A \rceil^4 (c_4 - 3c_{22}) + (|A|^4 + 2|A^\top A|^2)c_{22}.$$

Proof. Clearly,

$$E\xi = E\left(\frac{|Ax|^2}{|x|^2}\right) = E\left(\frac{1}{|x|^2} \sum_k \left(\sum_i a_{ki} x_i\right)^2\right) = E\left(\frac{1}{|x|^2} \sum_{k,i,j} a_{ki} x_i a_{kj} x_j\right),$$

and since $E(x_i x_j) = 0$ for $i \neq j$, it follows that

$$E\xi = \sum_{k,i} a_{ki}^2 E\left(\frac{x_i^2}{|x|^2}\right) = |A|^2 c_2,$$

as asserted. In the same vein,

$$E(\xi^2) = E\left(\frac{|Ax|^4}{|x|^4}\right) = E\left(\frac{1}{|x|^4} \sum a_{ij} x_j a_{ik} x_k a_{\ell p} x_p a_{\ell q} x_q\right),$$

the sum over the six indices i, j, k, ℓ, p, q. Again we only need to consider even powers, that is, x_j^4 and $x_j^2 x_k^2$ ($j \neq k$). We are therefore led to four cases:

$$j = k = p = q \quad \text{gives} \quad \sum a_{ij}^2 a_{\ell j}^2 c_4 = \sum_j \left(\sum_i a_{ij}^2\right)^2 c_4 = \lceil A \rceil^4 c_4,$$

$$j = k \neq p = q \quad \text{gives} \quad \sum a_{ij}^2 a_{\ell p}^2 c_{22} = |A|^4 c_{22},$$

$$j = p \neq k = q \quad \text{gives} \quad \sum a_{ij} a_{ik} a_{\ell j} a_{\ell k} c_{22} = |A^\top A|^2 c_{22},$$

$$j = q \neq k = p \quad \text{gives} \quad \sum a_{ij} a_{ik} a_{\ell k} a_{\ell j} c_{22} = |A^\top A|^2 c_{22}.$$

In the second and third cases we have to exclude the case $k = p$ and in the last case we must exclude $k = q$. This amounts to subtracting the term $\lceil A \rceil^4 c_{22}$ in each of these three cases. Thus,

$$E(\xi^2) = \lceil A \rceil^4 c_4 + (|A|^4 - \lceil A \rceil^4)c_{22} + 2(|A^\top A|^2 - \lceil A \rceil^4)c_{22},$$

which is equivalent to the asserted formula for $E(\xi^2)$. $\qquad\square$

Corollary 2.2. *Let $P^* \in (0, 1)$. If the number $\varepsilon > 0$ defined by*

$$\varepsilon^2 = \frac{\lceil A \rceil^4(c_4 - 3c_{22}) + |A|^4(c_{22} - c_2^2) + 2|A^\top A|^2 c_{22}}{P^*|A|^4 c_2^2} \tag{2}$$

is strictly smaller than 1, then

$$P\Big(\xi < (1 - \varepsilon)|A|^2 c_2\Big) \le P^*.$$

Proof. This follows from Theorem 2.1 along with Chebyshev's inequality:

$$P\Big(\xi < (1 - \varepsilon)E\xi\Big) \le P\Big(|\xi - E\xi| > \varepsilon E\xi\Big) \le \frac{(E\xi)^2 - E(\xi^2)}{\varepsilon^2 (E\xi)^2} = P^*. \qquad\square$$

Theorem 2.3. *If $x \in \mathbf{R}^N$ is drawn from the unit sphere or the unit ball with the uniform distribution or if the components of x are $N(0, \sigma)$ distributed, then*

$$c_2 = \frac{1}{N}, \quad c_{22} = \frac{1}{N(N + 2)}, \quad c_4 = \frac{3}{N(N + 2)}.$$

Proof. The expected values under consideration are integrals of the form

$$\int_G \frac{f(x_1^2 + \cdots + x_N^2)}{(x_1^2 + \cdots + x_N^2)^\lambda} x_1^{p_1 - 1} \ldots x_N^{p_N - 1}\, dx_1 \ldots dx_N.$$

For the uniform distribution on the unit ball $\mathbf{B}_N$, Liouville's formula

$$\int_{\mathbf{B}_N} \frac{x_1^{p_1 - 1} \ldots x_N^{p_N - 1}}{(x_1^2 + \cdots + x_N^2)^\lambda}\, dx_1 \ldots dx_N = \frac{2}{p_1 + \cdots + p_n - 2\lambda} \frac{\Gamma\left(\frac{p_1}{2}\right) \ldots \Gamma\left(\frac{p_N}{2}\right)}{\Gamma\left(\frac{p_1 + \cdots + p_N}{2}\right)},$$

for which see, e.g., [5, No. 676, 8(b)], yields the result. The case of the unit sphere can be reduced to the case of the unit ball by virtue of the identity

$$\frac{1}{|\mathbf{S}^{N-1}|}\int_{\mathbf{S}^{N-1}} x_1^{p_1} \ldots x_N^{p_N}\, d\sigma = \frac{1}{|\mathbf{B}_N|}\int_{\mathbf{B}_N} \frac{x_1^{p_1} \ldots x_N^{p_N}}{(x_1^2 + \cdots + x_N^2)^{(p_1 + \cdots + p_N)/2}}\, dx_1 \ldots dx_N,$$

which follows simply by introducing polar coordinates in the right integral. In the case of the $N(0, \sigma)$ distribution, we have to compute integrals of the form

$$\int_{\mathbf{R}^N} \frac{e^{-(x_1^2 + \cdots + x_N^2)/(2\sigma^2)}}{(x_1^2 + \cdots + x_n^2)^\lambda} x_1^{p_1 - 1} \ldots x_N^{p_N - 1}\, dx_1 \ldots dx_N. \tag{3}$$

Integral (3) is the limit of the integral over $r\mathbf{B}_N$ as $r \to \infty$. The integral over $r\mathbf{B}_N$ can be transformed into an integral over $\mathbf{B}_N$ by the substitution $x_j = rz_j$.

This integral can in turn be tackled by a more general formula of Liouville (see, for example, [5, No. 676, 8]):

$$\int_{\mathbf{B}_N} \varphi(z_1^2 + \cdots + z_N^2)\, z_1^{p_1-1} \ldots z_N^{p_N-1}\, dz_1 \ldots dz_N$$

$$= \frac{\Gamma\left(\frac{p_1}{2}\right)\ldots\Gamma\left(\frac{p_N}{2}\right)}{\Gamma\left(\frac{p_1+\cdots+p_N}{2}\right)} \int_0^1 \varphi(u)\, u^{(p_1+\cdots+p_N)/2-1}\, du.$$

Substituting $u = v/r^2$ one eventually gets that (3) equals

$$\frac{\Gamma\left(\frac{p_1}{2}\right)\ldots\Gamma\left(\frac{p_N}{2}\right)}{\Gamma\left(\frac{p_1+\cdots+p_N}{2}\right)} \lim_{r\to\infty} \int_0^{r^2} e^{-v/(2\sigma^2)}\, v^{(p_1+\cdots+p_N)/2-\lambda-1}\, dv$$

$$= \frac{\Gamma\left(\frac{p_1}{2}\right)\ldots\Gamma\left(\frac{p_N}{2}\right)}{\Gamma\left(\frac{p_1+\cdots+p_N}{2}\right)} \frac{\Gamma\left(\frac{p_1+\cdots+p_N}{2}-\lambda\right)}{(2\sigma^2)^{\lambda-(p_1+\cdots+p_N)/2}},$$

which can be used to prove the theorem for the $N(0,\sigma)$ distribution. It turns out that the final result is independent of σ. $\qquad\square$

3. The Toeplitz-plus-Hankel structure

We identify $M_n(\mathbf{R})$ with $\mathbf{R}^{n^2} =: \mathbf{R}^N$ by row stacking and abbreviate W_n and S_n to W and S, respectively. In this way the operator D_n defined by (1) becomes the $N \times N$ matrix

$$A := SW \otimes S^\top - S \otimes (WS)^\top = SW \otimes S - S \otimes SW$$

where $\otimes$ denotes the Kronecker product. The matrix A is block tridiagonal. The blocks on the main diagonal are $0, -SW, \ldots, -SW, 0$, the blocks on the superdiagonal are $0, S, \ldots, S$, and those on the subdiagonal are $S, \ldots, S, 0$. For example, if $n = 5$ then

$$A = \begin{pmatrix} 0 & 0 & 0 & 0 & 0 \\ S & -SW & S & 0 & 0 \\ 0 & S & -SW & S & 0 \\ 0 & 0 & S & -SW & S \\ 0 & 0 & 0 & 0 & 0 \end{pmatrix}.$$

It follows that

$$|A|^2 = 2(n-2)\cdot(n-2) + (n-2)\cdot 2(n-2) = 4(n-2)^2. \tag{4}$$

The matrix $A^\top A$ is block pentadiagonal. The second superdiagonal and the second subdiagonal are $\mathrm{diag}\left(S^2, \ldots, S^2\right)$, the first superdiagonal and the first subdiagonal are

$$\mathrm{diag}\left(-S^2W, -WS^2 - S^2W, \ldots, -WS^2 - S^2W, -WS^2\right),$$
$$\mathrm{diag}\left(-WS^2, -S^2W - WS^2, \ldots, -S^2W - WS^2, -S^2W\right),$$

respectively, and the main diagonal is

$$\mathrm{diag}\left(S^2, WS^2W + S^2, WS^2W + 2S^2, \ldots, WS^2W + 2S^2, WS^2W + S^2, S^2\right).$$

Thus, for $n = 5$, the matrix $A^\top A$ equals

$$\begin{pmatrix} S^2 & -S^2W & S^2 & 0 & 0 \\ -WS^2 & WS^2W + S^2 & -WS^2 - S^2W & S^2 & 0 \\ S^2 & -S^2W - WS^2 & WS^2W + 2S^2 & -WS^2 - S^2W & S^2 \\ 0 & S^2 & -S^2W - WS^2 & WS^2W + S^2 & -WS^2 \\ 0 & 0 & S^2 & -S^2W & S^2 \end{pmatrix}.$$

We obtain

$$\begin{aligned} |A^\top A|^2 &= [2(n-2) + 2] \cdot (n-2) + 2(n-3) \cdot [2(n-3) \cdot 2^2 + 4] + 4 \cdot 2(n-2) \\ &\quad + (n-4) \cdot [(n-2) \cdot 4^2 + 2(n-4)] + 2 \cdot [(n-2) \cdot 3^2 + 2(n-4)] \\ &= 36n^2 - 176n + 216. \end{aligned} \tag{5}$$

If $X \in M_n(\mathbf{R}) \cong \mathbf{R}^N$ has one of the three distributions of Theorem 2.3, then $c_4 = 3c_{22}$ and hence the term with $\lceil A \rceil^4$ in (2) vanishes. Combining (2), (4), (5), and Theorem 2.3 we arrive at the equality

$$\varepsilon^2 = \frac{5n^4 - 12n^3 - 42n^2 + 128n - 64}{P^* \cdot 2(n^2 + 2)(n-2)^4}. \tag{6}$$

Since $|A|^2 c_2 = 4(n-2)^2/n^2$, we deduce from Corollary 2.2 that the probability for accepting a matrix X with

$$\frac{|D_n X|^2}{|X|^2} < (1 - \varepsilon) \frac{4(n-2)^2}{n^2} \tag{7}$$

as a T+H matrix although it is not a T+H matrix is at most P^*.

To be more concrete, suppose that $n \geq 20$. Then the right-hand side of (6) is at most $0.0082/P^*$. Choosing $P_1^* = 0.05$ and $P_2^* = 0.01$ we get $\varepsilon_1 \leq 0.41$ and $\varepsilon_2 \leq 0.91$, respectively, and the right-hand side of (7) is at least

$$(1 - \varepsilon_1) \frac{4(20 - 2)^2}{20^2} \geq 1.91 \quad \text{and} \quad (1 - \varepsilon_2) \frac{4(20 - 2)^2}{20^2} \geq 0.29$$

in these two cases. Consequently, if X is a matrix of dimension $n \geq 20$ and $|D_n X|^2/|X|^2 < 1.91$ (resp. 0.29), we may accept X to be T+H. The probability for accepting such a matrix as T+H although it is not T+H does not exceed 5% (resp. 1%).

To state it in another way, suppose we randomly take 100 matrices X. Then at most about 5 (resp. 1) of them will satisfy the inequality $|D_n X|^2/|X|^2 < 1.91$ (resp. 0.29). We accept these matrices to be T+H. Although none of them may actually be T+H, we erroneously accepted a matrix as T+H in at most about 5% of all cases for the bound 1.91 and in at most about 1% of all cases for the bound 0.29.

4. General inhomogeneous linear equations

Let A and x be as in Section 2, but consider now the equation $Ax = b$ with a given right-hand side b. In addition to the numbers c_2, c_{22}, c_4 introduced in Section 2, we need the constants

$$c_0 = E\left(\frac{1}{|x|^2}\right), \quad c_0^* = E\left(\frac{1}{|x|^4}\right), \quad c_2^* = E\left(\frac{x_j^2}{|x|^4}\right).$$

Theorem 4.1. *For the random variable $\eta := |Ax - b|^2/|x|^2$ we have*

$$\begin{aligned}
E\eta &= |A|^2 c_2 + |b|^2 c_0, \\
E(\eta^2) &= \lceil A\rceil^4(c_4 - 3c_{22}) + (|A|^4 + 2|A^\top A|^2)c_{22} \\
&\quad + (2|A|^2|b|^2 + 4|A^\top b|^2)c_2^* + |b|^4 c_0^*.
\end{aligned}$$

Proof. Let $(\cdot, \cdot)$ be the usual scalar product in $\mathbf{R}^N$. Then

$$\begin{aligned}
E\eta &= E\left(\frac{1}{|x|^2}(Ax - b, Ax - b)\right) = E\left(\frac{1}{|x|^2}\left(|Ax|^2 - 2(Ax, b) + |b|^2\right)\right) \\
&= E\left(\frac{|Ax|^2}{|x|^2}\right) + |b|^2 E\left(\frac{1}{|x|^2}\right),
\end{aligned}$$

since $E((Ax, b)/|x|^2) = 0$ by our symmetry requirement. From Theorem 2.1 we therefore get $E\eta = |A|^2 c_2 + |b|^2 c_0$. Analogously,

$$\begin{aligned}
E(\eta^2) &= E\left(\frac{1}{|x|^4}\left(|Ax|^2 - 2(Ax, b) + |b|^2\right)^2\right) \\
&= E\Bigg(\frac{1}{|x|^4}\Big(|Ax|^4 - 4|Ax|^2(Ax, b) + 2|Ax|^2|b|^2 \\
&\qquad\qquad + 4(Ax, b)^2 - 4(Ax, b)|b|^2 + |b|^4\Big)\Bigg) \\
&= E\left(\frac{1}{|x|^4}\left(|Ax|^4 + 2|Ax|^2|b|^2 + 4(Ax, b)^2 + |b|^4\right)\right).
\end{aligned}$$

Theorem 2.1 gives us $E(|Ax|^4/|x|^4)$. The value $E(|Ax|^2/|x|^4)$ can be computed in the same way as $E(|Ax|^2/|x|^2)$ in the proof of Theorem 2.1, the only difference being that the c_2 now becomes c_2^*. Further,

$$\begin{aligned}
E\left(\frac{(Ax, b)^2}{|x|^4}\right) &= E\left(\frac{1}{|x|^4}\sum a_{ij}x_j b_i a_{k\ell}x_\ell b_k\right) \\
&= E\left(\frac{1}{|x|^4}\sum a_{ij}b_i a_{kj}b_k x_j^2\right) = \sum_j\left(\sum_i a_{ij}b_i\right)^2 c_2^* = |A^\top b|^2 c_2^*.
\end{aligned}$$

Finally, $E(|b|^4/|x|^4) = |b|^4 c_0^*$. Putting the things together we arrive at the asserted formula for $E(\eta^2)$. $\qquad\square$

Corollary 4.2. *Let $P^* \in (0,1)$ and define $\varepsilon > 0$ by*

$$\varepsilon^2 = \frac{1}{P^*(|A|^2 c_2 + |b|^2 c_0)^2}\left(\lceil A \rceil^4 (c_4 - 3c_{22}) + |A|^4 (c_{22} - c_2^2) + 2|A^\top A|^2 c_{22}\right.$$

$$\left. + 4|A^\top b|^2 c_2^* + 2|A|^2 |b|^2 (c_2^* - c_0 c_2) + |b|^4 (c_0^* - c_0^2)\right).$$

If $\varepsilon < 1$, then

$$P\Big(\eta < (1 - \varepsilon)(|A|^2 c_2 + |b|^2 c_0)\Big) \le P^*. \tag{8}$$

Proof. Proceed as in the proof of Corollary 2.2. $\square$

Theorem 4.3. *If x is drawn from $\mathbf{S}^{N-1}$ with the uniform distribution, then*

$$c_0 = 1, \quad c_0^* = 1, \quad c_2^* = \frac{1}{N};$$

if x is taken from $\mathbf{B}_N$ with the uniform distribution, then

$$c_0 = \frac{N}{N-2}, \quad c_0^* = \frac{N}{N-4}, \quad c_2^* = \frac{1}{N-2};$$

if the components of x are $N(0,\sigma)$ distributed, then

$$c_0 = \frac{1}{\sigma^2(N-2)}, \quad c_0^* = \frac{1}{\sigma^4(N-2)(N-4)}, \quad c_2^* = \frac{1}{\sigma^2 N(N-2)}.$$

Proof. The integrals one has to compute are all of the same form as those in the proof of Theorem 2.3. $\square$

5. Toeplitz-plus-Hankel matrices again

Let $X_0 \in M_n(\mathbf{R})$ be a fixed and known matrix and suppose $X \in M_n(\mathbf{R})$ is a perturbation to X_0. We want to know whether $X_0 + X$ is T+H. Consequently, we consider the equation $D_n(X_0 + X) = 0$, which reads $D_n X = -D_n X_0$. After identifying $M_n(\mathbf{R})$ and $\mathbf{R}^{n^2} =: \mathbf{R}^N$ we therefore arrive at an inhomogeneous linear system $Ax = b$ where A is as in Section 4 and b results from $-D_n X_0$ by row stacking.

Suppose X is taken at random and the vector x that emerges from X by row stacking has one of the three distributions listed in Theorem 4.3. We then can use Corollary 4.2 and Theorem 4.3 to associate with a given $P^* \in (0,1)$ a number $\varepsilon > 0$ such that (8) holds. The numbers ε and P^* are related by an equality of the form $\varepsilon^2 = K_n/P^*$. The constant K_n depends not only on n but also on X_0 (and the probability distribution under consideration). Assume we are given an X_0 for each n, which means that we actually deal with a sequence $X_{0,1}, X_{0,2}, X_{0,3}, \ldots$. A careful analysis shows that if there are constants $0 < C_1 < C_2 < \infty$ and $k \in \mathbf{Z}$ such that

$$C_1 \, n^k \le |D_n X_{0,n}|^2 \le C_2 \, n^k$$

for all n, then $K_n = O(1/n^2)$ and hence $\varepsilon < 1$ whenever n is large enough. We can even improve this estimate for K_n in some cases. For example, if x is from the unit sphere with the uniform distribution and $k \geq 1$, then $K_n = O(1/n^{k+2})$ and if x is from the unit ball with the uniform distribution, then $K_n = O(1/n^3)$ for $k = 1$ and $K_n = O(1/n^4)$ for $k \geq 2$.

References

[1] A. Böttcher, *On the problem of testing the structure of a matrix by displacement operations*, SIAM J. Numer. Analysis, to appear.

[2] A. Böttcher and S. Grudsky, *The norm of the product of a large matrix and a random vector*, Electronic Journal of Probability **8** (2003), Paper no. 7, pages 1–29.

[3] A. Böttcher and D. Wenzel, *How big can the commutator of two matrices be and how big is it typically?*, Linear Algebra Appl. **403** (2005), 216–228.

[4] G. Heinig and K. Rost, *Algebraic Methods for Toeplitz-Like Matrices and Operators*, Birkhäuser, Basel 1984.

[5] G.M. Fichtenholz, *Differential- und Integralrechnung*, Vol. III, Deutscher Verlag der Wissenschaften, Berlin 1977.

Albrecht Böttcher
Fakultät für Mathematik
TU Chemnitz
D-09107 Chemnitz, Germany
e-mail: `aboettch@mathematik.tu-chemnitz.de`

David Wenzel
Fakultät für Mathematik
TU Chemnitz
D-09107 Chemnitz, Germany
e-mail: `david.wenzel@s2000.tu-chemnitz.de`

Operator Theory:
Advances and Applications, Vol. 170, 53–74
© 2006 Birkhäuser Verlag Basel/Switzerland

Asymmetric Factorizations of Matrix Functions on the Real Line

L.P. Castro, R. Duduchava and F.-O. Speck

Dedicated to I.B. Simonenko on the occasion of his 70th birthday

Abstract. We indicate a criterion for some classes of continuous matrix functions on the real line with a jump at infinity to admit both, a classical right and an asymmetric factorization. It yields the existence of generalized inverses of matrix Wiener-Hopf plus Hankel operators and provides precise information about the asymptotic behavior of the factors at infinity and of the solutions to the corresponding equations at the origin.

Mathematics Subject Classification (2000). Primary 47A68; Secondary 15A23, 47B35, 15A21.

Keywords. Right factorization, asymmetric factorization, anti-symmetric factorization.

1. Introduction

In 1968, I.B. Simonenko published his celebrated paper *Some general questions in the theory of the Riemann boundary problem* [Si] that gave rise to intensive studies on Riemann problems, singular integral and Toeplitz operators, etc. including the concepts of generalized factorization [ClGo], Φ-factorization [LiSp] and Wiener-Hopf factorization [BöSi]. In that paper, I. Simonenko gave a rather general definition of factorization of matrices with measurable functions as entries. He proved equivalence of generalized factorization with the solvability of the corresponding systems of singular integral operators and gave many properties of generalized factorization. The paper [Si] continuous to influence the investigations almost four decades already.

Among the pioneering works on the subject one should mention contributions by T. Carleman, N. Wiener and H. Hopf, F. Gakhov, N. Muskhelishvili, M. Krein,

This article was started during the second author's visit to Instituto Superior Técnico, U.T.L., and Universidade de Aveiro, Portugal, in February–March 2005.

I. Gohberg, I. Simonenko and many others. See also [GoKaSp] for a survey on matrix-valued functions factorization.

Different types of matrix factorizations revealed to be a powerful tool for solving explicitly many problems, e.g. in mathematical physics. Recent work on applications in diffraction theory [CaSpTe1, CaSpTe3] initiated a detailed investigation of Wiener–Hopf plus Hankel operators in spaces of Bessel potentials and their theoretical background.

The present paper continues the investigation started in [CaSpTe2, CaSp]. Some related results on factorization of matrix symbols of pseudodifferential operators are exposed in [ChDu, Sh]. Corresponding work for the circle instead of $\mathbb{R}$ and the factorization theory for Toeplitz plus Hankel operators can be found in [Eh]. The present environment is designed for further applications in mathematical physics as started in [CaSpTe1].

Here we devote particular attention to factorization of matrix-valued functions with discontinuity at infinity, which plays a crucial role in solving some problems of mathematical physics. We establish a criterion for such matrix-valued functions on the real line admit, both, an asymmetric and a classical right factorization. It yields the existence of generalized inverses of *matrix convolution type operators with symmetry* [CaSpTe2] (or Wiener–Hopf plus/minus Hankel operators), and provides precise information about the asymptotic behavior of the factors at infinity, and of the solutions to the corresponding equations at the origin.

2. Classical factorization

Let $\mathscr{A}$ be a bounded matrix-valued function which belongs to the Zygmund space $\mathscr{Z}^{\mu}(\overline{\mathbb{R}})$ or to the algebra $\mathscr{H}_0^{\mu}(\overline{\mathbb{R}})$, $\mu > 0$ (see Appendix, § A.2) and is supposed to be elliptic:

$$\inf_{x \in \mathbb{R}} |\det \mathscr{A}(x)| > 0. \tag{2.1}$$

The limits $\mathscr{A}(+\infty)$ and $\mathscr{A}(-\infty)$ might differ (in contrast to the case $\mathscr{A} \in \mathscr{Z}^{\mu}(\mathbb{R})$ or $\mathscr{A} \in \mathscr{H}_0^{\mu}(\mathbb{R})$ when these limits coincide) and we consider the *Jordan normal decomposition* of the matrix

$$\mathscr{A}_{\infty} := [\mathscr{A}(+\infty)]^{-1}\mathscr{A}(-\infty) = \mathscr{K}\,\Lambda_{\mathscr{A}_{\infty}}B_{\mathscr{A}_{\infty}}(1)\,\mathscr{K}^{-1}. \tag{2.2}$$

Here $\Lambda_{\mathscr{A}_{\infty}}$ is a diagonal matrix of eigenvalues of $\mathscr{A}_{\infty}$, $B_{\mathscr{A}_{\infty}}(1)$ is upper triangular with entries 1 on the main diagonal and $\mathscr{K}$ is an elliptic (det $\mathscr{K} \neq 0$) transformation matrix (see Appendix, § A.1 for details).

Let $\lambda_1, \ldots, \lambda_{\ell}$ be all eigenvalues of the matrix $\mathscr{A}_{\infty}$ with the Riesz indices $m_1, \ldots, m_{\ell}$, respectively (i.e., λ_j defines m_j linearly independent associated vectors for $\mathscr{A}_{\infty}$; see [Ga]) and

$$\delta_j := \frac{1}{2\pi i}\log \lambda_j, \quad \gamma < \Re e\, \delta_j \leq \gamma + 1, \quad j = 1, \ldots, \ell \tag{2.3}$$

for some $\gamma \in \mathbb{R}$.

Theorem 2.1. *Let $m = 2, \ldots$, and $\mathscr{A} \in \mathscr{L}^2(\overline{\mathbb{R}})$ (or $\mathscr{A} \in \mathscr{H}_0^m(\overline{\mathbb{R}})$) be an elliptic $N \times N$ matrix-valued function. Then*

$$\mathscr{A}(x) = [\mathscr{A}_-(x)]^{-1}\Xi(x)\mathscr{A}_+(x), \tag{2.4}$$

$$\Xi(x) = \mathscr{A}(+\infty)\mathscr{K}\left(\frac{x-i}{x+i}\right)^{-\Delta+\varkappa} B_{\mathscr{A}_\infty}\left(\frac{1}{2\pi i}\log\frac{x-i}{x+i}\right)\mathscr{K}^{-1}.$$

Here:

(i) *The matrix-valued functions $\mathscr{A}_-^{\pm 1}$, $\mathscr{A}_+^{\pm 1}$ belong to $\mathscr{L}^1(\overline{\mathbb{R}})$ (belong to $\mathscr{H}_0^{m-1}(\overline{\mathbb{R}})$). The factors $\mathscr{A}_-^{\pm 1}(x-it)$ and $\mathscr{A}_+^{\pm 1}(x+it)$ have uniformly bounded analytic continuation for $t > 0$ and $\mathscr{A}_-(\pm\infty) = \mathscr{A}_+(\pm\infty) = I_N$, where I_N is the identity matrix of order N.*

(ii) *The numbers δ_j are defined in (2.3), the vector $\Delta := (\delta_1, \ldots, \delta_\ell)$ has length N (each δ_j occurs m_j times according to its algebraic multiplicity), $\varkappa = (\varkappa_1, \ldots, \varkappa_N) \in \mathbb{N}_0^N$ are integers (known as the partial indices of $\mathscr{A}$) and*

$$h^{\Delta+\varkappa} := \operatorname{diag}\{h^{\delta_1+\varkappa_1}, \ldots, h^{\delta_\ell+\varkappa_N}\} \quad for \quad h \in \mathbb{C}$$

is a diagonal matrix.

(iii) *$B_{\mathscr{A}_\infty}(z)$, $z \in \mathbb{C}$, is an upper triangular polynomial matrix-valued function related to the Jordan normal form of $\mathscr{A}_\infty$, thoroughly described in Appendix, §A.1.*

Remark 2.2. The factorization (2.4) depends on a real number $\gamma \in \mathbb{R}$ (cf. (2.3)) which will be fixed uniquely later and carries the information about the space where an operator with the symbol $\mathscr{A}$ is treated.

Remark 2.3. The factorization (2.4) can also be presented in the form:

$$\mathscr{A}(x) = [\mathscr{A}_-^0(x)]^{-1}\Xi_0(x)\mathscr{A}_+^0(x), \tag{2.5}$$

$$\Xi_0(x) = \left(\frac{x-i}{x+i}\right)^{-\Delta+\varkappa} B_{\mathscr{A}_\infty}\left(\frac{1}{2\pi i}\log\frac{x-i}{x+i}\right).$$

Here:

(i) the matrix-valued functions

$$\left[\mathscr{A}_-^0\right]^{\pm 1} = \left[\mathscr{K}^{-1}\mathscr{A}^{-1}(+\infty)\mathscr{A}_-\right]^{\pm}, \quad \left[\mathscr{A}_+^0\right]^{\pm 1} := \left[\mathscr{K}^{-1}\mathscr{A}_+\right]^{\pm 1}$$

belong to $\mathscr{L}^1(\overline{\mathbb{R}})$ (belong to $\mathscr{H}_0^{m-1}(\overline{\mathbb{R}})$). The factors $\left[\mathscr{A}_-^0\right]^{\pm 1}(x-it)$ and $\left[\mathscr{A}_+^0\right]^{\pm 1}(x+it)$ have uniformly bounded analytic continuation for $t > 0$ and $\mathscr{A}_+^0(\pm\infty) = \mathscr{K}^{-1}$.

(ii) The matrix $\Xi_0(x)$ is upper triangular and its factors have the properties (ii) and (iii) as described in the foregoing Theorem 2.1.

Proof of Theorem 2.1. Let

$$\mathscr{A}^*(x) = (x-i)^\Delta B_-(x)\mathscr{A}_1(x)B_+^{-1}(x)(x+i)^{-\Delta},$$
$$\mathscr{A}_1(x) = \mathscr{K}^{-1}\mathscr{A}^{-1}(+\infty)\mathscr{A}(x)\mathscr{K}, \tag{2.6}$$

where $B_\pm(x)$ are related to the Jordan normal form of $\mathscr{A}_\infty$ and are defined by (A.3). Due to (2.2) and (2.6), we have

$$
\begin{aligned}
\mathscr{A}_1(-\infty) &= \lim_{x \to -\infty} \mathscr{A}_1(x) = \mathscr{K}^{-1}[\mathscr{A}(+\infty)]^{-1}\mathscr{A}(-\infty)\mathscr{K} = \Lambda_{\mathscr{A}_\infty} B_{\mathscr{A}_\infty}(1)\,, \\
\mathscr{A}_1(+\infty) &= \lim_{x \to +\infty} \mathscr{A}_1(x) = I_N\,.
\end{aligned}
\tag{2.7}
$$

According to their definition, the matrix-valued functions $B_\pm(x)$ and $B_{\mathscr{A}_\infty}(\mathrm{x})$ are block-diagonal with blocks of upper triangular matrices of dimensions $m_1, \ldots, m_\ell$. The matrices $\Lambda_{\mathscr{A}_\infty}$ and $(x-i)^{\pm\Delta}$ are diagonal, with blocks of equal constants (functions) of the same dimension $m_1, \ldots, m_\ell$. Therefore all these matrices commute and have the following properties, cf. (A.1)–(A.4):

$$
\begin{aligned}
B_\pm \Lambda_{\mathscr{A}_\infty} &= \Lambda_{\mathscr{A}_\infty} B_\pm\,, & (x-i)^{\pm\Delta}\Lambda_{\mathscr{A}_\infty} &= \Lambda_{\mathscr{A}_\infty}(x-i)^{\pm\Delta}\,, \\
B_\pm B_{\mathscr{A}_\infty} &= B_{\mathscr{A}_\infty} B_\pm\,, & B_{\mathscr{A}_\infty}(-z) &= B_{\mathscr{A}_\infty}^{-1}(z)\,.
\end{aligned}
\tag{2.8}
$$

Based on (2.7) and on (2.8) the matrix-valued function $\mathscr{A}^*$ in (2.6) can be rewritten as follows

$$
\mathscr{A}^*(x) = \mathscr{A}_2^\pm(x) + \mathscr{A}_3^\pm(x)\,,
\tag{2.9}
$$

$$
\mathscr{A}_2^\pm(x) = (x-i)^\Delta B_-(x)[\mathscr{A}_1(x) - \mathscr{A}_1(\pm\infty)]B_+^{-1}(x)(x+i)^{-\Delta}\,,
\tag{2.10}
$$

$$
\begin{aligned}
\mathscr{A}_3^+(x) &= (x-i)^\Delta B_-(x)B_+^{-1}(x)(x+i)^{-\Delta} \\
&= \left(\frac{x-i}{x+i}\right)^\Delta B_{\mathscr{A}_\infty}\left(\frac{1}{2\pi i}\log\frac{x-i}{x+i}\right) \\
&= B_{\mathscr{A}_\infty}\left(\frac{1}{2\pi i}\log\frac{x-i}{x+i}\right)\left(\frac{x-i}{x+i}\right)^\Delta\,,
\end{aligned}
\tag{2.11}
$$

$$
\begin{aligned}
\mathscr{A}_3^-(x) &= (x-i)^\Delta B_-(x)\Lambda_{\mathscr{A}} B_{\mathscr{A}_\infty}(1)B_+^{-1}(x)(x+i)^{-\Delta} \\
&= \Lambda_{\mathscr{A}} B_{\mathscr{A}_\infty}(1)B_-(x)B_+^{-1}(x)(x-i)^\Delta(x+i)^{-\Delta} \\
&= \Lambda_{\mathscr{A}} B_{\mathscr{A}_\infty}(1)B_{\mathscr{A}_\infty}\left(\frac{1}{2\pi i}\log\frac{x-i}{x+i}\right)\left(\frac{x-i}{x+i}\right)^\Delta \\
&= \Lambda_{\mathscr{A}} B_{\mathscr{A}_\infty}(1)\mathscr{A}_3^+(x) = \mathscr{A}_3^+(x)B_{\mathscr{A}_\infty}(1)\Lambda_{\mathscr{A}}\,.
\end{aligned}
\tag{2.12}
$$

Further, due to the definition of the function $(x \pm i)^{\pm\Delta}$ (see (2.3), (2.6)),

$$
\left(\frac{x-i}{x+i}\right)^{\pm\Delta} = \begin{cases} I_N + \mathscr{O}(\langle x\rangle^{-1}) & \text{as} \quad x \to +\infty\,, \\ \Lambda_{\mathscr{A}}^{\mp 1} + \mathscr{O}(\langle x\rangle^{-1}) & \text{as} \quad x \to -\infty\,, \end{cases}
\tag{2.13}
$$

$$
B_{\mathscr{A}_\infty}\left(\frac{1}{2\pi i}\log\frac{x-i}{x+i}\right) = \begin{cases} B_{\mathscr{A}_\infty}(1) + \mathscr{O}(\langle x\rangle^{-1}) & \text{as} \quad x \to +\infty\,, \\ B_{\mathscr{A}_\infty}(0) = I_N + \mathscr{O}(\langle x\rangle^{-1}) & \text{as} \quad x \to -\infty\,, \end{cases}
\tag{2.14}
$$

where $\langle x \rangle := (1+|x|^2)^{1/2}$. From (2.11)–(2.14) it is clear that $\mathscr{A}_3^{\pm}(-\infty) = \mathscr{A}_3^{\pm}(+\infty)$ and

$$\mathscr{A}_3^{\pm} \in \mathscr{L}^{\sigma}(\mathbb{R}) \cap \mathscr{H}_0^{\sigma}(\mathbb{R}) \qquad \text{for all} \quad \sigma > 0. \tag{2.15}$$

Next we prove that

$$\mathscr{A}_2^{\pm} \in \mathscr{L}^{2-\delta_0-\varepsilon}(\mathbb{R}) \quad \left(\mathscr{A}_2^{\pm} \in \mathscr{H}_0^{m-\delta_0-\varepsilon}(\mathbb{R}), \text{ respectively}\right), \quad 0 < \delta_0 + 2\varepsilon < 1, \tag{2.16}$$

where $\varepsilon > 0$, $0 < \delta_0 + 2\varepsilon < 1$, is arbitrarily small and δ_0 is defined by the relations (see (2.3))

$$0 \le \delta_0 := \max_{j,q=1,\ldots,N} \{\Re e\,(\delta_j - \delta_q)\} < \delta_0 + 2\varepsilon < 1. \tag{2.17}$$

To this end we note that

$$\partial_x^k \mathscr{A}_1^0(x) = \mathscr{O}(\langle x \rangle^{-k-1}), \qquad \mathscr{A}_1^0(x) := \mathscr{A}_1(x) - \mathscr{A}_1(\pm\infty), \tag{2.18}$$

where $k = 2$ in the case of $\mathscr{A} \in \mathscr{L}^2(\overline{\mathbb{R}})$ and $k = m$ in the case of $\mathscr{A} \in \mathscr{H}_0^m(\overline{\mathbb{R}})$. A typical entry of $\mathscr{A}_2^{\pm}$ is

$$[\mathscr{A}_2^{\pm}]_{jl} = b_{jl}(x)[\mathscr{A}_1^0(x)]_{jl}, \quad j,l = 0,\ldots,N,$$

with

$$b_{jl}(x) := (x-i)^{\delta_p}(x+i)^{-\delta_q}[\log(x-i)]^r[\log(x-i)]^s = \mathscr{O}\left(\langle x \rangle^{\delta_p-\delta_q+\varepsilon}\right)$$

and Propositions A.1(i), A.1(iv) with (2.13) and (2.18) yield the claimed inclusion (2.16).

From (2.9), (2.15) and (2.16) we obtain

$$\mathscr{A}^* \in \mathscr{L}^{2-\delta_0-\varepsilon}(\mathbb{R}) \qquad \left(\mathscr{A}^* \in \mathscr{H}_0^{m-\delta_0-\varepsilon}(\mathbb{R}), \text{ respectively}\right). \tag{2.19}$$

Then, due to Proposition A.6, the elliptic matrix-valued function $\mathscr{A}^*$ admits a classical right factorization

$$\mathscr{A}^*(x) = [\mathscr{A}_-^*(x)]^{-1} \left(\frac{x-i}{x+i}\right)^{\varkappa} \mathscr{A}_+^*(x), \tag{2.20}$$

$$\varkappa = (\varkappa_1,\ldots,\varkappa_N) \in \mathbb{Z}^N, \quad \mathbb{Z} = \{0,\pm 1,\ldots\}$$

with factors $[\mathscr{A}_-^*(x)]^{\pm 1}$, $[\mathscr{A}_+^*(x)]^{\pm 1}$ belonging to the same algebras as $\mathscr{A}^*$ (cf. (2.19)). These factors have uniformly bounded analytic continuations into the half-planes $\Im m\, x < 0$ and $\Im m\, x > 0$, respectively.

Since the limits $\mathscr{A}_{\pm}^*(\infty)$, $\mathscr{A}^*(\infty)$ exist and $\mathscr{A}^*(\infty) = I_N$, from (2.20) there follows

$$[\mathscr{A}_-^*(\infty)]^{-1}\mathscr{A}_+^*(\infty) = \mathscr{A}^*(\infty) = I_N.$$

Therefore, without restricting generality we can assume that $\mathscr{A}_{\pm}^*(\infty) = I_N$. Then (cf. (2.20))

$$\partial_x^k \left[[\mathscr{A}_{\pm}^*(x)]^{\pm 1} - I_N\right] = \mathscr{O}\left(\langle x \rangle^{\delta_0+\varepsilon-k-1}\right) \quad \text{as} \quad x \to \infty, \tag{2.21}$$

where $k = 2$ in the case of $\mathscr{A} \in \mathscr{L}^2(\overline{\mathbb{R}})$ and $k = m$ in the case of $\mathscr{A} \in \mathscr{H}_0^m(\overline{\mathbb{R}})$.

58   L.P. Castro, R. Duduchava and F.-O. Speck

From (2.6) and (2.20) we find the components of the factorization (2.4):

$$\mathscr{A}_{\pm}(x) := \mathscr{M}_{\pm}(x \pm i)^{-\Delta} B_{\pm}^{-1}(x)\mathscr{A}_{\pm}^{*}(x) B_{\pm}(x)(x \pm i)^{\Delta}\mathscr{M}_{\pm}^{-1}$$

$$= I_N + \mathscr{M}_{\pm}(x \pm i)^{-\Delta} B_{\pm}^{-1}(x)[\mathscr{A}_{\pm}^{*}(x) - I_N]B_{\pm}(x)(x \pm i)^{\Delta}\mathscr{M}_{\pm}^{-1}, \quad (2.22)$$

$$\mathscr{M}_{+} := \mathscr{K}, \quad \mathscr{M}_{-} := \mathscr{A}(+\infty)\mathscr{K}.$$

The theorem will be proved if we succeed in verifying the inclusions

$$\langle\cdot\rangle^{k}\partial^{k}\mathscr{A}_{+}^{\pm 1}(\cdot), \quad \langle\cdot\rangle^{k}\partial^{k}\mathscr{A}_{-}^{\pm 1}(\cdot) \in \mathscr{L}^{1}(\mathbb{R}), \qquad (2.23)$$

where $k = 0$ in the case of $\mathscr{A} \in \mathscr{L}^{2}(\overline{\mathbb{R}})$ and $k = m - 2$ in the case of $\mathscr{A} \in \mathscr{H}_{0}^{m}(\overline{\mathbb{R}})$.
A typical entry of the matrix $\mathscr{A}_{+}^{\pm 1}(x) - I_N$ is

$$[\mathscr{A}_{+}^{\pm 1}(x) - I_N]_{jq} = (x + i)^{\delta_p - \delta_r} \sum_{l \leq q} c_{jql}\left[[\mathscr{A}_{+}^{*}(x)]_{jl}^{\pm 1} - \delta_{jl}\right] \log^{m_{lq}}(x + i),$$

$$\partial^{k}[\mathscr{A}_{+}^{\pm 1}(x) - I_N]_{jq} = \begin{cases} \mathscr{O}\left(\langle x\rangle^{\Re e(\delta_p - \delta_r)+\delta_0+2\varepsilon-k-1}\right) & \text{if} \quad \Re e\,\delta_p > \Re e\,\delta_r \\ \mathscr{O}\left(\langle x\rangle^{\delta_0+2\varepsilon-k-1}\right) & \text{if} \quad \Re e\,\delta_p \leq \Re e\,\delta_r \end{cases} \qquad (2.24)$$

(cf. (2.21)), where $m_{qq} = 0$ and δ_{jl} is the Kronecker's symbol. From (2.20) we have

$$\mathscr{A}_{+}^{*} - \mathscr{A}_{-}^{*} = \mathscr{A}^{*} - I_N + \left[I_N - \left(\frac{x - i}{x + i}\right)^{\varkappa}\right]\mathscr{A}_{+}^{*} + [\mathscr{A}_{-}^{*} - I_N][\mathscr{A}^{*} - I_N],$$

$$[\mathscr{A}_{+}^{*}]^{-1} - [\mathscr{A}_{-}^{*}]^{-1} = I_N - [\mathscr{A}^{*}]^{-1} + \left[\left(\frac{x - i}{x + i}\right)^{-\varkappa} - I_N\right][\mathscr{A}_{-}^{*}]^{-1} \qquad (2.25)$$

$$+ \left[[\mathscr{A}_{-}^{*}]^{-1} - I_N\right]\left[I_N - [\mathscr{A}^{*}]^{-1}\right],$$

and applying (2.19) and (2.21) we obtain

$$\partial_{x}^{k}\left[[\mathscr{A}_{+}^{*}]^{\pm 1}(x) - [\mathscr{A}_{-}^{*}]^{\pm 1}(x)\right]_{jl} = \mathscr{O}(\langle x\rangle^{-k-1}) + \partial_{x}^{k}\sum_{r=1}^{N}\{[\mathscr{A}_{-}^{*}]_{jr}^{\pm 1} - \delta_{jr}\}[\mathscr{A}^{*} - I_N]_{rl}$$

$$+ \mathscr{O}\left(\langle x\rangle^{\Re e\,(\delta_j - \delta_l)+\varepsilon-k-1}\right)$$

$$= \sum_{r=1}^{N}\mathscr{O}\left(\langle x\rangle^{\Re e\,(\delta_r - \delta_l)+\varepsilon-1-k}\right) = \mathscr{O}\left(\langle x\rangle^{\delta_l^+ +\varepsilon-1-k}\right)$$

where $\delta_j^+ := \max_{q}\{\Re e\,[\delta_q - \delta_j]\} = \Re e\,[\delta_{j_*} - \delta_j]$ for a certain $1 \leq j_* \leq n$ (note that
we have inserted $\partial_{x}^{l}\{[\mathscr{A}_{-}^{*}]_{jr}^{\pm 1}(x) - \delta_{jl}\} = \mathscr{O}(\langle x\rangle^{\delta_0+\varepsilon-l-1}) = \mathscr{O}(\langle x\rangle^{-l})$; cf. (2.21)).
According to Proposition A.1(i), $[[\mathscr{A}_{+}^{*}]^{\pm 1} - [\mathscr{A}_{-}^{*}]^{\pm 1}]_{jl} \in \widetilde{\mathscr{H}}^{k-\delta_l^+ -\varepsilon}(\mathbb{R})$ (we remind
that $k = 0$ in the case of $\mathscr{A} \in \mathscr{L}^{2}(\overline{\mathbb{R}})$ and $k = m - 2$ in the case of $\mathscr{A} \in \mathscr{H}_{0}^{m}(\overline{\mathbb{R}})$).
We will use the Hilbert transformation $H_{\mathbb{R}}$ (cf. (A.15)) to define the pro-
jections $P_{\mathbb{R}}^{\pm} = \frac{1}{2}(I \pm H_{\mathbb{R}})$ that eliminate functions, analytic in the half-planes

$\mp \Im m\, x < 0$ (see [ClGo, GoKr]), and are bounded in $\widetilde{\mathscr{H}}^{\mu}(\mathbb{R})$ (see Theorem A.5); hence

$$[\mathscr{A}_{\pm}^{*}]_{jl}^{\pm 1} = \pm P_{\mathbb{R}}^{\pm}\left[[\mathscr{A}_{+}^{*}]^{\pm 1} - [\mathscr{A}_{-}^{*}]^{\pm 1}\right]_{jl} \in \widetilde{\mathscr{H}}^{k-\delta_{l}^{+}-\varepsilon}(\mathbb{R})$$

and, therefore (see (A.8) and cf. (2.21)),

$$\partial_{x}^{k}\left[[\mathscr{A}_{\pm}^{*}]_{jl}^{\pm 1}(x) - I_{N}\right]_{jl} = \mathscr{O}\left(\langle x\rangle^{\delta_{l}^{+}+\varepsilon-k-1}\right). \tag{2.26}$$

Inserting the obtained asymptotic for $\left[[\mathscr{A}_{-}^{*}]_{jl}^{\pm 1}(x) - I_{N}\right]_{jl}$ into (2.25) and invoking (2.19) once again we get a more precise asymptotic behavior

$$\partial_{x}^{k}\left[[\mathscr{A}_{+}^{*}]^{\pm 1}(x) - [\mathscr{A}_{-}^{*}]^{\pm 1}(x)\right]_{jl} = \mathscr{O}\left(\langle x\rangle^{-k-1}\right)$$

$$+ \sum_{r=1}^{N}\mathscr{O}\left(\langle x\rangle^{\delta_{r}^{+}+2\varepsilon-1+\Re e\,(\delta_{r}-\delta_{l})+\varepsilon-k-1}\right) + \mathscr{O}\left(\langle x\rangle^{\Re e\,(\delta_{j}-\delta_{l})+\varepsilon-k-1}\right)$$

$$= \sum_{r=1}^{N}\mathscr{O}\left(\langle x\rangle^{\Re e\,(\delta_{r_{*}}-\delta_{l})+3\varepsilon-k-2}\right) + \mathscr{O}\left(\langle x\rangle^{\Re e\,(\delta_{j}-\delta_{l})+\varepsilon-k-1}\right)$$

$$= \mathscr{O}\left(\langle x\rangle^{\Re e\,(\delta_{j}-\delta_{l})+\varepsilon-k-1}\right),$$

where $\delta_{r_{*}} := \delta_{r} + \delta_{r}^{+}$. Thus, $\left[[\mathscr{A}_{+}^{*}]^{\pm 1} - [\mathscr{A}_{-}^{*}]^{\pm 1}\right]_{jl} \in \widetilde{\mathscr{H}}^{k-\Re e\,(\delta_{j}-\delta_{l})-\varepsilon}(\mathbb{R})$ and we conclude, as above, $[\mathscr{A}_{\pm}^{*}]_{jl}^{\pm 1} = \pm P_{\mathbb{R}}^{\pm}\left[[\mathscr{A}_{+}^{*}]^{\pm 1} - [\mathscr{A}_{-}^{*}]^{\pm 1}\right]_{jl} \in \widetilde{\mathscr{H}}^{k-\Re e\,(\delta_{j}-\delta_{l})-\varepsilon}(\mathbb{R})$. The latter yields (cf. (2.21))

$$\partial_{x}^{k}\left[[\mathscr{A}_{\pm}^{*}(x)]_{jl}^{\pm 1} - I_{N}\right]_{jl} = \mathscr{O}\left(\langle x\rangle^{\Re e\,(\delta_{j}-\delta_{l})+\varepsilon-k-1}\right).$$

By virtue of (2.24)

$$\partial_{x}^{k}[\mathscr{A}_{+}]_{jq}^{\pm 1}(x) = \mathscr{O}\left(\langle x\rangle^{\Re e\,(\delta_{q}-\delta_{j})+\Re e\,(\delta_{j}-\delta_{l})+2\varepsilon-k-1}\right) = \mathscr{O}(\langle x\rangle^{\theta-k-1})$$

since $\delta_{l} = \delta_{j}$ and since $\Re e\,(\delta_{q} - \delta_{j}) + 2\varepsilon = \theta < 1$. $\qquad\square$

For further purposes, we recall that a matrix $\mathscr{B}$ is called *normal* if it commutes with its own transposed matrix $\mathscr{B}^{\top}\mathscr{B} = \mathscr{B}\mathscr{B}^{\top}$ and $\mathscr{B}$ is called *positive definite* if

$$(\mathscr{B}\eta, \eta) \geq M|\eta|^{2} \qquad \forall\, \eta \in \mathbb{C}^{n}$$

with some constant $M > 0$.

Lemma 2.4. *If the matrix $\mathscr{A}_{\infty}$ in (2.2) is normal, then it is simple $\ell = N$ (i.e., each eigenvalue λ_{j} has algebraic multiplicity 1) and, therefore, $\mathscr{A}_{\infty}$ is diagonalizable:*

$$B_{\mathscr{A}_{\infty}}(x) \equiv I, \quad \mathscr{A}_{\infty} = \mathscr{K}\,\mathrm{diag}\,\{\lambda_{1},\ldots,\lambda_{N}\}\,\mathscr{K}^{*}, \tag{2.27}$$

$$\det\mathscr{K} \neq 0, \qquad \mathscr{K}^{-1} = \mathscr{K}^{*}.$$

If the matrices $\mathscr{A}(\pm\infty)$ are positive definite, then $\mathscr{A}_\infty$ in (2.2) is simple, the eigenvalues $\lambda_1,\ldots,\lambda_\ell$ are all real positive numbers and, therefore,

$$\mathfrak{Re}\,\delta_1 = \cdots = \mathfrak{Re}\,\delta_\ell \equiv 0^1\,. \tag{2.28}$$

Proof. For the first claim of the lemma we quote [La, Theorem 2.10.2].

The second assertion is proved in [DuSäWe, Lemma A.6] as follows. Since the matrices $\mathscr{A}(\pm\infty)$ are positive definite, the square roots $[\mathscr{A}(+\infty)]^{\pm 1/2}$ are well defined and the matrix

$$\begin{aligned}
\mathscr{A}_1(\omega) \;&:=\; [\mathscr{A}(+\infty)]^{1/2}\,\mathscr{A}_\infty(\omega)\,[\mathscr{A}(+\infty)]^{-1/2} \\
&=\; [\mathscr{A}(+\infty)]^{-1/2}\,\mathscr{A}(-\infty)\,[\mathscr{A}(+\infty)]^{-1/2}\,,
\end{aligned}$$

due to similarity, has the common eigenvalues, the common eigenvectors and the common Jordan chains of associated vectors with $\mathscr{A}_\infty$. On the other hand $\mathscr{A}_1$ is self-adjoint, i.e., is normal and has no associated vectors as noted above. Let $\eta,\ldots,\eta_N \in \mathbb{C}^N$ be eigenvectors corresponding to the eigenvalues $\lambda_1,\ldots,\lambda_N$; then

$$\mathscr{A}_\infty\eta_j = \lambda_j\eta_j, \quad j = 1,\ldots,N$$

and we get

$$\lambda_j = \frac{(\mathscr{A}_\infty(+\infty)\eta_j,\eta_j)}{(\mathscr{A}_\infty(-\infty)\eta_j,\eta_j)} > 0$$

because of the positive definiteness of $\mathscr{A}(\pm\infty)$. This implies (2.28). $\qquad\square$

3. Asymmetric and anti-symmetric factorizations

In this section we present two different kinds of factorizations of matrix-valued functions which display some symmetries in their structure. These factorizations are tightly connected with the theory of *convolution type operators with symmetry* [CaSpTe2]

$$T = r_+\mathscr{F}^{-1}\mathscr{B}\cdot\mathscr{F}\ell^c : [L^2(\mathbb{R}_+)]^N \to [L^2(\mathbb{R}_+)]^N\,, \tag{3.1}$$

and play a central role in the description of (generalized) invertibility properties of such operators (cf. [CaSp, CaSpTe2]). Here, the operator r_+ stands for the restriction to the positive half-line, $\mathscr{F}^{-1}$ and $\mathscr{F}$ are the inverse and direct Fourier transformations, $\mathscr{B}$ is a measurable $N \times N$ matrix-valued function, and ℓ^c denotes the even (ℓ^e) or odd (ℓ^o) extension as a continuous operator from $[L^2(\mathbb{R}_+)]^N$ into $[L^2(\mathbb{R})]^N$.

We shall also make use of $[L^2_\pm(\mathbb{R})]^{N\times N}$ to be the images of the space $[L^2(\mathbb{R})]^{N\times N}$ under the projections

$$P_\mathbb{R}^\pm = \frac{1}{2}(I \pm H_\mathbb{R})\,. \tag{3.2}$$

---

[1] The numbers δ_j in (2.3) and ν_j in [DuSäWe, (A.32)] are related as follows: $\delta_j = -i\nu_j$.

For a space $[X(\mathbb{R})]^{N \times N}$ and a weight function ρ the notation $[X(\mathbb{R}, \rho)]^{N \times N}$ will refer to the subspace of those elements $\mathscr{B}$ for which $\rho \mathscr{B} \in [X(\mathbb{R})]^{N \times N}$. In particular, we will make use of the subspaces

$$[L^{2,e}(\mathbb{R}, \rho)]^{N \times N} = \left\{ \mathscr{B} \in [L^2(\mathbb{R}, \rho)]^{N \times N} : \mathscr{B}(x) = \mathscr{B}(-x) \right\}$$

$$[L^{2,o}(\mathbb{R}, \rho)]^{N \times N} = \left\{ \mathscr{B} \in [L^2(\mathbb{R}, \rho)]^{N \times N} : \mathscr{B}(x) = -\mathscr{B}(-x) \right\} .$$

Definition 3.1. A matrix-valued elliptic function $\mathscr{B} \in \mathscr{G}[L^\infty(\mathbb{R})]^{N \times N}$ admits an *asymmetric generalized factorization with respect to L^2 and ℓ^e*, written as

$$\mathscr{B}(x) = \mathscr{B}_-(x) \left(\frac{x-i}{x+i} \right)^\kappa \mathscr{B}_e(x), \qquad x \in \mathbb{R}, \tag{3.3}$$

where $\kappa = (\kappa_1, \ldots, \kappa_N)$ and $\kappa_1, \ldots, \kappa_N \in \mathbb{Z}$ are integers, if:

(i) the factors belong to the following spaces

$$\mathscr{B}_- \in [L^2_-(\mathbb{R}, \lambda_-^{-2})]^{N \times N}, \qquad \mathscr{B}_-^{-1} \in [L^2_-(\mathbb{R}, \lambda_-^{-1})]^{N \times N}, \tag{3.4}$$

$$\mathscr{B}_e \in [L^{2,e}(\mathbb{R}, \lambda^{-1})]^{N \times N}, \qquad \mathscr{B}_e^{-1} \in [L^{2,e}(\mathbb{R}, \lambda^{-2})]^{N \times N}, \tag{3.5}$$

where $\lambda_-(\xi) = \xi - i$ and $\lambda(\xi) = (\xi^2 + 1)^{1/2}$, $\xi \in \mathbb{R}$;

(ii) the operator

$$V_e = A_e^{-1} \ell^e r_+ A_-^{-1}, \tag{3.6}$$

where

$$A_e = \mathscr{F}^{-1} \mathscr{B}_e \cdot \mathscr{F}, \tag{3.7}$$

$$A_- = \mathscr{F}^{-1} \mathscr{B}_- \cdot \mathscr{F}, \tag{3.8}$$

defined on a dense subspace of $[L^2(\mathbb{R})]^m$, has a bounded extension to $[L^2(\mathbb{R})]^m$.

The spaces of bounded rational functions without poles in the closed lower half-plane $\overline{\mathbb{C}_-} = \{\xi \in \mathbb{C} : \Im m\, \xi \leq 0\}$, or those which are even, are dense in the corresponding factor spaces (where the factors of $\mathscr{B}$ belong to) with respect to the weighted L^2 norm.

When all κ_j components of κ in (3.3) are zero, we will refer to the factorization as a *canonical asymmetric generalized factorization with respect to L^2 and ℓ^e* and so we shall use the word *canonical* in other similar factorizations.

Definition 3.2. We will say that a matrix-valued function $\mathscr{B} \in \mathscr{G}[L^\infty(\mathbb{R})]^{N \times N}$ admits an *asymmetric generalized factorization with respect to L^2 and ℓ^o*, if it is factorable in the form (3.3), with $\kappa = (\kappa_1, \ldots, \kappa_N)$, $\kappa_1, \ldots, \kappa_N \in \mathbb{Z}$,

$$\mathscr{B}_- \in [L^2(\mathbb{R}, \lambda_-^{-1})]^{N \times N}, \qquad \mathscr{B}_-^{-1} \in [L^2_-(\mathbb{R}, \lambda_-^{-2})]^{N \times N}, \tag{3.9}$$

$$\mathscr{B}_e \in [L^{2,e}(\mathbb{R}, \lambda_-^{-2})]^{N \times N}, \qquad \mathscr{B}_e^{-1} \in [L^{2,e}(\mathbb{R}, \lambda^{-1})]^{N \times N} \tag{3.10}$$

and if the operator

$$V_o = A_e^{-1} \ell^o r_+ A_-^{-1} \tag{3.11}$$

(cf. (3.7) and (3.8) for A_e and A_- respectively) defined on a dense subspace of $[L^2(\mathbb{R})]^N$, has a bounded extension to $[L^2(\mathbb{R})]^N$.

Given a matrix-valued function $\mathscr{A}$, on the real line, we will abbreviate by $\widetilde{\mathscr{A}}$ that one defined by

$$\widetilde{\mathscr{A}}(x) = \mathscr{A}(-x), \qquad x \in \mathbb{R}. \tag{3.12}$$

Definition 3.3. A matrix-valued function $\mathscr{C} \in \mathscr{G}[L^\infty(\mathbb{R})]^{N \times N}$ admits an *anti-symmetric generalized factorization with respect to L^2 and ℓ^e*

$$\mathscr{C}(x) = \mathscr{C}_-(x) \left(\frac{x-i}{x+i}\right)^{2\kappa} \widetilde{\mathscr{C}}_-^{-1}(x), \qquad x \in \mathbb{R}, \tag{3.13}$$

with integer-valued partial indices $\kappa = (\kappa_1, \ldots, \kappa_N)$, $\kappa_1, \ldots, \kappa_N \in \mathbb{Z}$, if:

(i) the factors belong to the following spaces

$$\mathscr{C}_- \in [L_-^2(\mathbb{R}, \lambda_-^{-2})]^{N \times N}, \qquad \mathscr{C}_-^{-1} \in [L_-^2(\mathbb{R}, \lambda_-^{-1})]^{N \times N}; \tag{3.14}$$

(ii) the operator

$$U_e = \widetilde{A}_- \ell^e r_+ A_-^{-1} \tag{3.15}$$

defined on a dense subset of $[L^2(\mathbb{R})]^N$ has a bounded extension to $[L^2(\mathbb{R})]^N$ (where $\widetilde{A}_- = \mathscr{F}^{-1} \widetilde{\mathscr{C}}_- \cdot \mathscr{F}$ and $A_- = \mathscr{F}^{-1} \mathscr{C}_- \cdot \mathscr{F}$).

Definition 3.4. We will say that a matrix-valued function $\mathscr{C} \in \mathscr{G}[L^\infty(\mathbb{R})]^{N \times N}$ admits an *anti-symmetric generalized factorization with respect to L^2 and ℓ^o*, if:

(i) $\mathscr{C}$ is decomposed as in (3.13) with integer-valued partial indices

$$\kappa = (\kappa_1, \ldots, \kappa_N), \quad \kappa_1, \ldots, \kappa_N \in \mathbb{Z};$$

(ii) the factors belong to the following spaces

$$\mathscr{C}_- \in [L_-^2(\mathbb{R}, \lambda_-^{-1})]^{N \times N}, \qquad \mathscr{C}_-^{-1} \in [L_-^2(\mathbb{R}, \lambda_-^{-2})]^{N \times N}; \tag{}$$

(iii) the operator

$$U_o = \widetilde{A}_- \ell^o r_+ A_-^{-1} \tag{3.16}$$

defined on a dense subset of $[L^2(\mathbb{R})]^N$ has a bounded extension to $[L^2(\mathbb{R})]^N$ (where $\widetilde{A}_- = \mathscr{F}^{-1} \widetilde{\mathscr{C}}_- \cdot \mathscr{F}$ and $A_- = \mathscr{F}^{-1} \mathscr{C}_- \cdot \mathscr{F}$).

In the next result we will explore a link between asymmetric and anti-symmetric generalized factorizations, which is useful for transferring results between the two types of factorizations.

Lemma 3.5. *Let $\mathscr{B} \in \mathscr{G}[L^\infty(\mathbb{R})]^{N \times N}$ and consider $\mathscr{C} = \mathscr{B}\widetilde{\mathscr{B}}^{-1}$.*

(i) *If $\mathscr{B}$ admits an asymmetric generalized factorization with respect to L^2 and ℓ^c,*

$$\mathscr{B}(x) = \mathscr{B}_-(x) \left(\frac{x-i}{x+i}\right)^\kappa \mathscr{B}_e(x), \qquad x \in \mathbb{R}, \tag{3.17}$$

then $\mathscr{C}$ admits an anti-symmetric generalized factorization with respect to L^2 and ℓ^c in the form

$$\mathscr{C}(x) = \mathscr{B}_-(x)\left(\frac{x-i}{x+i}\right)^{2\kappa} \widetilde{\mathscr{B}}_-^{-1}(x), \qquad x \in \mathbb{R}. \tag{3.18}$$

(ii) *If $\mathscr{C}$ admits an anti-symmetric generalized factorization with respect to L^2 and ℓ^c,*

$$\mathscr{C}(x) = \mathscr{C}_-(x)\left(\frac{x-i}{x+i}\right)^{2\kappa} \widetilde{\mathscr{C}}_-^{-1}(x), \qquad x \in \mathbb{R}, \tag{3.19}$$

then $\mathscr{B}$ admits an asymmetric generalized factorization with respect to L^2 and ℓ^c in the form

$$\mathscr{B}(x) = \mathscr{C}_-(x)\left(\frac{x-i}{x+i}\right)^{\kappa} \left(\left(\frac{x-i}{x+i}\right)^{-\kappa} \mathscr{C}_-^{-1}(x)\mathscr{B}(x)\right), \qquad x \in \mathbb{R}, \tag{3.20}$$

where $\left(\frac{x-i}{x+i}\right)^{-\kappa}\mathscr{C}_-^{-1}(x)\mathscr{B}(x)$ is the even factor (cf. (3.3)).

Proof. We will present the proof for $\ell^c = \ell^e$. The case $\ell^c = \ell^o$ runs analogously, with obvious changes.

(i) Assume that $\mathscr{B}$ has an asymmetric generalized factorization with respect to L^2 and ℓ^e

$$\mathscr{B}(x) = \mathscr{B}_-(x)\left(\frac{x-i}{x+i}\right)^{\kappa} \mathscr{B}_e(x), \qquad x \in \mathbb{R}, \tag{3.21}$$

with $\kappa_j \in \mathbb{Z}$, $j = 1,\ldots,N$, $\mathscr{B}_- \in [L^2_-(\mathbb{R}, \lambda_-^{-2})]^{N \times N}$, $\mathscr{B}_-^{-1} \in [L^2_-(\mathbb{R}, \lambda_-^{-1})]^{N \times N}$, $\mathscr{B}_e \in [L^{2,e}(\mathbb{R}, \lambda^{-1})]^{N \times N}$, $\mathscr{B}_e^{-1} \in [L^{2,e}(\mathbb{R}, \lambda^{-2})]^{N \times N}$ and with an operator

$$V_e = \mathscr{F}^{-1}\mathscr{B}_e^{-1} \cdot \mathscr{F}\ell^e r_+ \mathscr{F}^{-1}\mathscr{B}_-^{-1} \cdot \mathscr{F} \tag{3.22}$$

having a bounded extension to $[L^2(\mathbb{R})]^N$. We start by choosing the same "minus" factor $\mathscr{B}_-$ for the factorization of $\mathscr{C}$ and observe in addition that

$$\widetilde{\mathscr{B}}_-^{-1}(x) = \mathscr{B}_e^{-1}(x)\left(\frac{x-i}{x+i}\right)^{\kappa} \widetilde{\mathscr{B}}_-^{-1}(x) \tag{3.23}$$

holds since $\mathscr{B}_e$ is even. Therefore,

$$\mathscr{C}(x) = \mathscr{B}(x)\,\widetilde{\mathscr{B}}_-^{-1}(x) = \left(\mathscr{B}_-(x)\left(\frac{x-i}{x+i}\right)^{\kappa} \mathscr{B}_e(x)\right)\left(\mathscr{B}_e^{-1}(x)\left(\frac{x-i}{x+i}\right)^{\kappa} \widetilde{\mathscr{B}}_-^{-1}(x)\right)$$

$$= \mathscr{B}_-(x)\left(\frac{x-i}{x+i}\right)^{2\kappa} \widetilde{\mathscr{B}}_-^{-1}(x), \tag{3.24}$$

with

$$\mathscr{B}_- \in [L^2_-(\mathbb{R}, \lambda_-^{-2})]^{N \times N}, \qquad \mathscr{B}_-^{-1} \in [L^2_-(\mathbb{R}, \lambda_-^{-1})]^{N \times N}, \tag{3.25}$$

64   L.P. Castro, R. Duduchava and F.-O. Speck

or equivalently

$$\widetilde{\mathscr{B}}_- \in [L^2_+(\mathbb{R}, \lambda_+^{-2})]^{N \times N}, \qquad \widetilde{\mathscr{B}}_-^{-1} \in [L^2_+(\mathbb{R}, \lambda_+^{-1})]^{N \times N}, \quad (3.26)$$

where $\lambda_+(\xi) = \xi + i$, $\xi \in \mathbb{R}$.

The assumption of asymmetric generalized factorization entails that the operator

$$V = \mathscr{F}^{-1} \mathscr{B}_e^{-1} \cdot \mathscr{F} \ell^e r_+ \mathscr{F}^{-1} \mathscr{B}_-^{-1} \cdot \mathscr{F} \qquad (3.27)$$

is bounded in $[L^2(\mathbb{R})]^N$. As in the theory of generalized factorizations [Kr, § 9], this last condition (3.27) can be equivalently replaced by others. In particular, together with (3.23) we obtain that the operator

$$U_e = \mathscr{F}^{-1} \widetilde{\mathscr{B}}_- \cdot \mathscr{F} \ell^e r_+ \mathscr{F}^{-1} \mathscr{B}_-^{-1} \cdot \mathscr{F} \qquad (3.28)$$

is bounded in $[L^2(\mathbb{R})]^N$.

(ii) If $\mathscr{C}$ admits an anti-symmetric generalized factorization with respect to L^2 and ℓ^e,

$$\mathscr{C}(x) = \mathscr{B}(x)\, \widetilde{\mathscr{B}}^{-1}(x) = \mathscr{C}_-(x) \left(\frac{x-i}{x+i}\right)^{2\kappa} \widetilde{\mathscr{C}}_-^{-1}(x), \qquad x \in \mathbb{R}, \qquad (3.29)$$

then choosing

$$\mathscr{B}_e(x) = \left(\frac{x-i}{x+i}\right)^{-\kappa} \mathscr{C}_-^{-1}(x)\, \mathscr{B}(x) \qquad (3.30)$$

$$\mathscr{B}_-(x) = \mathscr{C}_-(x) \qquad (3.31)$$

it follows immediately that

$$\mathscr{B}(x) = \mathscr{B}_-(x) \left(\frac{x-i}{x+i}\right)^{\kappa} \mathscr{B}_e(x), \qquad x \in \mathbb{R}. \qquad (3.32)$$

In addition, due to (3.29), we have

$$\mathscr{C}_-^{-1}(x)\, \mathscr{B}(x)\, \widetilde{\mathscr{B}}^{-1}(x) = \left(\frac{x-i}{x+i}\right)^{2\kappa} \widetilde{\mathscr{C}}_-^{-1}(x) \qquad (3.33)$$

$$\widetilde{\mathscr{C}}_-^{-1}(x)\, \widetilde{\mathscr{B}}(x) = \left(\frac{x-i}{x+i}\right)^{-2\kappa} \mathscr{C}_-^{-1}(x)\, \mathscr{B}(x), \qquad (3.34)$$

and therefore (cf. (3.30) and the first identity in (3.29))

$$\widetilde{\mathscr{B}}_e(x) = \left(\frac{x-i}{x+i}\right)^{\kappa} \widetilde{\mathscr{C}}_-^{-1}(x)\, \widetilde{\mathscr{B}}(x) = \left(\frac{x-i}{x+i}\right)^{-\kappa} \mathscr{C}_-^{-1}(x)\mathscr{B}(x) = \mathscr{B}_e(x). \quad (3.35)$$

The obtained equality shows in particular that $\mathscr{B}_e$ is an even function.

Now, due to the anti-symmetric generalized factorization of $\mathscr{C}$, we already know that

$$\mathscr{B}_- = \mathscr{C}_- \in [L^2_-(\mathbb{R}, \lambda_-^{-2})]^{N \times N}, \qquad \mathscr{B}_-^{-1} = \mathscr{C}_-^{-1} \in [L^2_-(\mathbb{R}, \lambda_-^{-1})]^{N \times N}. \quad (3.36)$$

In combination with the inclusion $\mathscr{B} \in \mathscr{G}[L^\infty(\mathbb{R})]^{N \times N}$ and the property of the even function $\mathscr{B}_e$ in (3.30) leads to the further inclusions

$$\mathscr{B}_e \in [L^{2,e}(\mathbb{R}, \lambda^{-1})]^{N \times N}, \qquad \mathscr{B}_e^{-1} \in [L^2(\mathbb{R}, \lambda^{-2})]^{N \times N}. \tag{3.37}$$

Finally, similarly as in part (i), we obtain that the operator

$$V_e = \mathscr{F}^{-1}\mathscr{B}_e^{-1} \cdot \mathscr{F}\ell^e r_+ \mathscr{F}^{-1}\mathscr{B}_-^{-1} \cdot \mathscr{F}, \tag{3.38}$$

extended operator from a dense subspace, is bounded in $[L^2(\mathbb{R})]^N$. $\square$

Theorem 3.6. *Let $m = 2, \ldots$, $\mathscr{C} \in \mathscr{L}^2(\mathbb{R})$ (or $\mathscr{C} \in \mathscr{H}_0^m(\mathbb{R})$) be a $N \times N$ elliptic matrix-valued function and*

$$\mathscr{C}_\infty := [\mathscr{C}(+\infty)]^{-1}\mathscr{C}(-\infty). \tag{3.39}$$

Let $\lambda_1, \ldots, \lambda_\ell$ be all eigenvalues with Riesz indices $m_1, \ldots, m_\ell$ of the matrix $\mathscr{C}_\infty$, and consider the Jordan normal decomposition of $\mathscr{C}_\infty$,

$$\mathscr{C}_\infty = \mathscr{K} \Lambda_{\mathscr{C}_\infty} B_{\mathscr{C}_\infty}(1)\mathscr{K}^{-1} \tag{3.40}$$

(cf. the Appendix A.1 for details). Further, let

$$\delta_j := \frac{1}{2\pi i} \log \lambda_j, \quad \gamma < \Re\, \delta_j \leq \gamma + 1, \quad j = 1, \ldots, \ell \tag{3.41}$$

for some $\gamma \in \mathbb{R}$, and consider

$$\mathscr{C}^*(x) := (x - i)^\Delta B_-(x)\mathscr{C}_1(x)B_+^{-1}(x)(x + i)^{-\Delta}, \tag{3.42}$$
$$\mathscr{C}_1(x) := \mathscr{K}^{-1}\mathscr{C}^{-1}(+\infty)\mathscr{C}(x)\mathscr{K},$$

with $\Delta = (\delta_1, \ldots, \delta_\ell)$ having length N (where each δ_j occurs m_j times according to its algebraic multiplicity) and $B_\pm(x)$ are related to the Jordan normal form of $\mathscr{C}_\infty$ (cf. (A.3)).

If $\mathscr{C}^*$ admits an anti-symmetric factorization (within the classes mentioned lately),

$$\mathscr{C}^*(x) = \mathscr{C}_-^*(x) \left(\frac{x - i}{x + i}\right)^{2\varkappa} [\widetilde{\mathscr{C}_-^*}(x)]^{-1}, \tag{3.43}$$
$$\varkappa = (\varkappa_1, \ldots, \varkappa_N) \in \mathbb{Z}^N,$$

then the initial matrix $\mathscr{C}$ admits the factorization

$$\mathscr{C}(x) = \mathscr{C}_-(x) \Xi(x) [\widetilde{\mathscr{C}_-}(x)]^{-1}, \tag{3.44}$$

$$\Xi(x) = \mathscr{C}(+\infty)\mathscr{K} \left(\frac{x - i}{x + i}\right)^{-\Delta + 2\varkappa} B_{\mathscr{C}_\infty}^{-1} \left(\frac{1}{2\pi i} \log \frac{x - i}{x + i}\right) \mathscr{K}^{-1},$$

where the matrix-valued functions $\mathscr{C}_-^{\pm 1}$ belong to $\mathscr{L}^1(\mathbb{R})$ (or belong to $\mathscr{H}_0^{m-1}(\mathbb{R})$, respectively), and $\mathscr{C}_-^{\pm 1}(x - it)$ have uniformly bounded analytic continuation for $t > 0$.

Proof. The proof of Theorem 3.6 runs analogously to the proof of Theorem 2.1 with obvious modifications due to the different symmetry properties. $\square$

The foregoing result (together with Lemma 3.5) can be used in the description of the (generalized) inverses of convolution type operator with symmetry T (introduced in (3.1)), as described in [CaSpTe2, Theorem 3.2].

A. Appendix

In the Appendix we have collected related results which either are known and are applied in the foregoing sections, or might be useful for further considerations. In our exposition we follow mostly [ChDu, §§ 1.6–1.7].

A.1. Jordan decomposition

Let $\mathscr{B}$ be an elliptic $N \times N$ matrix (det $\mathscr{B} \neq 0$) and $\lambda_1, \ldots, \lambda_\ell$ be the eigenvalues of $\mathscr{B}$ with algebraic multiplicities $m_1, \ldots, m_\ell$, respectively. Hence the length of the chain of associated vectors with the eigenvalue λ_j is $\sum_{j=1}^{\ell} m_j = N$. Then $\mathscr{B}$ has the following decompositions

$$\mathscr{B} = \mathscr{K}_0 \mathscr{J}_\mathscr{B} \mathscr{K}_0^{-1} = \mathscr{K} \Lambda_\mathscr{B} B_\mathscr{B}(1) \mathscr{K}^{-1} \,,, \tag{A.1}$$

where $\mathscr{K}$ and $\mathscr{K}_0$ are some elliptic det $\mathscr{K}_0 \neq 0$, det $\mathscr{K} \neq 0$ transformation matrices, while the matrices $B_\mathscr{B}$ and $\mathscr{J}_\mathscr{B}$ are quasi-diagonal

$$\mathscr{J}_\mathscr{B} := \Lambda_\mathscr{B} + H_\mathscr{B} = \operatorname{diag}\{\lambda_1 I_{m_1} + H_{m_1}, \ldots, \lambda_\ell I_{m_\ell} + H_{m_\ell}\}$$
$$B_\mathscr{B}(x) := \operatorname{diag}\{B_{m_1}(x), \ldots, B_{m_\ell}(x)\}, \quad x \in \mathbb{C},$$
$$B_m(z) := \exp(z H_m), \quad z \in \mathbb{C},$$
$$\Lambda_\mathscr{B} := \operatorname{diag}\{\lambda_1 I_{m_1}, \ldots, \lambda_\ell I_{m_\ell}\},$$
$$H_\mathscr{B} := \operatorname{diag}\{H_{m_1}, \ldots, H_{m_\ell}\};$$

I_m is the identity and H_m is a nilpotent matrix that satisfies $H_m^m = 0$:

$$I_m := \begin{pmatrix} 1 & 0 & 0 & \cdots & 0 & 0 \\ 0 & 1 & 0 & \cdots & 0 & 0 \\ \cdot & \cdot & \cdot & \cdots & \cdot & \cdot \\ 0 & 0 & 0 & \cdots & 1 & 0 \\ 0 & 0 & 0 & \cdots & 0 & 1 \end{pmatrix}_{m \times m}, \quad H_m := \begin{pmatrix} 0 & 1 & 0 & \cdots & 0 & 0 \\ 0 & 0 & 1 & \cdots & 0 & 0 \\ \cdot & \cdot & \cdot & \cdots & \cdot & \cdot \\ 0 & 0 & 0 & \cdots & 0 & 1 \\ 0 & 0 & 0 & \cdots & 0 & 0 \end{pmatrix}_{m \times m}.$$

The first representation in (A.1) is known as the Jordan normal form and $\lambda I_m + H_m$ (for $\lambda \in \mathbb{C}$) is the Jordan cell of dimension m

$$\lambda I_m + H_m = \begin{pmatrix} \lambda & 1 & 0 & \cdots & 0 & 0 \\ 0 & \lambda & 1 & \cdots & 0 & 0 \\ \cdot & \cdot & \cdot & \cdots & \cdot & \cdot \\ 0 & 0 & 0 & \cdots & \lambda & 1 \\ 0 & 0 & 0 & \cdots & 0 & \lambda \end{pmatrix}_{m \times m}.$$

Since $B_m(z) = \exp(zH_m)$, $z \in \mathbb{C}$, and H_m is nilpotent, the exponent has a finite expansion

$$B_m(z) \quad := \quad \exp(zH_m) := I_n + \sum_{k=1}^{m-1} \frac{z^k}{k!} H_m^k$$

$$= \begin{pmatrix} 1 & \dfrac{z}{1!} & \dfrac{z^2}{2!} & \cdots & \dfrac{z^{m-2}}{(m-2)!} & \dfrac{z^{m-1}}{(m-1)!} \\ 0 & 1 & \dfrac{z}{1!} & \cdots & \dfrac{z^{m-3}}{(m-3)!} & \dfrac{z^{m-2}}{(m-2)!} \\ \cdot & \cdot & \cdot & \cdots & \cdot & \cdot \\ 0 & 0 & 0 & \cdots & 1 & \dfrac{z}{1!} \\ 0 & 0 & 0 & \cdots & 0 & 1 \end{pmatrix}_{m \times m} , \qquad z \in \mathbb{C}.$$

The sets $\{B_{\mathscr{B}}(z)\}_{z \in \mathbb{C}}$ and $\{B_m(z)\}_{z \in \mathbb{C}}$ are one parameter groups (see [Ar, §§ 14–23]) of matrix operators, and have the following properties:

$$B_{\mathscr{B}}(z_1 + z_2) = B_{\mathscr{B}}(z_1)B_{\mathscr{B}}(z_2),$$
$$B_{\mathscr{B}}(0) = I_N, \qquad B_{\mathscr{B}}(-z) = [B_{\mathscr{B}}(z)]^{-1}, \tag{A.2}$$
$$[B_{\mathscr{B}}(z)]^{\gamma} := \exp(z\gamma H_{\mathscr{B}}) = B_{\mathscr{B}}(\gamma z), \qquad z, \gamma \in \mathbb{C}.$$

According to the definition, e.g., in [Ga, § V.1]

$$b = \frac{1}{2\pi i} \log \mathscr{B} := \frac{1}{(2\pi)^2} \int_{\Gamma} [\mathscr{B} - zI]^{-1} \log z \, dz,$$

where I is the identity matrix, Γ is a closed contour, surrounding all eigenvalues $\lambda_1, \ldots, \lambda_\ell$ of $\mathscr{B}$ and leaving outside the negative real half-axes $\Re e\, z \le 0$. We assume $\log z := \log |z| + i \mathrm{Arg} z$, $-\pi < \mathrm{Arg} z < \pi$.

Here is the "purely algebraic" definition of the above-presented logarithm:

$$b = \frac{1}{2\pi i} \log \mathscr{B} \quad := \quad \frac{1}{2\pi i} \log \left[I - (I - \mathscr{B}) \right] = \sum_{k=0}^{\infty} \frac{1}{2\pi i k} \left[I - \mathscr{B} \right]^k$$

$$:= \sum_{k=0}^{\infty} \frac{1}{2\pi i k} \mathscr{K} \left[I - \Lambda_{\mathscr{B}} B_{\mathscr{B}}(1) \right]^k \mathscr{K}^{-1}$$

$$:= \frac{1}{2\pi i} \mathscr{K} \log \left[\Lambda_{\mathscr{B}} B_{\mathscr{B}}(1) \right] \mathscr{K}^{-1} = \mathscr{K} \left\{ \Delta + \frac{1}{2\pi i} H_{\mathscr{B}} \right\} \mathscr{K}^{-1},$$

$$\Delta := \frac{1}{2\pi i} \log \Lambda_{\mathscr{B}} = \mathrm{diag} \left\{ \underbrace{\delta_1, \ldots, \delta_1}_{m_1\text{-times}}, \ldots, \underbrace{\delta_\ell, \ldots, \delta_\ell}_{m_\ell\text{-times}} \right\}, \qquad \delta_j := \frac{1}{2\pi i} \log \lambda_j .$$

Introducing the notation

$$B_{\pm}(x) := B_{\mathscr{B}} \left(\frac{1}{2\pi i} \log(x \pm i) \right), \tag{A.3}$$

where the branch of the logarithm is fixed in the complex plane cut along the ray $\{z \in \mathbb{C} : \arg z = \gamma_0\}$, we find (cf. (A.2))

$$B_{\mathscr{B}}\left(\frac{1}{2\pi i} \log \frac{x-i}{x+i}\right) = B_-(x) B_+^{-1}(x)$$

$$= \begin{cases} [B_{\mathscr{B}}(1)]^{-1} + \mathcal{O}\left(|x-i|^{-1}\right) & \text{if} \quad x \to -\infty, \\ I_N + \mathcal{O}\left(|x+i|^{-1}\right) & \text{if} \quad x \to +\infty. \end{cases} \quad \text{(A.4)}$$

A.2. Hölder and Zygmund Spaces

We recall the definitions of some important spaces and expose their relevant properties to the present investigation.

For $s > 0$ the Zygmund space $\mathbb{Z}^s(\mathbb{R})$ is defined as the Banach space of functions with the finite norm

$$\left\|f\,|\,\mathbb{Z}^s(\mathbb{R})\right\| = \left\|f\,|\,C^m(\mathbb{R})\right\| + \sup_{h \neq 0}\left\{|h|^{-\nu}\left\|\Delta_h^2 \partial^m f\,|\,C(\mathbb{R})\right\|\right\},$$

$$s = m + \nu, \quad m \in \mathbb{N}_0, \quad 0 < \nu \leq 1,$$

where $\mathbb{N}_0 := \mathbb{N} \cup \{0\}$ and $\mathbb{N}$ denotes the set of all positive integers, $\Delta_h f(x) := f(x+h) - f(x)$, $\Delta_h^2 = \Delta_h \Delta_h$ and

$$\left\|f\,|\,C^m(\mathbb{R})\right\| = \sum_{k=0}^{m} \sup\{|\partial^k f(x)| : x \in \mathbb{R}\}.$$

For $s \in \mathbb{R}^+ \setminus \mathbb{N}$ the space $\mathbb{Z}^s(\mathbb{R})$ coincides with the Hölder space $C^s(\mathbb{R})$ (cf. [St, § V.4, Proposition 8]), which is endowed with the norm

$$\left\|f\,|\,C^s(\mathbb{R})\right\| = \left\|f\,|\,C^m(\mathbb{R})\right\| + \sup_{h \neq 0}\left\{|h|^{-\nu}\left\|\Delta_h \partial^m f\,|\,C(\mathbb{R})\right\|\right\},$$

$$s = m + \nu, \quad m \in \mathbb{N}_0, \quad 0 < \nu < 1.$$

For $s = m + \nu$, $m \in \mathbb{N}_0$, $0 < \nu \leq 1$, the space $H^s(\mathbb{R})$ of Hölder continuous functions on $\mathbb{R}$ consists of functions with the finite norm

$$\left\|\varphi\,|\,H^s(\mathbb{R})\right\| := \left\|\varphi\,|\,C^m(\mathbb{R})\right\| + \sup_{\substack{x,h \in \mathbb{R} \\ h \neq 0}} \frac{|\partial^m \varphi(x+h) - \partial^m \varphi(x)|}{\left|\dfrac{x+h}{x+h+i} - \dfrac{x}{x+i}\right|^\nu}. \quad \text{(A.5)}$$

This norm can also be written in the two following forms:

$$\left\|\varphi\,|\,H^s(\mathbb{R})\right\| = \left\|\varphi\,|\,C^m(\mathbb{R})\right\| + \sup_{\substack{x,h \in \mathbb{R} \\ h \neq 0}} \frac{|\partial^m \varphi(x+h) - \partial^m \varphi(x)|}{\left|\dfrac{1}{x+h+i} - \dfrac{1}{x+i}\right|^\nu}$$

$$= \left\|\varphi\,|\,C^m(\mathbb{R})\right\| + 2 \sup_{\substack{x,h \in \mathbb{R} \\ h \neq 0}} \frac{|\partial^m \varphi(x+h) - \partial^m \varphi(x)|}{\left|\dfrac{x+h-i}{x+h+i} - \dfrac{x-i}{x+i}\right|^\nu}. \quad \text{(A.6)}$$

Similarly $\mathscr{Z}^s(\mathbb{R})$ denotes the Zygmund space consisting of functions with the finite norm

$$\left\|\varphi\,|\,\mathscr{Z}^s(\mathbb{R})\right\| := \left\|\varphi\,|\,C^m(\mathbb{R})\right\| + \sup_{\substack{x,h\in\mathbb{R} \\ h\neq 0}} \frac{\left|\partial^m\varphi(x+h) + \partial^m\varphi(x-h) - 2\partial^m\varphi(x)\right|}{\left|\dfrac{x+h}{x+h+i} - \dfrac{x}{x+i}\right|^\nu}\,.$$

$$(A.7)$$

The space $H^\nu(\mathbb{R})$ differs from the above defined $C^\nu(\mathbb{R})$ since $\mathbb{R}$ is not compact; for compact curves Γ the spaces $H^\nu(\Gamma)$ and $C^\nu(\Gamma)$ are isomorphic.

For $s \in \mathbb{R}^+ \setminus \mathbb{N}$ the Zygmund space $\mathscr{Z}^s(\mathbb{R})$ coincides with the Hölder space $H^s(\mathbb{R})$ and differs for $s = 1, 2, \ldots$ ($\mathscr{Z}^s(\mathbb{R})$ contains $H^s(\mathbb{R})$ as a proper subspace; cf. [St, §V.4, Proposition 8] for details). The advantage of the Zygmund space $\mathscr{Z}^s(\mathbb{R})$ (compared with $H^s(\mathbb{R})$) is that the scale $\{\mathscr{Z}^s(\mathbb{R})\}_{s>0}$ allows interpolation (cf. [Tr]).

For a positive $\mu > 0$, $\mu = m + \nu$, $m \in \mathbb{N}$, $0 < \nu \leq 1$ we consider the following Banach algebra

$$\mathscr{H}^\mu(\mathbb{R}) := \left\{\varphi \in C^m(\mathbb{R}) : (x+i)^k \partial_x^k \varphi \in \mathscr{Z}^\nu(\mathbb{R}), k = 0, 1, \ldots, m\right\},$$

endowed with the norm

$$\left\|\varphi\,|\,\mathscr{H}^\mu(\mathbb{R})\right\| := \sum_{k=0}^m \left\|(x+i)^k \partial_x^k \varphi\,|\,\mathscr{Z}^\nu(\mathbb{R})\right\|\,.$$

If $\varphi \in \mathscr{H}^\mu(\mathbb{R})$ by sending in (A.6) $x \to 0$ we get

$$\partial_x^k[\varphi(h) - \varphi(\infty)] = \mathscr{O}\left(\langle h\rangle^{-\nu-k}\right)\,, \quad k = 0, 1, \ldots, m\,. \qquad (A.8)$$

Obviously,

$$g(-\infty) = \lim_{x\to-\infty} g(x) = \lim_{x\to+\infty} g(x) = g(+\infty) \qquad (A.9)$$

for all functions $g \in \mathscr{Z}^\mu(\mathbb{R})$ and $g \in \mathscr{H}^\mu(\mathbb{R})$ (cf. the definition of norms (A.5) and (A.7)). Therefore the Banach algebra $\mathscr{Z}^\mu(\overline{\mathbb{R}})$ of functions

$$\varphi(x) = \omega(x)\varphi_-(x) + [1 - \omega(x)]\varphi_+(x)\,, \qquad \varphi_\pm \in \mathscr{Z}^\mu(\mathbb{R})\,, \qquad (A.10)$$

where

$$\omega \in C^\infty(\mathbb{R})\,, \qquad \omega(x) = 1 \quad for \quad x < -1\,, \qquad \omega(x) = 0 \quad for \quad x > 1\,,$$

differs from the space $\mathscr{Z}^\mu(\mathbb{R})$ since the function $\varphi(x)$ in (A.10) has, in general, different limits:

$$\varphi(-\infty) = \varphi_-(-\infty)\,, \qquad \varphi(+\infty) = \varphi_+(+\infty)\,. \qquad (A.11)$$

The Banach algebra $\mathscr{H}^\mu(\overline{\mathbb{R}})$ is defined similarly.

For $0 < \nu \leq 1$ the spaces $\mathscr{Z}^\nu(\mathbb{R})$ and $\mathscr{Z}^\nu(\Gamma_0)$, where $\Gamma_0 = \{z \in \mathbb{C} : |z| = 1\}$ is the unit circle, are isomorphic:

$$\varpi_* : \mathscr{Z}^\nu(\mathbb{R}) \longrightarrow \mathscr{Z}^\nu(\Gamma_0)\,, \quad \varpi_*\varphi(z) := \varphi\left(i\frac{1+z}{1-z}\right)\,, \quad z \in \Gamma_0\,. \qquad (A.12)$$

The inverse isomorphism reads

$$\varpi_*^{-1}\psi(x) := \psi\left(\frac{x-i}{x+i}\right), \quad x \in \mathbb{R}.$$

In fact,

$$\left\|\varpi_*\varphi\,\big|\,\mathscr{L}^\nu(\Gamma_0)\right\| = \sup_{z \in \Gamma_0}\left|\varphi\left(i\frac{1+z}{1-z}\right)\right|$$

$$+ \sup_{\substack{z, z \pm h \in \Gamma_0 \\ |z_h - z| = |z_{-h} - z| \neq 0}} \frac{\left|\varphi\left(i\dfrac{1+z_h}{1-z_h}\right) + \varphi\left(i\dfrac{1+z_{-h}}{1-z_{-h}}\right) - 2\varphi\left(i\dfrac{1+z}{1-z}\right)\right|}{|z_h - z|^\nu}$$

$$= \sup_{x \in \mathbb{R}}|\varphi(x)| + \sup_{\substack{x, h \in \mathbb{R} \\ h \neq 0}} \frac{|\varphi(x+h) + \varpi(x-h) - 2\varphi(x)|}{\left|\dfrac{x+h-i}{x+h+i} - \dfrac{x-i}{x+i}\right|^\nu}$$

and, due to (A.6),

$$\left\|\varphi\,\big|\,\mathscr{L}^\nu(\mathbb{R})\right\| \leq \left\|\varpi_*\varphi\,\big|\,\mathscr{L}^\nu(\Gamma_0)\right\| \leq 2\left\|\varphi\,\big|\,\mathscr{L}^\nu(\mathbb{R})\right\|.$$

The next Proposition states certain inverse estimates to (A.8).

Proposition A.1. *Let* $0 < \nu \leq 1$, $m \in \mathbb{N}_0$.

(i) *If* $\varphi \in C^m(\mathbb{R})$ *and*

$$C_{k,\nu} := \sup_x \left|\,|x+i|^{k+\nu}\partial_x^k[\varphi(x) - \varphi(\infty)]\right| < \infty \quad \textit{for} \quad k = 0, 1, \ldots, m,$$

then $\varphi \in \mathscr{H}^{m-1+\nu}(\mathbb{R})$ *and* $\left\|\varphi\,\big|\,\mathscr{H}^{m-1+\nu}(\mathbb{R})\right\| \leq M \sum_{k=0}^m C_{k,\nu}$, *where* $M = const$
is independent of φ.

(ii) *If* $\varphi \in \widetilde{\mathscr{H}}^{m+\nu}(\mathbb{R})$ *and*

$$\partial_x^k b(x) = \mathcal{O}\left(\langle x \rangle^{-k}\right) \quad \textit{for} \quad k = 0, 1, \ldots, m, \tag{A.13}$$

then $b\varphi \in \widetilde{\mathscr{H}}^{m+\nu}(\mathbb{R})$.

(iii) *If* $\varphi \in \widetilde{\mathscr{L}}^\nu(\mathbb{R})$, $0 < \nu \leq 1$, *and if* (A.13) *holds, then* $b\varphi \in \widetilde{\mathscr{L}}^\nu(\mathbb{R})$.

(iv) *If* $\varphi \in \widetilde{\mathscr{L}}^\nu(\mathbb{R})$, $0 < \theta < \nu \leq 1$, *then* $(x+i)^\theta\varphi \in \widetilde{\mathscr{L}}^{\nu-\theta}(\mathbb{R})$.

Proof. For the proof of Proposition A.1.i and Proposition A.1.ii we refer to [ChDu, § 1.6]. Proposition A.1.iii and Proposition A.1.iv are proved by analogy to Proposition A.1.ii, based on similar assertions proved in [Mu, Chapt.1, § 6] for a smooth curve. $\qquad \square$

Remark A.2. As an example of the function $b(x)$ in (A.14) we can take $(x+i)^{i\mu}$, $\mu \in \mathbb{R}$.

Corollary A.3. *If* $0 < \mu_1 \leq \mu_2$, *the embedding* $\mathscr{H}^{\mu_2}(\mathbb{R}) \subset \mathscr{H}^{\mu_1}(\mathbb{R})$ *is continuous.*

Proof. The claim follows from the foregoing Proposition A.1 and from the asymptotic property (A.8). $\qquad \square$

Rational functions

$$r_\ell(x) = \sum_{|k| \le \ell} c_k \left(\frac{x-i}{x+i} \right)^k, \quad x \in \mathbb{R}, \quad c_k \in \mathbb{C} \tag{A.14}$$

belong to all $\mathscr{H}^\mu(\mathbb{R})$ (see Proposition A.1). Let $\widetilde{\mathscr{H}}^\mu(\mathbb{R})$ denote the sub-algebra of $\mathscr{H}^\mu(\mathbb{R})$ obtained by closing the algebra of rational functions (A.14). The algebra $\widetilde{\mathscr{H}}^\mu(\mathbb{R})$ is rationally dense by the definition in [BuGo] (see also [ClGo]).

In [Ta, §1.3.4] the sub-algebra $\widetilde{\mathscr{H}}^\mu(\mathbb{R})$ is characterized for $0 < \mu < 1$ as follows: $\varphi \in \widetilde{\mathscr{H}}^\mu(\mathbb{R})$ iff

$$\lim_{\varepsilon \to 0} \sup_{\substack{|x'-x| < \varepsilon \\ x' \ne x}} \frac{|\varphi(x') - \varphi(x)|}{\left| \dfrac{x'}{x'+i} - \dfrac{x}{x+i} \right|^\nu} = 0$$

uniformly for all $x \in \mathbb{R} \cup \{\infty\}$. Note that the same holds for all non-integer $\mu \in \mathbb{R}^+ \setminus \mathbb{N}_0$.

Proposition A.4. *If $0 < \mu = m + \nu < \mu' = m' + \nu'$, $m, m' \in \mathbb{N}_0$, $0 < \nu, \nu' < 1$, then the embedding $\mathscr{H}^{\mu'}(\mathbb{R}) \subset \widetilde{\mathscr{H}}^\mu(\mathbb{R})$ is continuous and dense.*
If $\varphi \in \widetilde{\mathscr{H}}^\mu(\mathbb{R})$ and

$$\partial_x^k b(x) = \mathcal{O}\left(|x+i|^{-k} \right) \quad \text{for} \quad k = 0, 1, \ldots, m,$$

then $b\varphi \in \widetilde{\mathscr{H}}^\mu(\mathbb{R})$.

Proof. For the proof we refer to [ChDu, §1.6]. $\qquad\square$

Let

$$\mathscr{L}_0^\nu(\mathbb{R}) := \{ \varphi \in \mathscr{L}^\nu(\mathbb{R}) \ : \ \varphi(\infty) = 0 \}.$$

Theorem A.5. *Let $\mu \in \mathbb{R}^+$. Then the Hilbert transform*

$$H_\mathbb{R}\varphi(x) := \frac{1}{\pi i} \int_{-\infty}^{\infty} \frac{\varphi(\tau)\, d\tau}{\tau - x} \tag{A.15}$$

is bounded in the spaces $\mathscr{L}^\nu(\mathbb{R})$, in $\mathscr{L}_0^\nu(\mathbb{R})$, in $\mathscr{H}^\nu(\mathbb{R})$ and in $\widetilde{\mathscr{H}}^\nu(\mathbb{R})$ for all $\nu \ge 0$.

Proof. Let us prove that $H_\mathbb{R}$ is a bounded operator in the space $\mathscr{L}_0^\nu(\mathbb{R})$. Then, due to the relations

$$\mathscr{L}^\nu(\mathbb{R}) = \{\text{const}\} \dotplus \mathscr{L}_0^\nu(\mathbb{R}), \quad H_\mathbb{R}c = 0 \text{ for } c = \text{const}, \tag{A.16}$$

$H_\mathbb{R}$ is bounded in $\mathscr{L}^\nu(\mathbb{R})$ as well. From (A.16) follows, in particular, that

$$H_\mathbb{R}\varphi = H_\mathbb{R}\varphi_0, \quad \varphi_0(x) := \varphi(x) - \varphi(\infty) \tag{A.17}$$

for arbitrary $\varphi \in \mathscr{L}^\nu(\mathbb{R})$. Then, integrating by parts,

$$\partial H_\mathbb{R}\varphi = \partial H_\mathbb{R}\varphi_0 = H_\mathbb{R}\partial\varphi_0 = H_\mathbb{R}\partial\varphi \tag{A.18}$$

which means that the Hilbert transform commutes with the derivative $\partial := d/dt$. Therefore it suffices to prove the theorem for $0 < \mu \leq 1$ and by (A.18) it is easily extensible to all $\mu > 0$.

Thus, we can assume $0 < \mu \leq 1$. The Cauchy singular integral operator

$$S_{\Gamma_0}\psi(z) := \frac{1}{\pi i} \int_{\Gamma_0} \frac{\psi(\zeta)d\zeta}{\zeta - z}$$

is bounded in $\mathscr{L}^\nu(\Gamma_0)$ for all $\nu > 0$. For $0 < \nu < 1$ this is known as *Privalov's Theorem* (see [GoKr, MiPr, Mu]), for $m < \nu < m+1$ (with $m = 1, 2, \ldots$) it follows by the property $\partial_z S_{\Gamma_0} = S_{\Gamma_0}\partial_\zeta$ (cf. a similar (A.18)), and for $\nu = m$ it is proved by interpolation (cf. [DuSp, CaDuSp, St]).

The operator $\varpi_*^{-1} S_{\Gamma_0} \varpi_*$, transformed by the isomorphism (A.12), acquires the form

$$\varpi_*^{-1} S_{\Gamma_0} \varpi_* \varphi(x) \;=\; \frac{1}{\pi i} \int_{-\infty}^{\infty} \frac{\varphi(\tau)}{\dfrac{\tau - i}{\tau + i} - \dfrac{x - i}{x + i}} \frac{2i\,d\tau}{(\tau + i)^2} = \frac{1}{\pi i} \int_{-\infty}^{\infty} \frac{x + i}{\tau + i} \frac{\varphi(\tau)d\tau}{\tau - x}$$

$$=\; H_{\mathbb{R}}\varphi(x) - K_1\varphi, \qquad\quad K_1\varphi := \frac{1}{\pi i} \int_{-\infty}^{\infty} \frac{\varphi(\tau)d\tau}{\tau + i}.$$

Since the one-dimensional operator K_1 is bounded in $\mathscr{L}^\mu(\mathbb{R}) \to \mathbb{C} \subset \mathscr{L}^\mu(\mathbb{R})$, the operator $H_{\mathbb{R}}$ is bounded in $\mathscr{L}^\mu(\mathbb{R})$ for all $0 < \mu \leq 1$.

Next we prove that $H_{\mathbb{R}}$ is bounded in $\mathscr{H}^\mu(\mathbb{R})$. For this we apply the integration by parts

$$(x + i)^k \partial_x^k H_{\mathbb{R}}\varphi = H_{\mathbb{R}}(y + i)^k \partial_y^k \varphi.$$

Applying the proved part of the theorem we proceed as follows:

$$\left\| H_{\mathbb{R}}\varphi \mid \mathscr{H}^\mu(\mathbb{R}) \right\| \;=\; \sum_{k=1}^{m} \left\| (x + i)^k \partial_x^k H_{\mathbb{R}}\varphi \mid \mathscr{L}^\nu(\mathbb{R}) \right\| = \sum_{k=1}^{m} \left\| H_{\mathbb{R}}(y + i)^k \partial_y^k \varphi \mid \mathscr{L}^\nu(\mathbb{R}) \right\|$$

$$\leq\; \| H_{\mathbb{R}} \| \sum_{k=1}^{m} \left\| (x + i)^k \partial_x^k \varphi \mid \mathscr{L}^\nu(\mathbb{R}) \right\| = \| H_{\mathbb{R}} \| \left\| \varphi \mid \mathscr{H}^\mu(\mathbb{R}) \right\|.$$

$H_{\mathbb{R}}$ is bounded in $\widetilde{\mathscr{H}^\mu}(\mathbb{R})$ because it is bounded in $\mathscr{H}^{\mu'}(\mathbb{R})$ for all $0 < \mu < \mu'$ (see Proposition A.4). $\qquad\qquad\qquad\qquad\qquad\qquad\qquad\qquad\square$

Proposition A.6. *Let* $\mu > 0$ *and* $\mathscr{A} \in \mathscr{L}^\mu(\mathbb{R})$ *(or* $\mathscr{A} \in \widetilde{\mathscr{H}^\mu}(\mathbb{R})$*) be an elliptic matrix-valued function. Then* $\mathscr{A}$ *admits the classical factorization*

$$\mathscr{A}(x) = [\mathscr{A}_-(x)]^{-1} \left(\frac{x - i}{x + i} \right)^\varkappa \mathscr{A}_+(x),$$

$$\varkappa = (\varkappa_1, \ldots, \varkappa_N) \in \mathbb{Z}^N, \quad \mathbb{Z} = \{0, \pm 1, \ldots\}$$

(A.19)

with factors $[\mathscr{A}_-(x)]^{\pm 1}$, $[\mathscr{A}_+(x)]^{\pm 1}$ in $\mathscr{L}^\mu(\mathbb{R})$ (or in $\widetilde{\mathscr{H}^\mu}(\mathbb{R})$, respectively) and have uniformly bounded analytic continuations into the half-planes $\Im\, x < 0$ and $\Im\, x > 0$, respectively.

Proof. For the proof we refer to [BuGo, ClGo]: for the space $\mathscr{L}^\mu(\mathbb{R})$ the proof in [BuGo, ClGo] is direct, while for the space $\widetilde{\mathscr{H}^\mu}(\mathbb{R})$ it follows from the general theorem on factorization in a rationally dense and decomposable Banach algebra (cf. Proposition A.4 and Theorem A.5). $\qquad\square$

Acknowledgment

The work was supported in part by "FCT-Portuguese Science Foundation", through the *Applied Mathematics Center* (Instituto Superior Técnico, U.T.L.) and Research Unit *Mathematics and Applications* (University of Aveiro).

References

[Ar]   V. Arnold, *Ordinary Differential Equations.* Springer-Verlag, Heidelberg 1992 (Russian original: 3rd edition, Nauka, Moscow 1984).

[BöSi]   A. Böttcher, B. Silbermann, *Analysis of Toeplitz Operators.* Springer-Verlag, Heidelberg 1990.

[BuGo]   M. Budjanu, I. Gohberg, General theorems on the factorization of matrix functions. II: Certain tests and their consequences. *Mat. Issled.* 3 (1968), 3–18.

[CaDuSp]   L.P. Castro, R. Duduchava, F.-O. Speck, Singular integral equations on piecewise smooth curves in spaces of smooth functions. *Operator Theory: Advances and Applications* **135**, 107–144. Birkhäuser-Verlag, Basel 2002.

[CaSp]   L.P. Castro, F.-O. Speck, Inversion of matrix convolution type operators with symmetry. *Port. Math. (N.S.)* **62** (2005), 193–216.

[CaSpTe1]   L.P. Castro, F.-O. Speck, F.S. Teixeira, On a class of wedge diffraction problems posted by Erhard Meister. In: *Operator Theoretical Methods and Applications to Mathematical Physics* (Eds. I. Gohberg et al.). *Operator Theory: Advances and Applications* **147**, 211–238. Birkhäuser-Verlag, Basel 2004.

[CaSpTe2]   L.P. Castro, F.-O. Speck, F.S. Teixeira, A direct approach to convolution type operators with symmetry. *Math. Nach.* **269-270** (2004), 73–85.

[CaSpTe3]   L.P. Castro, F.-O. Speck, F.S. Teixeira, Mixed boundary value problems for the Helmholtz equation in a quadrant. *Integr. Equ. Oper. Theory* **56** (2006), 1–44.

[ChDu]   O. Chkadua, R. Duduchava, Pseudodifferential equations on manifolds with boundary: Fredholm property and asymptotic. *Math. Nachr.* **222** (2001), 79–139.

[ClGo]   K. Clancey, I. Gohberg, *Factorization of Matrix Functions and Singular Integral Operators. Operator Theory: Advances and Applications* **3**. Birkhäuser-Verlag, Basel 1981.

[DuSp]   R. Duduchava, F.-O. Speck, Pseudo-differential operators on compact manifolds with Lipschitz boundary. *Math. Nachr.* **160** (1993), 149–191.

[DuSäWe] R. Duduchava, A.M. Sändig, W. Wendland, Interface cracks in anisotropic composites. *Math. Meth. Appl. Sciences* **22** (1999), 1413–1446.

[Eh] T. Ehrhardt, Invertibility theory for Toeplitz plus Hankel operators and singular integral operators with flip. *J. Funct. Anal.* **208** (2004), 64–106.

[Ga] F. Gantmacher, *Matrix Theory*. Nauka, Moscow, 1967.

[GoKr] I. Gohberg, N. Krupnik, *Introduction to the theory of one-dimensional singular integral operators*. Birkhäuser-Verlag, Basel 1992.

[GoKaSp] I. Gohberg, M.A. Kaashoek, I.M. Spitkovsky, *An overview of matrix factorization theory and operator applications*. In: *Factorization and Integrable Systems* (Eds. I. Gohberg et al.). Lecture notes of the summer school, Faro, Portugal, September 2000. *Operator Theory: Advances and Applications* **141**, Birkhäuser-Verlag, Basel 2003, p. 1–102.

[Kr] N.Ya. Krupnik, *Banach Algebras with Symbol and Singular Integral Operators*. Birkhäuser-Verlag, Basel 1987.

[La] P. Lancaster, *Theory of Matrices*. Academic Press, New York 1969.

[LiSp] G.S. Litvinchuk, I.M. Spitkovsky, *Factorization of Measurable Matrix Functions*. Birkhäuser-Verlag, Basel 1987.

[MiPr] S. Mikhlin, S. Prössdorf, *Singular Integral Operators*. Springer-Verlag, Heidelberg 1986.

[Mu] N. Muskhelishvili, *Singular Integral Equations*, Nordhoff, Groningen 1953. Last Russian edition: Nauka, Moscow 1968; Last English edition: Dover Publications, Inc., New York 1992.

[Sh] E. Shamir, Elliptic systems of singular integral equations. I: The half-space case. *Trans. Amer. Math. Soc.* **127** (1967), 107–124.

[Si] I.B. Simonenko, Some general questions in the theory of the Riemann boundary problem (Russian). Izv. Akad. Nauk SSSR Ser. Mat. **32** (1968), 1138–1146. English translation in Math. USSR, Izv. **2** (1968), 1091–1099.

[St] E. Stein, *Singular Integrals and Differentiability Properties of Functions*. Princeton Univ. Press, Princeton 1970.

[Ta] N. Tarkhanov, *The Cauchy Problem for Solutions of Elliptic Equations*. Akademie-Verlag, Berlin 1995.

[Tr] H. Triebel, *Interpolation Theory, Function Spaces, Differential Operators*. North-Holland, Amsterdam 1978.

L.P. Castro
Department of Mathematics, University of Aveiro
3810-193 Aveiro, Portugal
e-mail: lcastro@mat.ua.pt

R. Duduchava
A. Razmadze Mathematical Institute, Academy of Sciences of Georgia
1, M.Alexidze str., Tbilisi 93, Georgia
e-mail: dudu@rmi.acnet.ge

F.-O. Speck
Department of Mathematics, Instituto Superior Técnico, U.T.L.
Avenida Rovisco Pais, 1049–001 Lisboa, Portugal
e-mail: fspeck@math.ist.utl.pt

Operator Theory:
Advances and Applications, Vol. 170, 75–84
© 2006 Birkhäuser Verlag Basel/Switzerland

On the Structure of the Square of a $C_0(1)$ Operator

Ronald G. Douglas and Ciprian Foias

Dedicated to I.B. Simonenko on his seventieth birthday

Abstract. We use the structure theory for C_0 operators to determine when the square of a $C_0(1)$ operator is irreducible and when its lattices of invariant and hyperinvariant subspaces coincide.

Mathematics Subject Classification (2000). 47A15, 47A45.

Keywords. C_0 operators, invariant subspace lattice.

0. While the model theory for contraction operators (cf. [4]) is always a useful tool, it is particularly powerful when dealing with $C_0(1)$ operators. Recall that an operator T on a Hilbert space H is a $C_0(N)$-operator ($N = 1, 2, \ldots$) if $\|T\| \leq 1$, $T^n \to 0$ and, $T^{*n} \to 0$ (strongly) when $n \to \infty$ and $\operatorname{rank}(1 - T^*T) = N$. In particular, a $C_0(1)$ operator is unitarily equivalent to the compression of the unilateral shift operator S on the Hardy space H^2 to a subspace $H^2 \ominus mH^2$ for some inner function m in H^∞.

In this note we use the structure theory to determine when the lattices of invariant and hyperinvariant subspaces differ for the square T^2 of a $C_0(1)$ operator and the relationship of that to the reducibility of T^2. To accomplish this task we first determine very explicitly the characteristic operator function for T^2 and use the representation obtained to determine when the operator is irreducible. While every operator T in $C_0(1)$ is irreducible, it does not follow that T^2 is necessarily irreducible, that is, has no reducing subspaces. In particular, we characterize those T in $C_0(1)$ for which T^2 is irreducible but for which the lattices of invariant and hyperinvariant subspaces for T^2 are distinct.

Finally, we provide an example of an operator X of the form T^2, as above, on a four-dimensional Hilbert space for which the two lattices are distinct but X is irreducible. Moreover, we observe that no example exists on a three-dimensional space.

This work was prompted by a question to the first author from Ken Dykema (Sect. 2, [2]) concerning hyperinvariant subspaces in von Neumann algebras. He asked whether the lattices of invariant and hyperinvariant subspaces for an irreducible matrix must coincide. He provides an example in [2] on a six-dimensional Hilbert space showing that this is not the case.

We assume that the reader is familiar with the concepts and notation in [1] and [4].

1. Let $T \in C_0(1)$ on H, $\dim H \geq 2$. WLOG we can assume

$$T = P_H S|_H, \text{ where } H = H^2 \ominus mH^2,$$
$$(Sh)z = zh(z)(z \in D, h \in H^2), m \in H^\infty, m \text{ inner.} \tag{1}$$

Define

$$\Theta(\lambda) = \frac{1}{2} \begin{bmatrix} b(\lambda) & \lambda d(\lambda) \\ d(\lambda) & b(\lambda) \end{bmatrix} \quad (\lambda \in D), \tag{2}$$

where

$$b(\lambda) = m(\sqrt{\lambda}) + m(-\sqrt{\lambda}) \quad (\lambda \in \mathbb{D}) \quad \text{and} \tag{3a}$$

$$\begin{cases} d(\lambda) = \dfrac{m(\sqrt{\lambda}) - m(-\sqrt{\lambda})}{\sqrt{\lambda}} \quad (0 \neq \lambda \in \mathbb{D}) \\ d(0) = 2m'(0). \end{cases} \tag{3b}$$

Lemma 1. *The matrix function $\Theta(\cdot)$ is inner, pure and (up to a coincidence) the characteristic operator function of T^2.*

Proof. For $h \in H^2$ write

$$h(\lambda) = h_0(\lambda^2) + \lambda h_1(\lambda^2) \quad (\lambda \in \mathbb{D}). \tag{1.4a}$$

Clearly $h_0(\cdot), h_1(\cdot)$ $(= h_0(\lambda), h_1(\lambda), \lambda \in \mathbb{D})$ belong to H^2. Define $W \colon H^2 \mapsto H^2 \oplus H^2$ $(= H^2(\mathbb{C}^2))$ by

$$Wh = h_0 \oplus h_1, \text{ where } h \text{ is given by (1.4a).} \tag{1.4b}$$

Then W is unitary and

$$WS^2 = (S \oplus S)W. \tag{5}$$

Consequently,

$$WT^2 = WP_H S^2 = P_{WH} WS^2 = P_{WH}(S \oplus S)W; \tag{6}$$

moreover, since $S^2 mH^2 \subset mH^2$ we also have

$$(S \oplus S)WmH^2 = WS^2 mH^2 \subset WmH^2$$

and therefore

$$\begin{cases} P_{WH}(S \oplus S) = P_{WH}(S \oplus S)P_{WH} = WP_H W^*(S \oplus S)P_{WH} \\ \qquad = WP_H S^2 W^* P_{WH} = WT^2 P_H W^* \\ \qquad = W|_H T^2 (W|_H)^*. \end{cases} \tag{7}$$

These relationships show that $S \oplus S$ is an isometric lifting of $T_0 = P_{WH}(S \oplus S)|_{WH}$ and that this operator is unitarily equivalent to T^2. Moreover, since

$$\bigvee_{n=0}^{\infty} (S \oplus S)^n WH = H^2 \oplus H^2$$

is obvious, $S \oplus S$ is *the* minimal isometric lifting of $T = W|_H T^2 (W|_H)^*$.

Further,

$$\begin{aligned}
WmH^2 &= \{W(m_0(\lambda^2) + \lambda m_1(\lambda^2))(h_0(\lambda^2) + \lambda h_1(\lambda^2)): \ h \in H^2\} \\
&= \{W[(m_0 h_0)(\lambda^2) + \lambda^2(m_1 h_1)(\lambda^2) \\
&\qquad + \lambda(m_0 h_1 + m_1 h_0)(\lambda^2): \ h \in H^2\} \\
&= \{((m_0 h_0)(\lambda) + \lambda(m_1 h_1)(\lambda)) \oplus (m_0 h_1 + m_1 h_0)(\lambda): \ h \in H^2\} \\
&= \left\{ \begin{bmatrix} m_0 & \lambda m_1 \\ m_1 & m_0 \end{bmatrix} (h_0 \oplus h_1): \ h \in H^2 \right\} = \begin{bmatrix} m_0 & \lambda m_1 \\ m_1 & m_0 \end{bmatrix} H^2 \oplus H^2.
\end{aligned}$$

Note that the above computations also prove that

$$(Wm(S)W^*)(h_0 \oplus h_1) = \begin{bmatrix} m_0 & \lambda m_1 \\ m_1 & m_0 \end{bmatrix} h_0 \oplus h_1 \qquad (h_0 \oplus h_1 \in H^2 \oplus H^2). \qquad (8)$$

Since $m(S)$ is isometric, so is $Wm(S)W^*$, that is,

$$M(\lambda) \equiv \begin{bmatrix} m_0(\lambda) & \lambda m_1(\lambda) \\ m_1(\lambda) & m_0(\lambda) \end{bmatrix} \text{ is inner.} \qquad (9)$$

Consequently, T_0 is the compression of $S \oplus S$ to

$$WH = (H^2 \oplus H^2) \ominus M(H^2 \oplus H^2). \qquad (10)$$

Moreover, it is clear that

$$m_0(\lambda) = \frac{1}{2} b(\lambda), \quad m_1(\lambda) = \frac{1}{2} d(\lambda) \qquad (\lambda \in \mathbb{D})$$

so that the matrix $M(\cdot)$ defined by (9) is identical to the matrix $\Theta(\cdot)$ defined by (2).

Note that

$$\Theta(0) = \begin{bmatrix} m(0) & 0 \\ m'(0) & m(0) \end{bmatrix}$$

and

$$\Theta(0)^* \Theta(0) = \begin{bmatrix} |m(0)|^2 + |m'(0)|^2 & \overline{m'(0)} m(0) \\ \overline{m(0)} m'(0) & |m(0)|^2 \end{bmatrix}.$$

If $\Theta(\lambda)$ were not pure, then $\Theta(\lambda)^* \Theta(\lambda)$ would have the eigenvalue 1 and therefore the other eigenvalue must be $|m(0)|^4$. Taking traces we have

$$2|m(0)|^2 + |m'(0)|^2 = 1 + |m(0)|^4.$$

This implies that the modulus of the analytic function $\widetilde{m}(\lambda)$ defined by

$$\lambda \widetilde{m}(\lambda) = \frac{m(\lambda) - m(0)}{1 - \overline{m(0)} m(\lambda)} \qquad (\lambda \in \overline{\mathbb{D}}, \lambda \neq 0)$$

and

$$\widetilde{m}(0) = \frac{m'(0)}{1 - |m(0)|^2}$$

attains its maximum ($= 1$) at $\lambda = 0$. By virtue of the maximum principle, $\widetilde{m}(\lambda) = c = \text{constant}$, $|c| = 1$. Thus

$$m(\lambda) \equiv c \left(\frac{\lambda + \bar{c}m(0)}{1 + \lambda cm(0)} \right) \qquad (\lambda \in \overline{\mathbb{D}})$$

and

$$2 \leq \dim H = \dim(H^2 \ominus mH^2) = 1,$$

which is a contradiction.

We conclude that $\Theta(\cdot)$ is pure and, by virtue of (10) (recall $\Theta(\lambda) \equiv M(\lambda)$), that $\Theta(\cdot)$ is the characteristic operator function of T_0 and hence (up to a coincidence) also the characteristic operator function of T^2. This concludes the proof of the lemma. $\square$

Note that the preceding result also shows that T^2 is a $C_0(2)$ operator.

2.   Our next step is to characterize in terms of $\Theta(\lambda)$ the reducibility of T^2.

Lemma 2. *The operator T^2 is reducible if and only if there exist $Q_i = Q_i^* = Q_i^2$, $Q_i \in \mathcal{L}(\mathbb{C}^2)$ $(i = 1, 2)$ so that*

$$\Theta(\lambda)Q_2 = Q_1\Theta(\lambda) \qquad (\lambda \in \mathbb{D}) \tag{11}$$

and $0 \neq Q_i \neq I_{\mathbb{C}^2}$ $(i = 1, 2)$.

Proof. If Q_1, Q_2 as above exist, then (since rank $Q_1 = 1 = \text{rank } Q_2$) there exist unitary operators in $\mathcal{L}(\mathbb{C}^2)$ so that

$$W_1\Theta(\lambda)W_2 = \begin{bmatrix} \theta_1(\lambda) & 0 \\ 0 & \theta_2(\lambda) \end{bmatrix} \qquad (\lambda \in \mathbb{D}) \text{ for functions } \theta_1(\cdot), \theta_2(\cdot). \tag{12}$$

Indeed, if W_1 and W_2 are unitary operators in $\mathcal{L}(\mathbb{C}^2)$ such that

$$Q_1\mathbb{C}^2 = W_1^*(\mathbb{C} \oplus \{0\}), \quad Q_2\mathbb{C}^2 = W_2(\mathbb{C} \oplus \{0\}),$$

then

$$W_1\Theta(\lambda)W_2 \begin{bmatrix} 1 & 0 \\ 0 & 0 \end{bmatrix} - \begin{bmatrix} 1 & 0 \\ 0 & 0 \end{bmatrix} W_1\Theta(\lambda)W_2$$

$$= W_1 \left\{ \Theta(\lambda)W_2 \begin{bmatrix} 1 & 0 \\ 0 & 0 \end{bmatrix} W_2^* - W_1^* \begin{bmatrix} 1 & 0 \\ 0 & 0 \end{bmatrix} W_1\Theta(\lambda) \right\} W_2$$

$$= W_1(\Theta(\lambda)Q_2 - Q_1\Theta(\lambda)) = 0.$$

Thus $\mathbb{C} \oplus \{0\}$ (and hence also $\{0\} \oplus \mathbb{C}$) reduces $W_1\Theta(\lambda)W_2$ and consequently this operator has the form (12).

Clearly the θ_1, θ_2 in (12) are inner (and non-constant). Let

$$T_i = P_{H_i}S|_{H_i}, \text{ where } H_i = H^2 \ominus \theta_iH^2 \qquad (i = 1, 2). \tag{13}$$

Then the characteristic operator function of $T_1 \oplus T_2$ is the right-hand side of (12) which coincides with $\Theta(\lambda)$. Thus T^2 and $T_1 \oplus T_2$ are unitarily equivalent.

Conversely, if T^2 is reducible then T^2 is unitarily equivalent to the direct sum $T_1' \oplus T_2'$, where $T_i' = T^2|_{H_i}$ $(i = 1, 2)$, H_1, H_2 are reducing subspaces for T^2, and $H = H_1 \oplus H_2$. Clearly each $T_i' \in C_{00}$ and since the defect indices of the T_i's sum up to 2, it follows that each $T_i' \in C_0(1)$. Thus the characteristic operator function of $T_1' \oplus T_2'$ coincides with

$$\begin{bmatrix} \theta_1(\lambda) & 0 \\ 0 & \theta_2(\lambda) \end{bmatrix}, \tag{14}$$

where θ_i is the characteristic function of T_i' $(i = 1, 2)$. Again $\Theta(\lambda)$ is connected to (14) by a relation of the form (12), that is,

$$\Theta(\lambda) \equiv W_1^* \begin{bmatrix} \theta_1(\lambda) & 0 \\ 0 & \theta_2(\lambda) \end{bmatrix} W_2^*,$$

where W_1, W_2 are again unitary. Then

$$Q_1 = W_1^* \begin{bmatrix} 1 & 0 \\ 0 & 0 \end{bmatrix} W_1, \quad Q_2 = W_2 \begin{bmatrix} 1 & 0 \\ 0 & 0 \end{bmatrix} W_2^*$$

satisfy (11). $\qquad\square$

Remark. *Note that in (11), the orthogonal projections Q_1, Q_2 are of rank one. Such a projection Q is of the form*

$$Q = f \otimes f = \begin{bmatrix} |f_1|^2 & f_1 \bar{f_2} \\ f_2 \bar{f_1} & |f_2|^2 \end{bmatrix}, \tag{15}$$

where

$$f = f_1 \oplus f_2 \in \mathbb{C}^2, \quad \|f\| = 1.$$

Thus

$$Q = \begin{bmatrix} q & r\bar{\theta} \\ r\theta & 1 - q \end{bmatrix}, \quad \textit{where } 0 \leq q \leq 1, |\theta| = 1, r = (q(1 - q))^{1/2}. \tag{16}$$

3. In this paragraph we study the relation (11) using the representation (16) for $Q = Q_i$ $(i = 1, 2)$ and the form (2) of $\Theta(\lambda)$. Thus we have

$$\begin{bmatrix} b(\lambda) & \lambda d(\lambda) \\ d(\lambda) & b(\lambda) \end{bmatrix} \begin{bmatrix} q_2 & r_2 \bar{\theta}_2 \\ r_2 \theta_2 & 1 - q_2 \end{bmatrix} = \begin{bmatrix} q_1 & r_1 \bar{\theta}_1 \\ r_1 \theta_1 & 1 - q_1 \end{bmatrix} \begin{bmatrix} b(\lambda) & \lambda d(\lambda) \\ d(\lambda) & b(\lambda) \end{bmatrix}, \tag{17}$$

where

$$0 \leq q_1, q_2 \leq 1, |\theta_1| = |\theta_2| = 1, r_i = (q_i(1 - q_i))^{1/2} \quad (i = 1, 2). \tag{18}$$

We begin by noting that

$$|b(\lambda)|^2 + |d(\lambda)|^2 \not\equiv 0 \quad (\lambda \in \mathbb{D}), \tag{19}$$

since otherwise we would have $m(\lambda) \equiv 0$. In discussing (17) we will consider several cases:

Case I. If $b(\lambda) \equiv 0 \ (\lambda \in \mathbb{D})$, then (17) becomes:

$$\begin{bmatrix} \lambda d(\lambda) r_2 \theta_2 & \lambda d(\lambda)(1 - q_2) \\ d(\lambda) q_2 & d(\lambda) r_2 \bar{\theta}_2 \end{bmatrix} = \begin{bmatrix} r_1 \bar{\theta}_1 d(\lambda) & \lambda q_1 d(\lambda) \\ (1 - q_1) d(\lambda) & r_1 \theta_1 \lambda d(\lambda) \end{bmatrix}$$

which is possible if and only if $r_1 = 0 = r_2$ and $q_1 = 1 - q_2$. In this case T^2 is reducible.

Case II. If $d(\lambda) \equiv 0 \ (\lambda \in \mathbb{D})$, then

$$Q_2 = Q_1 = \text{ any } Q = Q^* = Q^2 \text{ with rank } Q = 1$$

and again T^2 is reducible.

Case III. If $b(\lambda) \not\equiv 0, d(\lambda) \not\equiv 0 \ (\lambda \in \mathbb{D})$, then (17) is equivalent to the equations

$$b(q_2 - q_1) = d(r_1 \bar{\theta}_1 - \lambda r_2 \theta_2), \quad b(r_2 \theta_2 - r_1 \theta_1) = \lambda d(q_1 + q_2 - 1)$$
$$d(q_1 + q_2 - 1) = b(r_1 \theta_1 - r_2 \theta_2), \qquad b(q_2 - q_1) = d(r_2 \bar{\theta}_2 - \lambda r_1 \bar{\theta}_1),$$

which in turn are equivalent to

$$\begin{cases} r_1 \theta_1 = r_2 \theta_2, \quad q_2 + q_1 = 1 \\ b(\lambda)(1 - 2q_1) \equiv d(\lambda)(\bar{\theta}_1 - \lambda \theta_1) r_1 \quad (\lambda \in \mathbb{D}). \end{cases} \tag{20}$$

In (20), $q_1 = 1/2$, if and only if $r_1 = 0$, i.e., $q_1 = 0$ or 1, a contradiction. Thus we can divide by $1 - 2q_1$ and (20) implies (with $\theta = \theta_1$)

$$\begin{cases} b(\lambda) \equiv d(\lambda)(\bar{\theta} - \lambda \theta)\rho \quad (\lambda \in \mathbb{D}) \\ \text{for some} \quad \rho \in \mathbb{R}, \rho \neq 0. \end{cases} \tag{21}$$

Conversely, if (21) holds, then setting

$$q_1 = \frac{1}{2} \pm \frac{1}{2} \frac{1}{(4\rho^2 + 1)^{1/2}} \ \text{(according to whether } \rho \lessgtr 0),$$

and $q_2 = 1 - q_1$, $\theta_2 = \theta_1 = \theta$, we obtain (20).

We now summarize our discussion in terms of $m(\cdot)$ (see (3a), (3b)), instead of $b(\cdot)$ and $d(\cdot)$, obtaining the following:

Lemma 3. *The operator T^2 is reducible if and only if one of the following conditions holds:*

$$m(-\lambda) \equiv -m(\lambda) \quad (\forall \lambda \in \mathbb{D}) \qquad (\textit{Case I above}); \tag{22}$$
$$m(-\lambda) \equiv m(\lambda) \quad (\forall \lambda \in \mathbb{D}) \qquad (\textit{Case II above}); \tag{23}$$

or there exist $\rho \in \mathbb{R}, \rho \neq 0$ and $\theta \in \mathbb{C}, |\theta| = 1$, such that the function

$$n(\lambda) \equiv m(\lambda)(\rho \theta \lambda^2 + \lambda - \rho \bar{\theta}) \qquad (\lambda \in \mathbb{D}) \tag{24a}$$

satisfies

$$n(\lambda) \equiv n(-\lambda) \qquad (\lambda \in \mathbb{D}) \qquad (\textit{Case III above}). \tag{24b}$$

4. We shall now give a more transparent form to conditions (24a), (24b) above. To this end note that

$$\rho\theta\lambda^2 + \lambda - \rho\bar\theta \equiv \rho\theta(\lambda - \delta_+\bar\theta)(\lambda - \delta_-\bar\theta),$$

where

$$\delta_\pm = \frac{-1 \pm \sqrt{4\rho^2 + 1}}{2\rho}. \tag{25}$$

Thus (with $\mu = \bar\theta\delta_+$), we have

$$\rho\theta\lambda^2 + \lambda - \rho\bar\theta = -\rho\delta_-(\lambda - \mu)(1 + \bar\mu\lambda). \tag{26}$$

Using this representation in(24a), condition (24b) becomes

$$m(\lambda)(\lambda - \mu)(1 + \bar\mu\lambda) \equiv m(-\lambda)(-\lambda - \mu)(1 - \lambda\bar\mu) \qquad (\lambda \in \mathbb{D}),$$

which can be written (since $0 < |\mu| < 1$) as

$$m(\lambda)\frac{\lambda - \mu}{1 - \bar\mu\lambda} \equiv m(-\lambda)\frac{(-\lambda) - \mu}{1 - \bar\mu(-\lambda)} \qquad (\lambda \in \mathbb{D}). \tag{27}$$

Thus $m(-\mu) = 0$ and therefore

$$m(\lambda) = p(\lambda)\frac{\lambda + \mu}{1 + \bar\mu\lambda} \qquad (\lambda \in \mathbb{D}), \tag{28}$$

where $p(\cdot) \in H^\infty$ is an (other) inner function. Obviously (27) is equivalent to

$$p(\lambda) \equiv p(-\lambda) \qquad (\lambda \in \mathbb{D}). \tag{29}$$

This discussion together with Lemma 3, readily yields the following

Theorem 1. *The operator T^2 is reducible iff either*

$$m(\lambda) = m(-\lambda) \qquad (\lambda \in \mathbb{D}) \tag{30}$$

or there exists a $\mu \in \mathbb{D}$ such that

$$m(\lambda) \equiv p(\lambda)\frac{\lambda + \mu}{1 + \bar\mu\lambda} \qquad (\lambda \in \mathbb{D}), \tag{31}$$

where $p(\cdot) \in H^\infty$ satisfies

$$p(\lambda) \equiv p(-\lambda) \qquad (\lambda \in \mathbb{D}). \tag{32}$$

Remark. *Case (22) is contained in the second alternative above when $\mu = 0$.*

5. In order to study the lattices $\mathrm{Lat}\{T^2\}$ and $\mathrm{Lat}\{T^2\}'$ we first bring together the following characterization of the $C_0(N)$ operators that are multiplicity free.

Proposition 1. *Let $\tilde T$ be a $C_0(N)$ operator. Then the following statements are equivalent.*

(1) $\tilde T$ is multiplicity free (that is, $\tilde T$ has a cyclic vector).
(2) $\mathrm{Lat}\{\tilde T\} = \mathrm{Lat}\{\tilde T\}'$.
(3) The minors of the characteristic matrix function of order $N - 1$ have no common inner divisor.

Proof. The equivalence of (1) and (3) is contained in the equivalence of (i) and (ii) in Theorem 2 in [3]. The implication (1) implies (2) is an easy corollary of the implication (i) implies (vi) of the same theorem and is contained in Corollary 2.14 in Chapter 3 of [1]. Finally, implication (3) implies (1) proceeds from the following lemma. $\qquad\square$

Lemma 4. *Let T be an C_0 operator on the Hilbert space $\mathcal{H}$ and f a maximal vector for T. Then f is cyclic for $\{T\}'$.*

Proof. Let $\mathcal{M}$ be the cyclic subspace for $\{T\}'$ generated by f and write $T \sim \left(\begin{smallmatrix} T' & X \\ 0 & T'' \end{smallmatrix}\right)$ for the decomposition $\mathcal{H} = \mathcal{M} \oplus \mathcal{M}^{\perp}$. Since $\mathcal{M}$ is hyperinvariant for T, it follows from Corollary 2.15 in Chapter 4 of [1], that the minimal functions satisfy $m_T = m_{T'} \cdot m_{T''}$. However, f maximal for T implies that $m_{T'} = m_T$ and hence $m_{T''} = 1$. Therefore, $\mathcal{M}^{\perp} = (0)$ or $\mathcal{M} = \mathcal{H}$ which completes the proof. $\qquad\square$

6. Our next aim is to characterize the case when the operator T^2 is multiplicity free. According to Proposition 1 that happens if and only if

$$b(\lambda), d(\lambda) \text{ and } \lambda d(\lambda)$$

have no common nontrivial inner divisor. Let $q(\lambda)$ be an inner divisor of $b(\lambda)$ and $d(\lambda)$, that is,

$$m(\sqrt{\lambda}) + m(-\sqrt{\lambda}) \equiv q(\lambda)r(\lambda) \tag{33a}$$

$$(\lambda \in \mathbb{D})$$

$$m(\sqrt{\lambda}) - m(-\sqrt{\lambda}) \equiv q(\lambda)\sqrt{\lambda}s(\lambda) \tag{33b}$$

for some $r, s \in H^{\infty}$. It follows that

$$m(\lambda) \equiv q(\lambda^2)(r(\lambda^2) - \lambda s(\lambda^2)), \tag{34}$$

that is, $m(\lambda)$ has an even inner divisor.

Conversely, if $m(\cdot)$ has an inner divisor (in H^{∞}) $p(\cdot)$ satisfying

$$p(\lambda) \equiv p(-\lambda), \tag{35}$$

then $q(\lambda) = p(\sqrt{\lambda}) = p(-\sqrt{\lambda})$ is in H^{∞} and inner. Thus $m(\lambda)$ can be represented as in (34) and clearly (34) implies (33a), (33b).

Thus we obtained the following:

Theorem 2. *The operator T^2 is multiplicity free iff the characteristic function $m(\lambda)$ for T has no nontrivial inner divisor $p(\lambda)$ in H^{∞} such that (see (35))*

$$p(\lambda) \equiv p(-\lambda) \qquad (\forall \lambda \in \mathbb{D}).$$

7. Our main result is now a direct consequence of Theorems 1 and 2 and Proposition 1, namely

Theorem 3. *Let $T \in C_0(1)$ satisfy:*

$$m_T(\lambda) \not\equiv m_T(-\lambda) \tag{A}$$

(B) *For $m_T(\lambda_0) = 0$, $\lambda_0 \in \mathbb{D}$, the function*

$$m_{T,\lambda_0}(\lambda) = m_T(\lambda) \Big/ \frac{\lambda - \lambda_0}{1 - \bar{\lambda}_0 \lambda} \qquad (\lambda \in \mathbb{D})$$

is not even, that is,

$$m_{T,\lambda_0}(\lambda) \not\equiv m_{T,\lambda_0}(-\lambda).$$

(C) *There exists a nontrivial inner divisor $p(\lambda)$ (in H^∞) of $m_T(\lambda)$ such that*

$$p(\lambda) \equiv p(-\lambda).$$

Then

$$T^2 \text{ is irreducible,} \tag{D}$$

and

$$\operatorname{Lat} T^2 \neq \operatorname{Lat}\{T^2\}'. \tag{E}$$

8. Remarks

1) Let

$$m_T(\lambda) = \frac{\lambda^2 - \lambda_1}{1 - \bar{\lambda}_1 \lambda^2} \left(\frac{\lambda - \lambda_2}{1 - \bar{\lambda}_2 \lambda} \right)^2 \qquad (\lambda \in \mathbb{D}), \tag{36}$$

where $\lambda_1, \lambda_2 \in \mathbb{D}$, $\lambda_2^2 \neq \lambda_1$. Then m fulfills the conditions (A), (B), (C) in Theorem 3, T^2 satisfies (D) and (E) above and $\dim H = 4$.

2) Using elementary arguments on a standard form for a 3×3 matrix T, one can show that if $\operatorname{Lat}\{T\} \neq \operatorname{Lat}\{T\}'$, then T is reducible. Hence, the example given in (1) is on the lowest-dimensional space possible.

3) Let m_T be singular, that is,

$$m_T(\lambda) = \exp\left[-\frac{1}{2\pi} \int_0^\pi \frac{e^{it} + \lambda}{e^{it} - \lambda} d\mu(e^{it}) \right]$$

with μ a singular measure on $\partial\mathbb{D} = \{e^{it} : 0 \le t < 2\pi\}$. Assume that there exists a Borel set $\Omega \subset \partial\mathbb{D}$ so that

$$\mu(\Omega) = \mu(\partial\mathbb{D}), \quad \mu(\{-\lambda : \lambda \in \Omega\}) = 0.$$

(e.g., $\mu = \delta_1$, the point mass at 1). Then

$$\operatorname{Lat}\{T^2\} = \operatorname{Lat}\{T^2\}' = \operatorname{Lat}\{T\}. \tag{37}$$

Indeed, in this case (C) above does not hold.

References

[1] Hari Bercovici, *Operator Theory and Arithmetic in H^∞*, Amer. Math. Soc., Providence, RI., 1988.

[2] Ken Dykema, Hyperinvariant subspaces for some B-circular operators, Math. Ann. (to appear).

[3] Bela Sz.-Nagy and Ciprian Foias, Opérateurs sans multiplicité, Acta Sci. Math. (Szeged) 30 (1969), 1–18.

[4] Bela Sz.-Nagy and Ciprian Foias, *Harmonic Analysis of Operators on Hilbert Space*, North Holland, 1970.

Ronald G. Douglas
Department of Mathematics
Texas A&M University
College Station, TX 77843-3368, USA
e-mail: `rdouglas@math.tamu.edu`

Ciprian Foias
Department of Mathematics
Texas A&M University
College Station, TX 77843-3368, USA

Operator Theory:
Advances and Applications, Vol. 170, 85–100
© 2006 Birkhäuser Verlag Basel/Switzerland

On the Connection Between the Indices of a Block Operator Matrix and of its Determinant

Israel Feldman, Nahum Krupnik and Alexander Markus

Dedicated to Igor Simonenko with respect and friendship

Abstract. We consider a finite block operator matrix $\mathcal{A}$ in a Hilbert space. If the entries of $\mathcal{A}$ commute modulo the compact operators, then $\mathcal{A}$ is a Fredholm operator if and only if $\det \mathcal{A}$ is a Fredholm operator, but in general ind $\mathcal{A} \neq$ ind $\det \mathcal{A}$. On the other hand, if the commutators of the entries of $\mathcal{A}$ are trace class operators then ind $\mathcal{A} =$ ind $\det \mathcal{A}$. We obtain formulas for the difference ind $\mathcal{A} -$ ind $\det \mathcal{A}$ provided the entries of $\mathcal{A}$ commute modulo some von Neumann–Schatten ideal. Then we indicate some ideals larger than the ideal of trace class operators for which the mentioned statement about the equality ind $\mathcal{A} =$ ind $\det \mathcal{A}$ remains true.

Mathematics Subject Classification (2000). Primary 47A53; Secondary 47B10, 49J55.

Keywords. Block operator matrices, Fredholm operators, index, trace, operator ideals, von Neumann–Schatten ideals, commutators.

1. Introduction

1. Let $\mathcal{L}(H)$ be the set of all linear bounded operators in an infinite-dimensional separable Hilbert space H and $\mathcal{K}(H)$ be the subset of all compact operators.

If $A \in \mathcal{K}(H)$ then A^*A is a compact non-negative operator. Let

$$\lambda_1(A^*A) \geq \lambda_2(A^*A) \geq \cdots$$

be the sequence of eigenvalues of A^*A where each non-zero eigenvalue is repeated as many times as the value of its multiplicity. By definition, the *s-numbers* of A are the numbers $s_n(A) = (\lambda_n(A^*A))^{1/2}$ $(n = 1, 2, \ldots)$. The term *"ideal"* always means a non-trivial (i.e., different from $\{0\}$ and $\mathcal{L}(H)$) two-sided ideal of the algebra $\mathcal{L}(H)$. Important well-known classes of ideals are the *von Neumann–Schatten* ideals $\mathcal{S}_p$ $(p > 0)$. By definition

$$\mathcal{S}_p = \{A \in \mathcal{K}(H) : \sum_{n=1}^{\infty} s_n^p(A) < \infty\}.$$

If $p = 1$ we obtain the ideal of *trace class operators* $\mathcal{S}_1$, and if $p = 2$ we obtain the ideal of *Hilbert–Schmidt operators* $\mathcal{S}_2$.

An operator $A \in \mathcal{L}(H)$ is called a *Fredholm* operator if its range is closed and the null spaces $\mathrm{Ker} A$ of A and $\mathrm{Ker} A^*$ of A^* are finite-dimensional. The number

$$\mathrm{ind}\ A := \dim \mathrm{Ker} A - \dim \mathrm{Ker} A^*$$

is said to be the *index* of A.

For two operators $A, B \in \mathcal{L}(H)$ we denote by $[A, B]$ their *commutator*, $[A, B] := AB - BA$.

If H^n is the orthogonal sum of n copies of the spaces H, then any operator $\mathcal{A} \in \mathcal{L}(H^n)$ can be represented in the form of an $n \times n$ block operator matrix:

$$\mathcal{A} = [A_{jk}]_{j,k=1}^n \quad (A_{jk} \in \mathcal{L}(H)). \tag{1.1}$$

Suppose that the entries A_{jk} commute modulo $\mathcal{K}(H)$, i.e., all commutators $[A_{jk}, A_{j'k'}]$ are compact operators. Define the determinant $\det \mathcal{A}$ in the usual way as a sum of products of its block matrices. If we are interested only in the Fredholm properties of $\det \mathcal{A}$ then the order of the factors A_{jk} in each term (product) of $\det \mathcal{A}$ does not matter since each product differs from any of its reordered products by a compact operator.

It has been known for a long time that $\mathcal{A}$ is a Fredholm operator if and only if $\det \mathcal{A}$ is. This result was proved in [K1] (see also [K2, Theorem 2.1]), and another approach to its proof can be found in [Ha, Problem 55]. On the other hand it is known that under the above-mentioned conditions the equality

$$\mathrm{ind}\ \mathcal{A} = \mathrm{ind}\ \det \mathcal{A} \tag{1.2}$$

does not hold in general (see Section 2). It was proved in [MF] that equality (1.2) holds under the stronger condition that the entries A_{jk} commute modulo $\mathcal{S}_1$. In [FM] it was conjectured that here the class $\mathcal{S}_1$ is in some sense sharp.

2. The main purposes of this paper are the following: (a) to obtain formulas for the difference $\mathrm{ind}\ \mathcal{A} - \mathrm{ind}\ \det \mathcal{A}$ in the case when the entries of $\mathcal{A}$ commute modulo some $\mathcal{S}_p$ $(p > 1)$; (b) to show that in the mentioned result [MF] regarding the validity of equality (1.2) the set (ideal) $\mathcal{S}_1$ can be replaced by some larger ideals of the algebra $\mathcal{L}(H)$.

An outline of the contents of the paper is as follows. In the next section we analyse one known example when equality (1.2) does not hold. We show that in this example the commutators of the entries of $\mathcal{A}$ satisfy the conditions

$$s_n\left([A_{jk}, A_{j'k'}]\right) = O\left(1/\sqrt{n}\right) \quad (n \to \infty).$$

We prove also that under these conditions the parameters $\mathrm{ind}\ \mathcal{A}$ and $\mathrm{ind}\ \det \mathcal{A}$ can take any two prescribed integers.

In Section 3 some known results are formulated, and among them one recent result [DFWW] which plays a decisive role in this paper (see Theorem 3.1).

In Section 4 we consider the case of 2×2 block operator matrices. We do this separately from the general case for two reasons. First, it is easier to explain the main ideas and methods on this simplest model. Second, the formulas for ind $\mathcal{A}$ − ind det $\mathcal{A}$ obtained in Section 4 are not direct corollaries of the general formulas from Section 5, and their independent deductions seem us more natural. We mention here one of these formulas. If

$$\mathcal{A} = \begin{bmatrix} A & B \\ C & D \end{bmatrix}$$

and all blocks A, B, C, D commute modulo $\mathcal{S}_2$, then

$$\text{ind } \mathcal{A} - \text{ind } \det \mathcal{A} = \text{tr } F(MA, MB, C, D).$$

Here $\text{tr} X$ is the trace of the operator X (see Section 3 for definition), M is an inverse of $\det \mathcal{A} = AD - BC$ modulo $\mathcal{S}_2$ and $F(X_1, X_2, X_3, X_4)$ is so-called standard polynomial (see Section 4).

Section 5 contains the main results. For operator matrix (1.1) under the conditions

$$[A_{jk}, A_{j'k'}] \in \mathcal{S}_p$$

we obtain for $p = 2, 3, \ldots$ some formulas for the difference ind $\mathcal{A}$ − ind det $\mathcal{A}$. Using these formulas, we prove that in the above-mentioned result [MF] about equality (1.2) ideal $\mathcal{S}_1$ can be replaced by some wider ideal. Namely any ideal J such that

$$J \subset \bigcup_{p>0} \mathcal{S}_p \quad \text{and} \quad \text{diag}\, [1/n] \notin J$$

is suitable.

2. Counterexamples

Here we discuss some examples of 2×2 block operator matrices such that their entries commute modulo $\mathcal{K}(H)$ but equality (1.2) does not hold. The following Example 2.1 in essence is contained in [V]. Another example of this kind can be obtained from [SS].

Example 2.1. Let $H = L^2(S^2)$ where S^2 is the two-dimensional sphere. There exist singular integral operators $A, B, C, D \in \mathcal{L}(H)$ with continuous symbols such that

$$\mathcal{A} = \begin{bmatrix} A & B \\ C & D \end{bmatrix} \tag{2.1}$$

is a Fredholm operator and ind $\mathcal{A} = 1$. This follows, e.g., from [MP, p.378]; see also [S2, Lemma 2 and Theorem 1] where explicit construction of such an operator $\mathcal{A}$ is proposed. Operator $\det \mathcal{A} = AD - BC$ also is a Fredholm operator but ind $\det \mathcal{A} = 0$ [MP, Ch. XIII, Theorem 3.2]. It is known also that all commutators of the operators A, B, C, D are compact [MP, Ch,XIII, Theorem 2.2]. We obtain some estimates for s-numbers of the commutators.

Since both numbers ind $\mathcal{A}$ and ind det $\mathcal{A}$ depend continuously on $\mathcal{A}$ we may assume that the symbol of $\mathcal{A}$ is infinitely smooth on the set of all unit vectors tangent to S^2, and hence the commutators map $L^2(S^2)$ into $W_1^2(S^2)$ (see [S1, Theorem 3]). If an operator $T \in \mathcal{L}(L^2(S^2))$ maps $L^2(S^2)$ into $W_1^2(S^2)$ then

$$s_n(T) = O\left(1/\sqrt{n}\right). \tag{2.2}$$

Indeed, T can be represented in the form $T = JR$ where R maps $L^2(S^2)$ into $W_1^2(S^2)$ and J is the embedding of $W_1^2(S^2)$ into $L^2(S^2)$. Obviously, $s_n(T) \le \|R\| s_n(J)$.

On the other hand $s_n(J) = O\left(1/\sqrt{n}\right)$ (see, e.g., [T, Theorem 4.10.1(a)]), and we obtain (2.2).

So we have an example of a Fredholm operator (2.1) such that ind $\mathcal{A} = 1$, ind det $\mathcal{A} = 0$ and any commutator Q of the entries of $\mathcal{A}$ satisfies $s_n(Q) = O\left(1/\sqrt{n}\right)$.

Now we show how to use Example 2.1 to construct examples with arbitrary values of ind $\mathcal{A}$ and ind det $\mathcal{A}$. This method was suggested by A.V. Kozak.

Theorem 2.2. *Let H be an infinite-dimensional separable Hilbert space. For arbitrary integers p and q there exists a 2×2 block operator matrix $\mathcal{A}$ with entries from $\mathcal{L}(H)$ such that all commutators Q of these entries satisfy*

$$s_n(Q) = O\left(1/\sqrt{n}\right), \tag{2.3}$$

$\mathcal{A}$ is a Fredholm operator, ind $\mathcal{A} = p$ *and* ind det $\mathcal{A} = q$.

Proof. Let $H_0 = \mathcal{L}\left(L^2(S^2)\right)$ and let

$$\mathcal{A} = \begin{bmatrix} A & B \\ C & D \end{bmatrix}$$

be the operator constructed in Example 2.1. For arbitrary $n > 1$ denote

$$A^{(n)} = \operatorname{diag}[A, \ldots, A] \in \mathcal{L}\left(H_0^n\right),$$

$$\mathcal{A}^{(n)} = \operatorname{diag}[\mathcal{A}, \ldots, \mathcal{A}] \in \mathcal{L}\left((H_0^2)^n\right) = \mathcal{L}\left((H_0^n)^2\right).$$

Obviously, operator $\mathcal{A}^{(n)}$ is unitarily equivalent to the operator

$$\mathcal{B}_n = \begin{bmatrix} A^{(n)} & B^{(n)} \\ C^{(n)} & D^{(n)} \end{bmatrix} \in \mathcal{L}\left((H_0^n)^2\right)$$

and

$$\det \mathcal{B}_n = A^{(n)} D^{(n)} - B^{(n)} C^{(n)} = (\det \mathcal{A})^{(n)}.$$

Hence

$$\operatorname{ind} \mathcal{B}_n = \operatorname{ind} \mathcal{A}^{(n)} = n, \quad \operatorname{ind} \det \mathcal{B}_n = 0.$$

Further,

$$\operatorname{ind} \mathcal{B}_n^* = -n, \quad \operatorname{ind} \det \mathcal{B}_n^* = 0.$$

So, for an arbitrary integer k there exists a Fredholm operator

$$\tilde{\mathcal{A}} = \begin{bmatrix} \tilde{A} & \tilde{B} \\ \tilde{C} & \tilde{D} \end{bmatrix} \in \mathcal{L}(\tilde{H}^2)$$

such that

$$\text{ind } \tilde{\mathcal{A}} = k, \quad \text{ind } \det \tilde{\mathcal{A}} = 0$$

and that condition (2.3) holds for any commutator Q of the entries of $\tilde{\mathcal{A}}$.

For an arbitrary integer m we fix a Fredholm operator $S \in \mathcal{L}(\tilde{H})$ such that $\text{ind} S = m$. Define

$$\hat{\mathcal{A}} = \begin{bmatrix} \hat{A} & \hat{B} \\ \hat{C} & \hat{D} \end{bmatrix} \in \mathcal{L}((\tilde{H}^2)^2),$$

where

$$\hat{A} = \begin{bmatrix} \tilde{A} & 0 \\ 0 & S \end{bmatrix}, \; \hat{B} = \begin{bmatrix} \tilde{B} & 0 \\ 0 & 0 \end{bmatrix}, \; \hat{C} = \begin{bmatrix} \tilde{C} & 0 \\ 0 & 0 \end{bmatrix}, \; \hat{D} = \begin{bmatrix} \tilde{D} & 0 \\ 0 & I \end{bmatrix}.$$

Obviously, all commutators Q of the entries of $\hat{\mathcal{A}}$ satisfy condition (2.3).

It is easy to see that

$$\det \hat{\mathcal{A}} = \begin{bmatrix} \det \tilde{\mathcal{A}} & 0 \\ 0 & S \end{bmatrix},$$

and hence $\text{ind } \det \hat{\mathcal{A}} = \text{ind} S = m$. On the other hand, $\hat{\mathcal{A}}$ is unitarily equivalent to the operator

$$R = \begin{bmatrix} \tilde{\mathcal{A}} & 0 \\ 0 & \tilde{S} \end{bmatrix}, \quad \text{where } \tilde{S} = \begin{bmatrix} S & 0 \\ 0 & I \end{bmatrix}$$

and therefore

$$\text{ind } \hat{\mathcal{A}} = \text{ind } \tilde{\mathcal{A}} + \text{ind } S = k + m.$$

Since k and m are arbitrary integers, m and $k + m$ are arbitrary integers as well. $\qquad\square$

3. Preliminaries

1. The *trace* of an operator $A \in \mathcal{S}_1$ is defined by the equality

$$\text{tr } A = \sum_{k=1}^{\infty} (Ae_k, e_k) \tag{3.1}$$

where $\{e_k\}_1^{\infty}$ is an orthonormal basis of H. It is well known that series (3.1) converges absolutely and its sum does not depend on the choice of the basis $\{e_k\}_1^{\infty}$. The trace has a number of interesting properties (see [GK, Ch. III, Section 8.1]). It is important for us that tr is a linear functional on the vector space $\mathcal{S}_1$ and that

$$\text{tr } [T, A] = 0, \quad (T \in \mathcal{S}_1, \; A \in \mathcal{L}(H)). \tag{3.2}$$

This equality holds also for $T \in \mathcal{S}_p$, $A \in \mathcal{S}_q$ $(p > 1, 1/p + 1/q = 1)$.

Consider the compact operator

$$\sum_{n=1}^{\infty} \frac{1}{n}(\cdot, e_n)e_n,\tag{3.3}$$

where $\{e_k\}_1^{\infty}$ is an orthonormal basis of H. Obviously, if operator (3.3) does not belong to an ideal J, then the same is true for the operator

$$\sum_{n=1}^{\infty} \frac{1}{n}(\cdot, f_n)f_n,$$

where $\{f_k\}_1^{\infty}$ is any other orthonormal basis of H. In this case we will write:

$$\operatorname{diag}[1/n] \notin J.$$

A decisive role in this paper is played by the following result [DFWW].

Theorem 3.1 *Let J be an ideal such that*

$$\operatorname{diag}[1/n] \notin J.\tag{3.4}$$

If a trace class operator A admits a representation

$$A = \sum_{k=1}^{m} [T_k, B_k],$$

where $T_k \in J$, $B_k \in \mathcal{L}(H)$ $(k = 1, 2, \ldots, m)$, then $\operatorname{tr} A = 0$.

Of course, the ideal $\mathcal{S}_1$ has the property (3.4), and hence Theorem 3.1 gives a very interesting generalization of the simple property (3.2).

2. It is well known that an operator $A \in \mathcal{L}(H)$ is a Fredholm operator if and only if A is invertible modulo $\mathcal{K}(H)$, and that the ideal $\mathcal{K}(H)$ can be replaced here by the ideal $\mathcal{F}(H)$ of all finite-dimensional operators [GGK, pp. 191, 192]. Since an arbitrary ideal J satisfies $\mathcal{F}(H) \subset J \subset \mathcal{K}(H)$ (see, e.g., [GK, Ch. III, Theorem 1.1]), we can also replace $\mathcal{K}(H)$ by any other ideal, in particular by some $\mathcal{S}_p$.

We will systematically use the following well-known trace formula for the index [H, Proposition 19.1.14].

Theorem 3.2. *Let $A, M \in \mathcal{L}(H)$ and*

$$I - MA, \ I - AM \in \mathcal{S}_p$$

for some positive integer p. Then

$$\operatorname{ind} A = \operatorname{tr}\left((I - MA)^p - (I - AM)^p\right).$$

3. Consider in more detail the notion of the determinant of an operator $\mathcal{A} \in \mathcal{L}(H^n)$. If this operator is represented in the form of a block operator matrix

$$\mathcal{A} = [A_{jk}]_{j,k}^{n} \quad (A_{jk} \in \mathcal{L}(H))\tag{3.5}$$

we define

$$\det \mathcal{A} := \sum_{\sigma} (\operatorname{sgn} \sigma) A_{1\sigma_1} A_{2\sigma_2} \cdots A_{n\sigma_n}\tag{3.6}$$

where σ runs through the symmetric group S_n and sgn σ denotes the sign of the permutation σ. Note that $\det \mathcal{A}$ is an operator in H. Its definition depends on $\mathcal{A}$ but also on the given decomposition of the "big space" as an orthogonal sum of n isomorphic summands and on the choice of the isomorphisms between these summands, but we suppose that the decomposition and the isomorphisms are fixed.

The main topic of this paper is the study of the index of a block operator matrix $\mathcal{A}$ under the condition that the entries commute modulo some ideal J. The following lemma shows how to reduce this problem to a more convenient case when the operator $\det \mathcal{A} - I$ also belongs to the ideal J. We will systematically use this lemma in the next sections.

Lemma 3.3. *Let $\mathcal{A}$ be a Fredholm operator in H^n defined by (3.5) and all entries of $\mathcal{A}$ commute modulo ideal J. Let M be an inverse of $\det \mathcal{A}$ modulo J and*

$$\mathcal{B} := \operatorname{diag}[M, I, \dots, I]\, \mathcal{A} \quad (\in \mathcal{L}(H^n)). \tag{3.7}$$

Then all entries of $\mathcal{B}$ commute modulo J, $\det \mathcal{B} - I \in J$ and

$$\operatorname{ind} \mathcal{A} - \operatorname{ind}\, \det \mathcal{A} = \operatorname{ind} \mathcal{B}.$$

Proof. Since $\det \mathcal{A}$ is also a Fredholm operator, it has an inverse M modulo J. Obviously, $\det \mathcal{A}$ commutes with each A_{jk} modulo J. It follows that modulo J

$$M A_{jk} - A_{jk} M = M A_{jk} (I - \det \mathcal{A}\, M) + (M\, \det \mathcal{A} - I)\, A_{jk} M,$$

and hence M commutes with each A_{jk} modulo J. This implies that the entries of $\mathcal{B}$ commute modulo J. By definition (3.7), $\det \mathcal{B} = M \det \mathcal{A}$, and hence, $\det \mathcal{B} - I \in J$. Equality (3.7) implies also that

$$\operatorname{ind} \mathcal{B} = \operatorname{ind} M + \operatorname{ind} \mathcal{A} = -\operatorname{ind}\, \det \mathcal{A} + \operatorname{ind} \mathcal{A}.$$

The lemma is proved. $\qquad\qquad\qquad\qquad\qquad\qquad\qquad\qquad\qquad\qquad\square$

4. 2×2 block operator matrices

1. Here we consider in detail the simplest case $n = 2$. Using Lemma 3.3 we start with the simplified situation when not only the commutators of the entries of a 2×2 operator matrix $\mathcal{B}$ but also the difference $I - \det \mathcal{B}$ belongs to an ideal J. We use the following notations

$$\mathcal{B} = \begin{bmatrix} A_0 & B_0 \\ C_0 & D_0 \end{bmatrix} \quad (A_0, B_0, C_0, D_0 \in \mathcal{L}(H)), \tag{4.1}$$

$$T := I - A_0 D_0 + B_0 C_0, \quad P := I - D_0 A_0 + C_0 B_0,$$
$$Q := I - D_0 A_0 + B_0 C_0, \quad R := I - A_0 D_0 + C_0 B_0. \tag{4.2}$$

Obviously

$$Q - P = T - R = [B_0, C_0] \tag{4.3}$$

and

$$Q - T = P - R = [A_0, D_0]. \tag{4.4}$$

We will use the notion of *standard polynomial* (see [R] or [K2, Section 20]):

$$F(X_1, \ldots, X_n) := \sum_{\sigma \in S_n} (\operatorname{sgn} \sigma) X_{\sigma_1} \cdots X_{\sigma_n}$$

where S_n is the symmetric group. According to the definition (3.6)

$$F(X_1, \ldots, X_n) = \det \begin{bmatrix} X_1 & X_2 & \cdots & X_n \\ X_1 & X_2 & \cdots & X_n \\ \cdots & \cdots & \cdots & \cdots \\ X_1 & X_2 & \cdots & X_n \end{bmatrix}.$$

We are especially interested in the case $n = 4$, and we will need the equality

$$F(X_1, X_2, X_3, X_4) = [X_1, X_2][X_3, X_4] - [X_1, X_3][X_2, X_4]$$

$$+ [X_1, X_4][X_2, X_3] + [X_3, X_4][X_1, X_2] - [X_2, X_4][X_1, X_3] + [X_2, X_3][X_1, X_4]. \quad (4.5)$$

This equality is readily verified by the removal of all parentheses on the right-hand side. Notice that by (4.5) $F(X_1, X_2, X_3, X_4) \in \mathcal{S}_1$ if all commutators $[X_j, X_k] \in \mathcal{S}_2$.

Lemma 4.1. *Let $\mathcal{B}$ be a Fredholm operator such that its entries commute modulo $\mathcal{S}_2$ and $I - \det \mathcal{B} \in \mathcal{S}_2$. Then*

$$\operatorname{ind} \mathcal{B} = \operatorname{tr} F(A_0, B_0, C_0, D_0). \quad (4.6)$$

Proof. Define

$$\mathcal{R} = \begin{bmatrix} D_0 & -B_0 \\ -C_0 & A_0 \end{bmatrix}. \quad (4.7)$$

Then

$$I - \mathcal{R}\mathcal{B} = \begin{bmatrix} Q & [B_0, D_0] \\ -[A_0, C_0] & R \end{bmatrix} \quad (4.8)$$

and

$$I - \mathcal{B}\mathcal{R} = \begin{bmatrix} T & [A_0, B_0] \\ -[C_0, D_0] & P \end{bmatrix}. \quad (4.9)$$

By the condition of the lemma, $I - \mathcal{B}\mathcal{R} \in \mathcal{S}_2$ and $I - \mathcal{R}\mathcal{B} \in \mathcal{S}_2$. (The identity operator in H^2, as well as in H^n below, also will be denoted by I.) Using (3.2) for $p = 2$ we have

$$\operatorname{ind} \mathcal{B} = \operatorname{tr} \left((I - \mathcal{R}\mathcal{B})^2 - (I - \mathcal{B}\mathcal{R})^2 \right).$$

It is well known (and easy to check) that the trace of an operator matrix from $\mathcal{S}_1$ equals the trace of the sum of its diagonal blocks. Simple calculation shows that

$$\operatorname{ind} \mathcal{B} = \operatorname{tr} \left(Q^2 - [B_0, D_0][A_0, C_0] + R^2 - [A_0, C_0][B_0, D_0] \right.$$

$$\left. - T^2 + [A_0, B_0][C_0, D_0] - P^2 + [C_0, D_0][A_0, B_0] \right). \quad (4.10)$$

Since property (3.2) holds for $T, A \in \mathcal{S}_2$, and since $Q, R, T, P, \in \mathcal{S}_2$,

$$\operatorname{tr} QT = \operatorname{tr} TQ, \quad \operatorname{tr} PR = \operatorname{tr} RP, \quad (4.11)$$

and

$$\operatorname{tr} \left([A_0, D_0][B_0, C_0] \right) = \operatorname{tr} \left([B_0, C_0][A_0, D_0] \right). \quad (4.12)$$

In view of (4,3), (4.4), (4.11) and (4.12)

$$\mathrm{tr}\,(Q^2 + R^2 - T^2 - P^2) = \mathrm{tr}\,((Q - T)(Q + T) - (P - R)(P + R))$$
$$= \mathrm{tr}\,([A_0, D_0](Q - P + T - R)) = 2\mathrm{tr}\,([A_0, D_0][B_0, C_0])$$
$$= \mathrm{tr}\,([A_0, D_0][B_0, C_0] + [B_0, C_0][A_0, D_0]).$$

The last equality and (4.10) imply that

$$\mathrm{ind}\,\mathcal{B} = \mathrm{tr}\big([A_0, B_0][C_0, D_0] - [A_0, C_0][B_0, D_0] + [A_0, D_0][B_0, C_0]$$

$$+[C_0, D_0][A_0, B_0] - [B_0, D_0][A_0, C_0] + [B_0, C_0][A_0, D_0]\big), \qquad (4.13)$$

and hence, (4.5) implies (4.6). $\qquad\qquad\square$

Theorem 4.2. *Let*

$$\mathcal{A} = \left[\begin{array}{cc} A & B \\ C & D \end{array} \right] \qquad\qquad (4.14)$$

be a Fredholm operator, and all its entries commute modulo $\mathcal{S}_2$. Let M be an inverse of $\det \mathcal{A} = AD - BC$ modulo $\mathcal{S}_2$. Then

$$\mathrm{ind}\,\mathcal{A} = \mathrm{ind}\,\det \mathcal{A} + \mathrm{tr}\,F(MA, MB, C, D). \qquad\qquad (4.15)$$

This theorem follows from Lemmas 3.3 and 4.1. $\qquad\qquad\square$

We do not know any example where under the condition of Theorem 4.2 the last term in (4.15) differs from zero (in Example 2.1 we only know that the commutators belong to $\mathcal{S}_p$ for $p > 2$). Now we prove that this term equals zero under some additional condition. We start with some additional remarks.

It follows from (4.12) and similar equalities

$$\mathrm{tr}\,([B_0, D_0][A_0, C_0]) = \mathrm{tr}\,([A_0, C_0][B_0, D_0]),$$
$$\mathrm{tr}\,([C_0, D_0][A_0, B_0]) = \mathrm{tr}\,([A_0, B_0][C_0, D_0])$$

that under the conditions of Lemma 4.1 we can rewrite (4.13) in a reduced form

$$\mathrm{ind}\,\mathcal{B} = 2\mathrm{tr}\,([C_0, D_0][A_0, B_0] - [A_0, C_0][B_0, D_0] + [A_0, D_0][B_0, C_0]). \qquad (4.16)$$

Direct calculations show that

$$[C_0, D_0][A_0, B_0] - [A_0, C_0][B_0, D_0] + [A_0, D_0][B_0, C_0]$$

$$= [[D_0, C_0]B_0, A_0] - [[A_0, C_0]B_0, D_0] + [[A_0, D_0]B_0, C_0]. \qquad (4.17)$$

Now we are able to prove the following result

Theorem 4.3. *Let J be an ideal such that*

$$J \subset \mathcal{S}_2 \quad \text{and} \quad \mathrm{diag}\,[1/n] \notin J.$$

If the operator $\mathcal{A}$ defined by (4.14) is a Fredholm operator and all its entries commute modulo J then

$$\mathrm{ind}\,\mathcal{A} = \mathrm{ind}\,\det \mathcal{A}. \qquad\qquad (4.18)$$

Proof. Let M be an inverse of $\det \mathcal{A}$ modulo J and

$$\mathcal{B} = \left[\begin{array}{cc} M & 0 \\ 0 & I \end{array} \right] \mathcal{A}. \qquad\qquad (4.19)$$

Then operator $\mathcal{B}$ satisfies the conditions of Lemma 4.1, and by (4.16), (4.17)

$$\text{ind } \mathcal{B} = 2\text{tr}\left([[D,C]MB,MA] - [[MA,C]MB,D] + [[MA,D]MB,C]\right). \quad (4.20)$$

Since the commutators $[D,C], [MA,C]$ and $[MA,D]$ belong to J, the right-hand side of (4.20) equals zero by Theorem 3.1, i.e., ind $\mathcal{B} = 0$. Now we obtain (4.18) from Lemma 3.3 (or simply from (4.19)). $\qquad\square$

Remark 1. If the entries of $\mathcal{A}$ commute modulo $\mathcal{S}_1$, the proof of (4.18) is immediate. Indeed, from Theorem 3.2 for $p = 1$ we immediately obtain (see (4.8), (4.9), (4.3), (4.4))

$$\text{ind } \mathcal{B} = \text{tr } (\mathcal{B}\mathcal{R} - \mathcal{R}\mathcal{B}) = \text{tr } (Q + R - T - P) = 0.$$

Recall that the result under discussion was proved in [MF] (for the general case of $n \times n$ operator matrices).

2. Now we suppose that the entries of the block operator matrix $\mathcal{A}$ (or $\mathcal{B}$) commute modulo $\mathcal{S}_3$. We use below notations (4.1), (4.2) and (4.14).

Lemma 4.4. *Let $\mathcal{B}$ be a Fredholm operator such that its entries commute modulo $\mathcal{S}_3$ and $I - \det \mathcal{B} \in \mathcal{S}_3$. Then*

$$\begin{aligned}
\text{ind}\mathcal{B} = 3\text{tr } \big(TF(A_0, B_0, C_0, D_0) &\qquad\qquad (4.21)\\
+ [A_0, D_0]([B_0, C_0][A_0, D_0] - [B_0, D_0][A_0, C_0] + [C_0, D_0][A_0, B_0])&\\
- [B_0, C_0]([A_0, D_0][B_0, C_0] - [A_0, C_0][B_0, D_0] + [C_0, D_0][A_0, B_0])\big).&
\end{aligned}$$

Proof. We define the operator $\mathcal{R}$ by equality (4.7). Since $I - \mathcal{R}\mathcal{B}, \ I - \mathcal{B}\mathcal{R} \in \mathcal{S}_3$ we can use Theorem 3.2 for $p = 3$:

$$\text{ind } \mathcal{B} = \text{tr }\left((I - \mathcal{R}\mathcal{B})^3 - (I - \mathcal{B}\mathcal{R})^3\right).$$

It is not difficult to show using (4.8) that

$$\begin{aligned}
\text{tr}(I - \mathcal{R}\mathcal{B})^3 &= \text{tr } \big(Q^3 - Q[B_0,D_0][A_0,C_0] - [B_0,D_0][A_0,C_0]Q - [B_0,D_0]R[A_0,C_0]\\
&\qquad - [A_0,C_0]Q[B_0,D_0] - [A_0,C_0][B_0,D_0]R + R^3 - R[A_0,C_0][B_0,D_0]\big)\\
&= \text{tr } \big(Q^3 + R^3 - 3Q[B_0,D_0][A_0,C_0] - 3R[A_0,C_0][B_0,D_0]\big).
\end{aligned}$$

Analogously, using (4.9), we obtain

$$\text{tr } (I - \mathcal{B}\mathcal{R})^3 = \text{tr } \big(T^3 + P^3 - 3T[A_0, B_0][C_0, D_0] - 3P[C_0, D_0][A_0, B_0]\big).$$

Hence

$$\begin{aligned}
\text{ind}\mathcal{B} = \text{tr } \big(Q^3 + R^3 - T^3 - P^3\big)&\\
+ 3\text{tr } \big(T[A_0, B_0][C_0, D_0] + P[C_0, D_0][A_0, B_0]&\\
- Q[B_0, D_0][A_0, C_0] - R[A_0, C_0][B_0, D_0]\big).&\qquad (4.22)
\end{aligned}$$

It is easy to see that

$$\mathrm{tr}\left(Q^3 + R^3 - T^3 - P^3\right)$$
$$= \mathrm{tr}\left((Q-T)^3 + 3QT(Q-T) - (P-R)^3 - 3PR(P-R)\right)$$
$$= 3\mathrm{tr}\left((QT - PR)[A_0, D_0]\right). \tag{4.23}$$

By (4.3),(4.4)

$$Q = T + [A_0, D_0], \ \ R = T + [C_0, B_0], \ P = T + [A_0, D_0] + [C_0, B_0]$$

and we obtain

$$\mathrm{tr}\left(Q^3 + R^3 - T^3 - P^3\right)$$
$$= 3\mathrm{tr}\left([A_0, D_0](QT - PR)\right)$$
$$= 3\mathrm{tr}\left([A_0, D_0]([A_0, D_0][B_0, C_0] - [B_0, C_0]^2 + T[B_0, C_0] + [B_0, C_0]T)\right)$$
$$= 3\mathrm{tr}\left([A_0, D_0]^2[B_0, C_0] - [A_0, D_0][B_0, C_0]^2\right.$$
$$\left. + T([B_0, C_0][A_0, D_0] + [A_0, D_0][B_0, C_0])\right) \tag{4.24}$$

Some simple calculations show that (4.22)–(4.24) and (4.5) imply (4.21). $\qquad\square$

Lemmas 3.3 and 4.4 imply

Theorem 4.5. *Let $\mathcal{A}$ be Fredholm operator and its entries commute modulo $\mathcal{S}_3$. Let M be an inverse of* det $\mathcal{A} = AD - BC$ *modulo $\mathcal{S}_3$. Then*

$$\mathrm{ind}\ \mathcal{A} - \mathrm{ind}\ \det \mathcal{A} = 3\mathrm{tr}\big((I - M\det \mathcal{A})F(MA, MB, C, D)$$

$$+[MA, D]\,([MB, C][MA, D] - [MB, D][MA, C] + [C, D][MA, MB])$$

$$-[MB, C]([MA, D][MB, C] - [MA, C][MB, D] + [C, D][MA, MB])\big). \tag{4.25}$$

Remark 2. Theorem 4.5 allows us to show that Theorem 4.3 holds under the condition $J \subset \mathcal{S}_3$ (instead of $\mathcal{S}_2$), but we will prove this in Section 5 for arbitrary $\mathcal{S}_p$ and for $n \times n$ operator matrices.

Remark 3. Unlike Theorem 4.2, we know from Section 2 that under the conditions of Theorem 4.5 the right-hand side of equality (4.25) can be different from zero.

5. The general case

In this section we use the following notion of adjoint matrix for an $n \times n$ block operator matrix $\mathcal{B} = [B_{jk}]_{j,k=1}^n$. Let M_{jk} be the $(n-1) \times (n-1)$ block operator matrix obtained from $\mathcal{B}$ by deleting the jth row and kth column. Denote

$$R_{jk} = (-1)^{j+k} \det M_{kj}.$$

The operator matrix

$$\mathcal{R} = [R_{jk}]_{j,k=1}^n$$

is called the *adjoint* matrix of $\mathcal{B}$.

Lemma 5.1. *Let $\mathcal{B}$ be a Fredholm operator and let the entries of $\mathcal{B}$ commute modulo $\mathcal{S}_p$ for some positive integer p. Assume that $I - \det \mathcal{B} \in \mathcal{S}_p$. If $\mathcal{R}$ is the adjoint matrix of $\mathcal{B}$ then*

$$\operatorname{ind} \mathcal{B} = \operatorname{tr} \sum_{m=1}^{p} (-1)^m \binom{p}{m} D_m, \tag{5.1}$$

where

$$D_m = \sum_{j,i_1,\ldots,i_{2m-1}=1}^{n} \left[R_{j i_{2m-1}} B_{i_{2m-1} i_{2m-2}} \cdots R_{i_2 i_1}, B_{i_1 j} \right]. \tag{5.2}$$

Proof. It is easy to check that

$$I - \mathcal{R}\mathcal{B} \in \mathcal{S}_p, \quad I - \mathcal{B}\mathcal{R} \in \mathcal{S}_p,$$

and

$$\operatorname{ind} \mathcal{B} = \operatorname{tr} \left((I - \mathcal{R}\mathcal{B})^p - (I - \mathcal{B}\mathcal{R})^p \right)$$

$$= \operatorname{tr} \sum_{m=1}^{p} (-1)^m \binom{p}{m} \left((\mathcal{R}\mathcal{B})^m - (\mathcal{B}\mathcal{R})^m \right). \tag{5.3}$$

Direct calculations show that the sum of diagonal blocks of the operator matrix $(\mathcal{R}\mathcal{B})^m - (\mathcal{B}\mathcal{R})^m$ coincides with the operator in the right-hand side of (5.2). Hence (5.3) implies (5.1). $\qquad\square$

The next statement follows from Lemma 3.3.

Theorem 5.2. *Let $\mathcal{A}$ be a Fredholm operator and let the entries of $\mathcal{A}$ commute modulo $\mathcal{S}_p$ for some positive integer p. If M is an inverse of $\det \mathcal{A}$ modulo $\mathcal{S}_p$ and*

$$\mathcal{B} = \operatorname{diag}[M, I, \ldots, I]\mathcal{A}$$

then

$$\operatorname{ind} \mathcal{A} = \operatorname{ind} \det \mathcal{A} + \operatorname{ind} \mathcal{B},$$

and $\operatorname{ind} \mathcal{B}$ *can be calculated by formula (5.1).*

Now we indicate some conditions when $\operatorname{ind} \mathcal{B} = 0$, and hence, $\operatorname{ind} \mathcal{A} = \operatorname{ind} \det \mathcal{A}$. The main technical difficulties are overcome in the next statement.

Lemma 5.3. *Let $\mathcal{B}$ be a Fredholm operator such that its entries commute modulo an ideal J and $I - \det \mathcal{B} \in J$. Then for any positive integer m the operator D_m defined by (5.2) admits representation as a finite sum*

$$\sum [T_k, A_k] \quad (T_k \in J, \ A_k \in \mathcal{L}(H)). \tag{5.4}$$

Proof. We prove the lemma by induction. First we suppose that it holds for D_m and prove that it holds for D_{m+1}. Denote

$$P_{jk} = \sum_{i=1}^{n} R_{ji} B_{ik} - \delta_{jk} I.$$

It follows from the operator matrix analogue of Cramer's rule [GGK, p. 194] that $P_{jk} \in J$ $(j, k = 1, 2, \ldots, n)$. Since

$$D_{m+1} = \sum_{j, i_1, \ldots, i_{2m}=1}^{n} \left[P_{ji_{2m}} R_{i_{2m}i_{2m-1}} B_{i_{2m-1}i_{2m-2}} \cdots R_{i_2 i_1}, B_{i_1 j} \right] + D_m$$

and $P_{jk} \in J$, we obtain the statement of the lemma for D_{m+1}.

It remains to prove the statement for D_1. By definition

$$D_1 = \sum_{i,j=1}^{n} [R_{ij}, B_{ji}]$$

and

$$R_{ij} = (-1)^{i+j} \sum_{\sigma'} (\operatorname{sgn} \sigma') B_{1\sigma'_1} \cdots B_{j-1,\sigma'_{j-1}} B_{j+1,\sigma'_{j+1}} \cdots B_{n\sigma'_n}$$

where the summation is over all bijective mappings

$$\sigma' : \{1, \ldots, j-1, j+1, \ldots, n\} \longrightarrow \{1, \ldots, i-1, i+1, \ldots, n\}.$$

It is easy to check that

$$D_1 = \sum_{\sigma \in S_n} (\operatorname{sgn} \sigma) \Big(\sum_{j=2}^{n-1} B_{1\sigma_1} \cdots B_{j-1,\sigma_{j-1}} B_{j+1,\sigma_{j+1}} \cdots B_{n\sigma_n} B_{j\sigma_j}$$

$$+ B_{2\sigma_2} \cdots B_{n\sigma_n} B_{1\sigma_1} - \sum_{j=2}^{n-1} B_{j\sigma_j} B_{1\sigma_1} \cdots B_{j-1,\sigma_{j-1}} B_{j+1,\sigma_{j+1}} \cdots B_{n\sigma_n}$$

$$- B_{n\sigma_n} B_{1\sigma_1} \cdots B_{n-1,\sigma_{n-1}} \Big). \tag{5.5}$$

We have to prove that the expression in the right-hand side of (5.5) admits a representation in the form (5.4). It is enough to prove this for the operator

$$S := \sum_{j=2}^{n-1} S_1 \cdots S_{j-1}S_{j+1} \cdots S_n S_j + S_2 \cdots S_n S_1$$

$$- \sum_{j=2}^{n-1} S_j S_1 \cdots S_{j-1}S_{j+1} \cdots S_n - S_n S_1 \cdots S_{n-1}$$

where $S_j := B_{j\sigma_j}$. Denote

$$A_j = S_1 \cdots S_{j-1} \ (1 < j \leq n), \quad C_j = S_{j+1} \cdots S_n \ (1 \leq j < n),$$

$$W_j = [[A_j, C_j], S_j] \ (1 < j < n).$$

Since $A_j S_j = A_{j+1}$ and $S_j C_j = S_{j-1}$, we have

$$W_j = A_j C_j S_j - C_j A_{j+1} - S_j A_j C_j + C_{j-1} A_j.$$

It is easy to see that

$$\sum_{j=2}^{n-1} (C_{j-1}A_j - C_j A_{j+1}) = C_1 A_2 - C_{n-1} A_n = S_2 \cdots S_n S_1 - S_n S_1 \cdots S_{n-1}$$

and therefore

$$S = \sum_{j=2}^{n-1} W_j = \sum_{j=2}^{n-1} [[A_j, C_j], S_j].$$ (5.6)

Since the operators B_{jk} commute modulo J,

$$[A_j, C_j] = \left[B_{1\sigma_1} \cdots B_{j-1,\sigma_{j-1}}, B_{j+1,\sigma_{j+1}} \cdots B_{n\sigma_n} \right] \in J.$$

Therefore equality (5.6) shows that S, and hence D_1, admit representation in the form (5.4) $\qquad\square$

Lemma 5.4. *Let J be an ideal such that*

1) $J \subset \bigcup_{p>0} S_p$ *and* 2) $\operatorname{diag}[1/n] \notin J$.

If $\mathcal{B}$ is a Fredholm operator, its entries commute modulo J and $I - \det \mathcal{B} \in J$ then

$$\operatorname{ind} \mathcal{B} = 0.$$ (5.7)

Proof. Since the number of all commutators $[B_{jk}, B_{j'k'}]$ is finite, we may suppose that $J \in S_p$ for some (positive integer) p. Let operators D_m be defined by (5.2). It follows from Lemma 5.3 that the operator

$$\sum_{m=1}^{p} (-1)^m \binom{p}{m} D_m$$

admits representation in the form (5.4). Since this operator belongs to S_1 (see the proof of Lemma 5.1), Theorem 3.1 and equality (5.1) imply (5.7). $\qquad\square$

Note that condition 2) of Lemma 5.4. does not imply condition 1).

Theorem 5.5. *Suppose the ideal J satisfies the conditions of Lemma 5.4. If $\mathcal{A}$ is a Fredholm operator and its entries commute modulo J then*

$$\operatorname{ind} \mathcal{A} = \operatorname{ind} \det \mathcal{A}.$$ (5.8)

This theorem follows from Lemmas 3.3 and 5.4.

Corollary 5.6. *Let $\mathcal{A}$ be a Fredholm operator such that every commutator of its entries satisfies*

$$s_n\left([A_{jk}, A_{j'k'}]\right) = o(1/n) \ (n \to \infty).$$ (5.9)

Then $\operatorname{ind} \mathcal{A} = \operatorname{ind} \det \mathcal{A}$.

Proof. Consider the ideal

$$J = \{ T \in K(H) : s_n(T) = o(1/n) \}$$

and use Theorem 5.5. $\qquad\square$

Condition (5.9) can be weakened, namely it can be replaced by condition

$$\sum_{m=1}^{n} s_m\left([A_{jk}, A_{j'k'}]\right) = o(\ln n) \ (n \to \infty).$$ (5.10)

We reformulate condition (5.10) in a slightly different form (see Corollary 5.7 below).

Define

$$S_\Omega = \{T \in \mathcal{K}(H) : \sum_{k=1}^{n} s_k(T) = O\left(\sum_{k=1}^{n} 1/k\right)\}, \tag{5.11}$$

$$S_\Omega^{(0)} = \{T \in \mathcal{K}(H) : \sum_{k=1}^{n} s_k(T) = o\left(\sum_{k=1}^{n} 1/k\right)\}. \tag{5.12}$$

These two sets, the so-called *symmetrically-normed ideals*, are ideals and also Banach spaces with respect to the norm

$$\|T\|_\Omega = \sup_n \frac{\sum_{k=1}^{n} s_k(T)}{\sum_{k=1}^{n} 1/k}.$$

Ideals (5.11) and (5.12) play a special role in some problems of the theory of non-selfadjoint operators. Ideal S_Ω is the adjoint to the well-known Matsaev ideal S_ω, and $S_\Omega^{(0)}$ is the pre-adjoint of S_ω (for details see [GK, Ch. III, Section 15]). It is obvious that $S_\Omega^{(0)} \subset S_\Omega \subset S_p$ for each $p > 1$.

Corollary 5.7. *Let A be a Fredholm operator with*

$$[A_{jk}, A_{j'k'}] \in S_\Omega^{(0)}$$

for arbitrary j, k, j', k'. Then ind $A =$ ind det A.

This corollary shows that Conjecture 1 from [FM] is false. Unfortunately, we know nothing new about Conjecture 2 which concerns the ideals S_p. So the gap between the condition $s_n([A_{jk}, A_{j'k'}]) = o(1/n)$ which guarantees equality (5.8) and condition $s_n([A_{jk}, A_{j'k'}]) = O(1/\sqrt{n})$ which allows counterexamples still remains large.

We note in conclusion that some other conditions sufficient for the validity of (5.8) have been found in [K2, Section 3] and [VT].

References

[DFWW] K. Dykema, T. Figiel, G. Weiss, M. Wodzicki, Commutator structure of operator ideals, *Adv. Math.* **185** (2004), 1–79.

[FM] I. Feldman, A. Markus, On the connection between the indices of an operator matrix and its determinant, *Lecture Notes in Math.* **1573** (1994), 248–249.

[GGK] I. Gohberg, S. Goldberg, M.A. Kaashoek, *Classes of Linear Operators,* vol. 1, Birkhäuser Verlag, Basel – Boston – Berlin, 1990.

[GK] I. Gohberg, M. Krein, *Introduction to the Theory of Linear Nonselfadjoint Operators,* Amer. Math. Soc., Providence, R.I., 1969.

[Ha] P. Halmos, *A Hilbert Space Problem Book,* Van Nostrand, Princeton, 1967.

[H] L. Hörmander, *The Analysis of Linear Partial Differential Operators,* vol. 3, Springer-Verlag, Berlin, 1985.

[K1] N. Krupnik, On the normal solvability and the index of singular integral operators, *Uchenye Zap. Kishinev. Gos. Univ.,* **82** (1965), 3–7 (Russian).

[K2]   N. Krupnik, *Banach Algebras with Symbol and Singular Integral Operators*, Birkhäuser Verlag, Basel – Boston, 1987.

[MF]   A. Markus, I. Feldman, Index of an operator matrix, *Funct. Anal. Appl.* **11**(1977), 149–151.

[MP]   S.G. Mikhlin, S. Prössdorf, *Singular Integral Operators,* Springer-Verlag, Berlin, 1986.

[R]   L.H. Rowen, *Polynomial Identities in Ring Theory,* Academic Press, New York, 1980.

[S1]   R.T. Seeley, Singular integrals on compact manifolds, *Amer. J. Math.* **81** (1959), 658–690.

[S2]   R.T. Selley, Integro-differential operators on vector bundles, *Trans. Amer. Math. Soc.* **117** (1965), 167–204.

[SS]   V.N. Semenjuta, I.B. Simonenko, Computation of the index of multidimensional discrete convolutions, *Mat. Issled.* **4,** vyp. 4(14) (1969), 134–141 (Russian).

[T]   H. Triebel, *Interpolation Theory, Function Spaces, Differential Operators,* North Holland, Amsterdam, 1978.

[VT]   N.L. Vasilevski, R. Trujillo, On Φ_R-operators in matrix algebras of operators, *Soviet Math. Dokl.* **20** (1979), 406–409.

[V]   A.I. Volpert, On the index of boundary problem for a system of harmonic functions with three independent variables, *Soviet Math. Dokl.* **1** (1960), 791–793.

Israel Feldman
Dept. of Mathematics
Bar-Ilan University
Ramat-Gan, 52900, Israel
e-mail: `feldmni@math.biu.ac.il`

Nahum Krupnik
Dept. of Mathematics
Bar-Ilan University
Ramat-Gan, 52900, Israel
e-mail: `nkrupnik@accessv.com`

Alexander Markus
Dept. of Mathematics
Ben-Gurion University of the Negev
Beer-Sheva, 84105, Israel
e-mail: `markus@cs.bgu.ac.il`

Operator Theory:
Advances and Applications, Vol. 170, 101–106
© 2006 Birkhäuser Verlag Basel/Switzerland

Quasi-commutativity of Entire Matrix Functions and the Continuous Analogue of the Resultant

I. Gohberg, M.A. Kaashoek and L. Lerer

Dedicated to I.B. Simonenko, with respect and admiration,
on the occasion of his 70th birthday.

Abstract. This paper is an addition to the paper [3], where it was proved that the theorem about the null space of the classical Sylvester resultant matrix also holds for its continuous analogue for entire matrix function provided that a certain so-called quasi-commutativity condition is fulfilled. In the present paper we show that this quasi-commutativity condition is not only sufficient but also necessary.

Mathematics Subject Classification (2000). Primary 47B35, 45E10, 30D20; Secondary 47A56.

Keywords. Entire matrix functions, quasi-commutativity, resultant operator, continuous analogue of the resultant.

First, let us recall the definition of the continuous analogue of the resultant for entire matrix functions. Let ω be a fixed positive number, and let b and d be $n \times n$ matrix functions, $b \in L_1^{n \times n}[-\omega, 0]$ and $d \in L_1^{n \times n}[0, \omega]$. Consider the operator R, acting on $L_1^n[-\omega, \omega]$, defined by

$$(Rf)(t) = \begin{cases} f(t) + \displaystyle\int_{-\omega}^{\omega} d(t-s)f(s)\,ds, & 0 \le t \le \omega, \\[2ex] f(t) + \displaystyle\int_{-\omega}^{\omega} b(t-s)f(s)\,ds, & -\omega \le t < 0. \end{cases} \tag{1}$$

Here we follow the convention that $d(t)$ and $b(t)$ are zero whenever t does not belong to $[0, \omega]$ or $[-\omega, 0]$, respectively. The operator R is called the *resultant*

The third author gratefully acknowledges the support of the Glasberg-Klein Research Fund at the Technion.

102   I. Gohberg, M.A. Kaashoek and L. Lerer

operator associated to the entire matrix functions $\mathcal{B}$ and $\mathcal{D}$ given by

$$\mathcal{B}(\lambda) = I + \int_{-\omega}^{0} e^{i\lambda t} b(t)\, dt, \quad \mathcal{D}(\lambda) = I_n + \int_{0}^{\omega} e^{i\lambda t} d(t)\, dt. \tag{2}$$

For that reason we shall write $R(\mathcal{B}, \mathcal{D})$ instead of R.

Note that the right-hand side of (1) has the same form as the classical Sylvester resultant matrix, provided one replaces finite Toeplitz matrices by convolution operators over a finite interval. Moreover, as has been shown in [1], analogous to the resultant matrix, in the scalar case ($n = 1$) the dimension of the null space of $R(\mathcal{B}, \mathcal{D})$ is precisely equal to the total common multiplicity $\nu(\mathcal{B}, \mathcal{D})$ of the common zeros of the entire matrix functions $\mathcal{B}$ and $\mathcal{D}$. In fact (see [1]), if $\lambda_1, \ldots, \lambda_\ell$ are the distinct common zeros of the entire scalar functions $\mathcal{B}$ and $\mathcal{D}$, and ν_j is the common multiplicity of λ_j as a common zero of $\mathcal{B}$ and $\mathcal{D}$, then the functions

$$\psi_{j\,k}(t) = t^k e^{-i\lambda_j t}, \quad k = 0, \ldots, \nu_j - 1, \ j = 1, \ldots, \ell$$

form a basis of $\mathrm{Ker}\, R(\mathcal{B}, \mathcal{D})$. The fact that the number of common zeros of the entire functions $\mathcal{B}$ and $\mathcal{D}$ is finite follows from

$$\lim_{\Im\lambda \leq 0, \, |\lambda| \to \infty} \mathcal{B}(\lambda) = I_n, \qquad \lim_{\Im\lambda \geq 0, \, |\lambda| \to \infty} \mathcal{D}(\lambda) = I_n, \tag{3}$$

In the matrix case, the number $\nu(\mathcal{B}, \mathcal{D})$ is well defined too, and stands for the total common multiplicity of the common eigenvalues of $\mathcal{B}$ and $\mathcal{D}$. The precise definition is more complex than for the scalar case and requires (see, e.g., Section 2 in [3]) the notion of eigenvalues and common Jordan changes. As before, from (3) it follows again that $\nu(\mathcal{B}, \mathcal{D})$ is finite, and one can show that

$$\dim \mathrm{Ker}\, R(\mathcal{B}, \mathcal{D}) \geq \nu(\mathcal{B}, \mathcal{D}). \tag{4}$$

More precisely, one can prove (see [3]) the following result. If $\lambda_1, \ldots, \lambda_\ell$ are the distinct common eigenvalues of $\mathcal{B}$ and $\mathcal{D}$, and if for each λ_ν the $\mathbb{C}^n$-vectors

$$x_{1,0}^{\nu}, \ldots, x_{1,\,r_1^{(\nu)}-1}^{\nu}, x_{2,0}^{\nu}, \ldots, x_{2,\,r_2^{(\nu)}-1}^{\nu}, \ldots, x_{p_\nu,0}^{\nu}, \ldots, x_{p_\nu,\,r_{p_\nu}^{(\nu)}-1}^{\nu}$$

form a canonical set of common Jordan chains of $\mathcal{B}$ and $\mathcal{D}$ at λ_ν, then the functions

$$\psi_{j,k}^{\nu}(t) = e^{-i\lambda_\nu t} \sum_{\mu=0}^{k} \frac{(-it)^{k-\mu}}{(k-\mu)!} x_{j,\mu}^{\nu}, \quad k = 0, \ldots, r_j^{(\nu)} - 1, \tag{5}$$

$$j = 1, \ldots, \quad p_\nu, \nu = 1, \ldots, \ell,$$

form a linear independent subset of $\mathrm{Ker}\, R(\mathcal{B}, \mathcal{D})$.

For $n = 2$ a simple example in [1] shows $\mathrm{Ker}\, R(\mathcal{B}, \mathcal{D})$ can contain non-smooth functions. Since those in (5) are all smooth, it follows that the inequality in (4) can be strict. Hence, for the matrix case, the question arises under what conditions on $\mathcal{B}$ and $\mathcal{D}$ do we have equality in (4). The next theorem answers this question.

Theorem 1. *Let $R(\mathcal{B}, \mathcal{D})$ be the resultant operator corresponding to the entire $n \times n$ matrix functions $\mathcal{B}$ and $\mathcal{D}$ in (2). Then*

$$\dim \mathrm{Ker}\, R(\mathcal{B}, \mathcal{D}) = \nu(\mathcal{B}, \mathcal{D}). \tag{6}$$

if and only if there exist entire $n \times n$ matrix functions A and C of the form

$$A(\lambda) = I_n + \int_0^\omega e^{i\lambda s} a(s)\, ds, \quad C(\lambda) = I_n + \int_{-\omega}^0 e^{i\lambda s} c(s)\, ds, \tag{7}$$

where $a \in L_1^{n\times n}[0,\omega]$ and $c \in L_1^{n\times n}[-\omega,0]$, such that

$$A(\lambda)B(\lambda) = C(\lambda)D(\lambda), \quad \lambda \in \mathbb{C}. \tag{8}$$

In the commutative case, when $B(\lambda)$ and $D(\lambda)$ commute for each $\lambda \in \mathbb{C}$, one can take $A = D$ and $C = B$, and then (8) is automatically fulfilled. Thus the above theorem covers the result for the scalar case in [1].

We shall say that B and D are *quasi-commutative* whenever one can find A and C as in (7) such that (8) holds. The fact that quasi-commutativity of B and D is a sufficient condition for the equality (6) to hold has been proved in [3]. Here we shall establish the necessity of this condition.

To deal with the necessity of the quasi-commutativity condition we consider the following entire matrix function equation:

$$X(\lambda)B(\lambda) + Y(\lambda)D(\lambda) = G(\lambda), \quad \lambda \in \mathbb{C}. \tag{9}$$

The right-hand side is assumed to be known, and of the form

$$G(\lambda) = \int_{-\omega}^\omega e^{i\lambda t} g(t)\, dt, \quad \text{with } g \in L_1^{m\times n}[-\omega,\omega]. \tag{10}$$

Thus g is a $m \times n$ matrix function of which the entries have their support in and are integrable on $[-\omega,\omega]$. The problem is to find entire $m \times n$ matrix functions X and Y,

$$X(\lambda) = \int_0^\omega e^{i\lambda t} x(t)\, dt, \quad Y(\lambda) = \int_{-\omega}^0 e^{i\lambda t} y(t)\, dt, \tag{11}$$

with $x \in L_1^{m\times n}[0,\omega]$ and $y \in L_1^{m\times n}[-\omega,0]$, such that (9) is satisfied. When such functions X and Y exist, we shall say that equation (9) is *solvable*. To establish the necessity of the condition (6) we need the following result.

Theorem 2. *Assume* $\dim \mathrm{Ker}\,(B,D) = \nu(B,D)$. *Then equation (9) is solvable if and only if each common Jordan chain of B and D is also a Jordan chain of G with respect to the same eigenvalue.*

Let us show how Theorem 2 can be used to prove that the property of quasi-commutativity property (8) is a necessary condition for (6) to hold.

Assume $\dim \mathrm{Ker}\,(B,D) = \nu(B,D)$. We have to find entire matrix functions A and C,

$$A(\lambda) = I_n + A(\lambda), \qquad A(\lambda) = \int_0^\omega e^{i\lambda t} a(t)\, dt, \tag{12}$$

$$C(\lambda) = I_n + C(\lambda), \qquad C(\lambda) = \int_{-\omega}^0 e^{i\lambda t} c(t)\, dt, \tag{13}$$

with $a \in L_1^{n \times n}[0, \omega]$ and $c \in L_1^{n \times n}[-\omega, 0]$, such that (8) holds. Using the special form of $\mathcal{A}$ and $\mathcal{C}$ in (12) and (13), we see that (8) can be rewritten in the following equivalent form

$$A(\lambda)\mathcal{B}(\lambda) - C(\lambda)\mathcal{D}(\lambda) = -\mathcal{B}(\lambda) + \mathcal{D}(\lambda), \quad \lambda \in \mathbb{C}. \tag{14}$$

Notice that (14) is an equation of the form (9) with $X = A$, $Y = -C$, and

$$G(\lambda) = -\mathcal{B}(\lambda) + \mathcal{D}(\lambda), \quad g(t) = \begin{cases} d(t), & 0 < t \le \omega, \\ -b(t), & -\omega \le t < 0. \end{cases}$$

Obviously, each common Jordan chain of $\mathcal{B}$ and $\mathcal{D}$ is a common Jordan chain of $-\mathcal{B} + \mathcal{D}$. Thus we can apply Theorem 2 to show that equation (14) is solvable with A and C of the desired form. Thus, if (6) holds, then $\mathcal{B}$, $\mathcal{D}$ has the quasi-commutativity property.

It remains to prove Theorem 2. For this purpose we need the following lemma.

Lemma 3. *With $m = 1$ equation (9) is solvable for G given by (10) if and only if*

$$\int_{-\omega}^{\omega} g(t)\varphi(-t) = 0, \quad \varphi \in \operatorname{Ker} R(\mathcal{B}, \mathcal{D}). \tag{15}$$

Proof. Since $m = 1$, the function $G(\lambda)$ and the unknowns $X(\lambda)$ and $Y(\lambda)$ are one row matrix functions. By taking inverse Fourier transforms in (9) we see that (9) is equivalent to

$$x(t) + y(t) + \int_0^{\omega} x(s)b(t - s)\, ds + \int_{-\omega}^0 y(s)d(t - s)\, ds = g(t), \quad -\omega \le t \le \omega. \tag{16}$$

Here, as before, we follow the convention that a function f on an interval $[\alpha, \beta]$ is considered as function on $\mathbb{R}$ by defining $f(t) = 0$ for $t \in \mathbb{R}\backslash[\alpha, \beta]$.

Let S be the operator on $L^{1 \times n}[-\omega, \omega]$ defined by

$$(Sh)(t) = h(t) + \int_0^{\omega} h(s)b(t - s)\, ds + \int_{-\omega}^0 h(s)d(t - s)\, ds, \quad -\omega \le t \le \omega. \tag{17}$$

By comparing (16) and (17) we see that equation (16) is equivalent to

$$Sh = g, \quad h(t) = \begin{cases} x(t) & \text{when } 0 < t \le \omega, \\ y(t) & \text{when } -\omega \le t < 0. \end{cases} \tag{18}$$

Next, let us compute the Banach space adjoint S' of S relative to the pairing $\langle \cdot, \cdot \rangle$ given by

$$\langle f, h \rangle = \int_{-\omega}^{\omega} h(t)f(t)\, dt, \quad h \in L_1^{1 \times n}[-\omega, \omega], \quad f \in L_\infty^{n \times 1}[-\omega, \omega].$$

As usual we write $L^n_\infty[-\omega,\omega]$ for the space $L^{n\times 1}_\infty[-\omega,\omega]$. A straightforward calculation shows that $S' = T$, where T is the operator on $L^n_\infty[-\omega,\omega]$ defined by

$$(Tf)(t) = \begin{cases} f(t) + \displaystyle\int_{-\omega}^{\omega} b(r-t)f(r)\,dr, & 0 \le t \le \omega, \\[3mm] f(t) + \displaystyle\int_{-\omega}^{\omega} d(r-t)f(r)\,dr, & -\omega \le t < 0. \end{cases}$$

Another straightforward calculation yields $T = JR_\infty(\mathcal{B},\mathcal{D})J$, where J is the flip over operator on $L^n_\infty[-\omega,\omega]$ given by $(Jf)(t) = f(-t)$, and $R_\infty(\mathcal{B},\mathcal{D})$ is the restriction of $R(\mathcal{B},\mathcal{D})$ to $L^n_\infty[-\omega,\omega]$,

$$R_\infty(\mathcal{B},\mathcal{D}) = R(\mathcal{B},\mathcal{D})|L^n_\infty[-\omega,\omega],$$

which maps $L^n_\infty[-\omega,\omega]$ into itself. Note that both $R(\mathcal{B},\mathcal{D})$ and $R_\infty(\mathcal{B},\mathcal{D})$ are operators of the form identity plus compact. Since $L^n_\infty[-\omega,\omega]$ is continuously and densely embedded in $L^n_1[-\omega,\omega]$, it follows (see Theorem V.3.4 in [2]) that $\operatorname{Ker} R_\infty(\mathcal{B},\mathcal{D}) = \operatorname{Ker} R(\mathcal{B},\mathcal{D})$. From $S' = T$ and the intertwining relation $T = JR_\infty(\mathcal{B},\mathcal{D})J$, we conclude that

$$S'J = JR_\infty(\mathcal{B},\mathcal{D}). \tag{19}$$

Next we use that S is also of the form $I + K$, where K is compact. According to the Fredholm alternative this implies that the equation $Sh = g$ is solvable in $L^{n\times n}_1[-\omega,\omega]$ if and only if g is perpendicular in the Banach space sense to the kernel of S'. From (19) we see that

$$\operatorname{Ker} S' = J\operatorname{Ker} R_\infty(\mathcal{B},\mathcal{D}) = J\operatorname{Ker} R(\mathcal{B},\mathcal{D}).$$

Using the definition of J and the equivalence of the equations (9), (16) and (18), we conclude that for (9) to be solvable it is necessary and sufficient that (15) holds. $\qquad\square$

Proof of Theorem 2. If equation (9) is solvable, then clearly any common Jordan chain of $\mathcal{B}$ and $\mathcal{D}$ is a Jordan chain of G. So we have to prove the reverse implication. For this purpose we need (6).

Assume $\dim \operatorname{Ker} R(\mathcal{B},\mathcal{D}) = \nu(\mathcal{B},\mathcal{D})$, and let each common Jordan chain of $\mathcal{B}$ and $\mathcal{D}$ be a Jordan chain of G with respect to the same eigenvalue. To solve (9) we may without loss of generality assume that $m = 1$. Thus, as in Lemma 3, the function $G(\lambda)$ and the unknowns $X(\lambda)$ and $Y(\lambda)$ are one row matrix functions.

Let $x_0, x_1, \ldots, x_{r-1}$ be any ordered sequence of vectors in $\mathbb{C}^n$, and consider the associated functions

$$\widetilde{x}_k(t) = e^{-i\lambda_0 t}\sum_{\nu=0}^{k}\frac{(-it)^\nu}{\nu!}x_{k-\nu}, \qquad k = 0,\ldots,r-1. \tag{20}$$

Here λ_0 is some complex number. Since $\dim \operatorname{Ker}(\mathcal{B},\mathcal{D}) = \nu(\mathcal{B},\mathcal{D})$ holds, we know that the set (5) forms a basis for $\operatorname{Ker}(\mathcal{B},\mathcal{D})$, and hence (cf., Lemma 4.1 in [3]) it follows that (15) holds if and only if

$$\int_{-\omega}^{\omega} g(t)\widetilde{x}_k(-t)\,dt = 0, \quad k = 0,\ldots,r-1, \tag{21}$$

where $x_0,\ldots,x_{r-1}$ is any common Jordan chain of $\mathcal{B}$ and $\mathcal{D}$ with corresponding eigenvalue λ_0.

Next, fix a common Jordan chain $x_0,\ldots,x_{r-1}$ of $\mathcal{B}$ and $\mathcal{D}$, and let λ_0 be the corresponding eigenvalue. Let us compute the left-hand side of (21) with $\widetilde{x}_k$ being given by (20). We have

$$\int_{-\omega}^{\omega} g(t)J\widetilde{x}_k(-t)\,dt = \sum_{\nu=0}^{k} \frac{1}{\nu!} \int_{-\omega}^{\omega} e^{i\lambda_0 t}(it)^\nu g(t)x_{k-\nu} = \sum_{\nu=0}^{k} \frac{1}{\nu!} G^{(\nu)}(\lambda_0)x_{k-\nu}.$$

Thus (21) holds if and only if $\sum_{\nu=0}^{k} \frac{1}{\nu!} G^{(\nu)}(\lambda_0)x_{k-\nu} = 0$ for $k = 0,\ldots,r-1$. By definition, the latter happens if and only if $x_0,\ldots,x_{r-1}$ is a Jordan chain of G at λ_0. According to our hypothesis any common Jordan chain of $\mathcal{B}$ and $\mathcal{D}$ is a Jordan chain of G with respect to the same eigenvalue. Thus (21) holds and the theorem is proved. $\qquad\square$

The discrete analogue of Theorem 1 for regular matrix polynomials also holds, and will be the topic of a further publication.

References

[1] I. Gohberg and G. Heinig, The resultant matrix and its generalizations, II. Continual analog of resultant matrix, *Acta Math. Acad. Sci. Hungar* **28** (1976), 198–209 [in Russian].

[2] I. Gohberg and N. Krupnik, *Introduction to the theory of one dimensional singular integral operators*, Ştiinţa, Kishinev, 1973.

[3] I. Gohberg, M.A. Kaashoek, and L. Lerer, The continuous analogue of the resultant and related convolution operators, in: *Proceedings IWOTA 2004*, to appear.

I. Gohberg
School of Mathematical Sciences
Raymond and Beverly Faculty of Exact Sciences, Tel-Aviv University
Ramat Aviv 69978, Israel
e-mail: `gohberg@math.tau.ac.il`

M.A. Kaashoek
Afdeling Wiskunde, Faculteit der Exacte Wetenschappen, Vrije Universiteit
De Boelelaan 1081a, 1081 HV Amsterdam, The Netherlands
e-mail: `ma.kaashoek@few.vu.nl`

L. Lerer
Department of Mathematics, Technion – Israel Institute of Technology
Haifa 32000, Israel
e-mail: `llerer@techunix.technion.ac.il`

Operator Theory:
Advances and Applications, Vol. 170, 107–135
© 2006 Birkhäuser Verlag Basel/Switzerland

Double Barrier Options Under Lévy Processes

Sergei M. Grudsky

To I.B. Simonenko on the occasion of his 70th birthday

Abstract. In this paper the problem of determination of the no arbitrage price of double barrier options in the case of stock prices is modelled on Lévy processes is considered. Under the assumption of existence of the Equivalent Martingale Measure this problem is reduced to the convolution equation on a finite interval with symbol generated by the characteristic function of the Lévy process. We work out a theory of unique solvability of the getting equation and stability of the solution under relatively small perturbations.

Mathematics Subject Classification (2000). Primary 47620; Secondary 60G35, 60G51.

Keywords. Barrier options, Lévy processes, Toeplitz operator.

1. Introduction

The problem of the determination of the price of a double barrier option in case when the stock price is modelled by geometric Brownian motion (classical hypothesis) is considered in [1]–[7]. The articles [4]–[7] are devoted to an approach connected with a solution of the Black-Scholes (partial) differential equation on a strip of finite width. But it should be noted that for many cases geometric Brownian motion is not an adequate model for stock price. Therefore in recent years many investigators have used Lévy processes as models for logarithmic stock price. In this way European options ([8]–[19]), perpetual American options ([15], [20]–[21]) and barrier options ([15], [22], [23]) were considered.

In this paper we consider double barrier options under Lévy processes. Following the monograph [15] we use the generalized Black-Scholes equation approach. That is we reduce the original option problem to a partial pseudodifferential equation of the type

$$\frac{\partial u(x,t)}{\partial t} - L_x u(x,t) = 0 \tag{1.1}$$

Author acknowledges financial support by CONACYT project 046936-F.

where the pseudodifferential, more exactly, convolution operator L_x (acting on the variable x) is generated by the characteristic exponent of the Lévy process $X_t := \ln S_t$ (here S_t is stock price). In the case of a double barrier the equation (1.1) is considered in the region

$$E = (x_1, x_2) \times (-\infty, T)$$

(where T is expiry date) and can be reduced (with the help of the Laplace transform in the variable $\tau = T - t$) to a convolution equation on an interval (x_1, x_2). We apply the Matrix Riemann Boundary Value Problem method worked out in the papers [24]–[27] for the investigation of the convolution equation. In this way we prove unique solvability of the problem and stability of the solution under relatively small perturbations.

2. Auxiliary material

In this section we introduce necessary definitions and formulate some well known results (see [15], [28], [29]) There are many kinds of double barrier option problems. We consider (in some sense) the basic problem which can be called Up-Down-And-Out barrier option. Other double barrier option problems can be reduced to this problem and (or) to single barrier problems.

Let S_t be stock price at the instant of time t, and $\varphi : (0, \infty) \to [0, \infty)$ be a measurable function.

Definition 2.1. *An Up-Down-And-Out barrier option is an agreement between two persons (Writer and Holder) at time instant t according to which Writer is obliged to pay to Holder the amount $\varphi(S_T)$ at the future instant of time T (expiry date) if and only if during the option life (between t and T), S_t is always within the interval (S_1, S_2) (here $0 < S_1 < S_2$ are some levels, i.e., barriers, of the stock price).*

Note that if there exists some instant of time $t_{1,2} \leq T$ such that $S_{t_1} \geq S_2$ or $S_{t_2} \leq S_1$ then the option expires worthless.

Consider a market model which consists of a bond with a constant riskless rate of return $r > 0$, and of a stock with price $S_t = \exp\{X_t\}$ where X_t is a Lévy process.

Let $(\Omega, \mathcal{F}, \mathbf{P})$ be a probability space where Ω is the space of elementary events and $\mathcal{F}$ is a σ-algebra of subsets of Ω.

Definition 2.2 ([28]–[29]). *An $\mathcal{F}$-adapted process X_t is called a Lévy process if the following conditions hold:*

1. *$X_0 = 0$ a.e.*
2. *X_t has stationary increment, that is, for arbitrary $t > s > 0$ the distribution of $(X_t - X_s)$ coincides with the distribution of X_{t-s}.*

3. X_t has independent increments, that is, for arbitrary $0 \le t_1 < t_2 < \ldots < t_n$, the random variables

$$X_{t_1}, X_{t_2} - X_{t_1}, \ldots, X_{t_n} - X_{t_{n-1}}$$

are independent.

4. For each $w \in \Omega$ the function $X_t = X_t(w)$ is right-continuous on $(0, \infty)$ and there exists a left limit at all $t \ge 0$.

5. X_t is stochastically continuous, that is for every $t > 0$ and $\varepsilon > 0$

$$\lim_{s \to t} \mathbf{P}[|X_t - X_s| > \varepsilon] = 0.$$

If X_t is a Lévy process, then according to the Lévy-Khintchine formula ([28]–[29])

$$E_{\mathbf{P}}[e^{i\xi X_t}] = e^{-t\psi^{\mathbf{P}}(\xi)}, \quad \xi \in \mathbb{R}, \tag{2.1}$$

where the function $\psi^{\mathbf{P}}(\xi)$ has the representation

$$\psi^{\mathbf{P}}(\xi) = \frac{1}{2}\sigma^2\xi^2 - i\mu\xi - \int_{-\infty}^{\infty} (e^{iu\xi} - 1 - i\xi u I_{(-1,1)}(u))\Pi(du) \tag{2.2}$$

with $\sigma \ge 0$, $\mu \in \mathbb{R}$, and Π is a measure on $\mathbb{R}$ satisfying the condition

$$\left| \int_{-\infty}^{\infty} \frac{u^2}{1 + u^2} \Pi(du) \right| < \infty, \tag{2.3}$$

$$I_{(-1,1)}(u) = \left\{ \begin{array}{ll} 1, & |u| < 1; \\ 0, & |u| \ge 1. \end{array} \right.$$

The expectation of exponent $E_{\mathbf{P}}[e^{i\xi X_t}]$ is called the characteristic function, the function $\psi^{\mathbf{P}}(\xi)$ is called the characteristic exponent of X_t (under the probability measure $\mathbf{P}$), the triplet (a, γ, Π) is called the generating triplet of X_t.

We will consider an arbitrage free market (see, for example, [30]). From results of [31] it follows that no-arbitrage pricing of options is possible if there exists an Equivalent Martingale Measure (EMM) $\mathbf{Q}$.

Let (Ω, F, F_t, P) be a probability space with right continuous filtration $F_t(\subset F)$.

Let $\mathbf{P}_t$, $\mathbf{Q}_t$ be restriction measures $\mathbf{P}\big|_{\mathcal{F}_t}$ and $\mathbf{Q}\big|_{\mathcal{F}_t}$ respectively.

Let $Z_t = \frac{d\mathbf{Q}_t}{d\mathbf{P}_t}$ be the density of $\mathbf{Q}_t$ with respect to $\mathbf{P}_t$.

If $0 < Z_t < \infty$ a.e. then the measures $\mathbf{P}$ and $\mathbf{Q}$ are called locally equivalent.

Definition 2.3. *A measure $\mathbf{Q}$ locally equivalent with respect to the measure $\mathbf{P}$ is called an Equivalent Martingale Measure (EMM) if the process $S_t^* = e^{-rt}S_t$ is a $\mathbf{Q}$-martingale (more exactly: is a $(\Omega, \mathcal{F}_t, \mathbf{Q})$-martingale).*

The process S_t^* is called the discounted stock price.

Let $V(S_t, t)$ be the option price for time $t \le T$, and let $V^*(S_t, t) := e^{-rt}V(S_t, t)$ be discounted option price. Under the measure $\mathbf{Q}$ all discounted price processes (such that the prices are $\mathbf{Q}$-integrable) are assumed to be martingales. By virtue of this assumption prices of certain securities whose price at some future date T are given random walk can be expressed by the help of conditional expectation. We write the conditional expectation formula for our case.

Let $x_{1,2} = \ln S_{1,2}$ and let $\eta := \eta(w)$ be the hitting time of the set $\mathbb{R} \setminus (x_1, x_2)$

$$\eta(w) := \inf\{t \geq 0 | X_t(w) \in \mathbb{R} \setminus (x_1, x_2)\}.$$

Then for the Up-Down-And-Out barrier option at expiry date $t = T$ we have

$$V(e^{X_T}, T) = \varphi(e^{X_T}) I_{\{\eta > T\}}$$

where I_A is the characteristic function of the set $A \subset \Omega$. Denote $\mathcal{U}(X_t, t) := V(e^{X_t}, t)$ and $g(X_T) = \varphi(e^{X_T})$. Then we have

$$e^{-rt}\mathcal{U}(x, t) = E_{\mathbf{Q}} \left[e^{-rT} g(X_T) I_{\{\eta > T\}} \bigg|_{\mathcal{F}_t} \right] \tag{2.4}$$

where the right-hand side is the conditional expectation under the measure $\mathbf{Q}$ with respect to the σ-algebra $\mathcal{F}_t$ with $X_t = x$.

Thus the existence of EMM is an important question in option theory. If X_t is neither Brownian motion nor a Poisson process then typically that EMM is not unique. Moreover there often exist infinitely many different EMMs. We formulate in this connection the main result of the article [32].

Suppose that X_t is a Lévy process with characteristic triplet $(0, \mu, \Pi)$ (for a similar result for processes with triplet (σ, μ, Π) for $\sigma > 0$ see [9]).

Let μ_r denote the class of measures $\mathbf{Q}$ locally equivalent to $\mathbf{P}$ under which $e^{-rt}S_t$ is a martingale and X_t is a Lévy process under the measure $\mathbf{Q}$.

Let $Y_{\mu,r}(\Pi(dx))$ denote the class of function $y : \mathbb{R} \to (0, +\infty)$ such that

$$\int_{-\infty}^{\infty} (\sqrt{y(x)} - 1)^2 \Pi(dx) + \int_{\{x > 1\}} (e^x - 1)y(x)\Pi(dx) < \infty$$

and

$$\mu - r + \int_{-\infty}^{\infty} ((e^x - 1)y(x) - x I_{[-1,1]}(x))\Pi(dx) = 0.$$

Theorem 2.1 ([32]).

  a) *If $Y_{\mu,r}(\Pi(dx)) = \emptyset$, then $\mu_r = \emptyset$.*

  b) *If $Y_{\mu,r}(\Pi(dx)) \neq \emptyset$, then μ_r is non-empty and for each $y \in Y_{\mu,r}(\Pi(dx))$ there is a measure $\mathbf{Q} \in \mu_r$ under which X_t is again a Lévy process with generating triplet $(0, \mu', \Pi')$, where*

$$\mu' = \mu + \int_{-1}^{1} x(y(x) - 1)\Pi(dx) \qquad and \qquad \Pi'(A) = \int_A y(x)\Pi(dx).$$

Conversely, if $\mathbf{Q} \in \mu_r$ is the measure under which X_t is a Lévy process, then its generating triplet is $(0, \mu', \Pi')$ where μ' and Π' are given by the above expressions with some $y \in Y_{\mu,r}$.

c) *Let y and $\mathbf{Q}$ be as in* b). *Then the characteristic exponents of X_t under $\mathbf{P}$ and $\mathbf{Q}$ are related by*

$$\psi^{\mathbf{Q}}(\xi) = \psi^{\mathbf{P}}(\xi) + \int_{-\infty}^{\infty} (1 - e^{ix\xi})(y(x) - 1)\Pi(dx).$$

Thus in the case of Lévy processes typically an EMM exists and is not unique. Thus there exists the problem of the choice of EMM. This problem is not trivial. For a discussion about such choice, see [15] (pp. 97–98).

From now on assume that an EMM $\mathbf{Q}$ is chosen and that X_t is a Lévy process under the measure $\mathbf{Q}$.

Definition 2.4. *We will say that the Lévy process X_t satisfies the (ACP)-condition (see [15] p. 59) if the function*

$$(U^r f)(x) := E_{\mathbf{Q}}\left[\int_0^{\infty} e^{-rt} f(X_t)dt \,\middle|\, X_0 = x\right]$$

is continuous for every $f \in L_{\infty}(R)$.

Some sufficient conditions for (ACP)-condition are given (for example) in [15] (Theorem 2.11 and Lemma 2.4).

We will consider everywhere below Lévy processes that satisfy the (ACP)-condition.

Let $g \in L_{\infty}(x_1, x_2)$, the set of all essentially bounded functions on (x_1, x_2) and let the process X_t satisfy the (ACP)-condition. Then according to Theorem 2.13 of [15] the function $\mathcal{U}(x, t)$ defined by (2.4) is a bounded solution of the following partial pseudodifferential problem:

$$\frac{\partial \mathcal{U}(x, t)}{\partial t} - (r - L_x^{\mathbf{Q}})\mathcal{U}(x, t) = 0, \ x \in (x_1, x_2), \ t < T, \tag{2.5}$$

$$\mathcal{U}(x, T) = g(x), \quad x \in (x_1, x_2), \tag{2.6}$$

$$\mathcal{U}(x, t) = 0, \quad x \in R\backslash(x_1, x_2), \quad t < T. \tag{2.7}$$

Here the pseudodifferential operator $L_x^{\mathbf{Q}}$ (acting on the variable x) is given by the formula

$$(L_x^{\mathbf{Q}} f)(x) = (\mathcal{F}^{-1}(-\psi^{\mathbf{Q}}(\cdot))\mathcal{F}f)(x), \tag{2.8}$$

where $\psi^{\mathbf{Q}}(\xi)$ is the characteristic exponent of the Lévy process X_t under the EMM $\mathbf{Q}$ and the Fourier transform is given (for $f \in L_1(\mathbb{R})$) as follows,

$$(\mathcal{F}f)(\xi) = \int_{-\infty}^{\infty} e^{-i\xi x} f(x)dx. \tag{2.9}$$

Without loss of generality suppose that $x_1 = 0$ and $x_2 = a > 0$. Making the change of variable $\tau = T - t$, $u(x, \tau) = \mathcal{U}(x, t)$ we obtain the following problem,

$$\frac{\partial u(x, \tau)}{\partial \tau} + (r - L_x^{\mathbf{Q}})u(x, \tau) = 0, \ (x, \tau) \in (0, a) \times (0, \infty), \tag{2.10}$$

$$u(x, 0) = g(x), \quad x \in (0, a), \tag{2.11}$$

$$u(x, \tau) = 0, \quad x \in R\backslash(0, a), \quad \tau \in (0, \infty). \tag{2.12}$$

Equation (2.10) is understood in the sense of generalized functions:

$$(u, (-\frac{\partial}{\partial \tau} + r - \tilde{L}_x^{\mathbf{Q}})w) = 0 \tag{2.13}$$

for all $w \in S(R \times R)$ such that supp $w \subseteq (0, a) \times (0, \infty)$, where $S(R \times R)$ is the space of infinitely differentiable functions vanishing at infinity faster any negative power of $(x^2 + t^2)^{1/2}$ together with all derivatives. Here $u \in S'(R \times R)$ the set of all continuous linear functionals (distributions) on $S(R \times R)$ and $\tilde{L}_x^{\mathbf{Q}} := F^{-1}(-\psi^{\mathbf{Q}}(-\xi))F$. (For details see [15].)

3. Convolution equation and classes of symbols

Introduce the Laplace transform (LT) by variable τ and denote

$$v(x, w) := (\mathcal{L}u)(x, w) = \int_0^\infty e^{-w\tau} u(x, \tau) d\tau. \tag{3.1}$$

Applying integration by parts we obtain

$$\left(\mathcal{L}\frac{\partial u}{\partial t}\right)(x, w) = u(x, \tau)e^{-w\tau}\Big|_0^\infty + w \int_0^\infty e^{-w\tau} u(x, \tau) d\tau$$

$$= -u(x, 0) + wv(x, w) = -g(x) + wv(x, w).$$

Thus we pass from problem (2.10)–(2.12) to the following problem

$$(-L_x^{\mathbf{Q}} + r + w)v(x, w) = g(x), \quad x \in (0, a), \tag{3.2}$$

$$v(x, w) = 0 \quad x \in R\backslash(0, a). \tag{3.3}$$

We interpret the problem (3.2)–(3.3) as an operator equation considered in some H^s-spaces.

We introduce the corresponding notation. For $s \in R$ denote by $H^s(R)$ the space of distributions $f(\in S'(R))$ with finite norm defined by

$$\|f\|_{H^s}^2 = \int_{\mathbb{R}} |(\mathcal{F}f)(\xi)|^2 (1 + |\xi|^2)^s d\xi.$$

Let U be an open subset of $\mathbb{R}$. Then denote by $H^s(U)$ the subspace of $H^s(R)$ consisting of distributions with supp $f \in \overline{U}$.

Introduce the set $C_0^\infty(U)$ of all functions f having all derivatives, with supp $f \in U$. It is well known that the closure $C_0^\infty(U)$ by norm of $H^s(R)$ coincides with $H^s(U)$. Suppose that $v(\cdot, w) \in H^{s_1}(0, a)$ and $g \in H^{s_2}(0, a)$. For such a function $v(\cdot, w)$ the condition (3.3) holds automatically.

Thus we can rewrite the problem (3.2)–(3.3) as the following equation,

$$P_{(0,a)}(\mathcal{F}^{-1}(\psi^{\mathbf{Q}}(\xi) + r + w)\mathcal{F})v(x, w) = g(x) \tag{3.4}$$

where $P_{(0,a)}$ is the operator of restriction to the interval $(0, a)$, $v(\cdot, w) \in H^{s_1}(0, a)$ and $g \in H^{s_2}(0, a)$. It should be noted that (3.4) is the convolution equation on the finite interval $(0, a)$ with the symbol $a(\xi, w) := \psi^{\mathbf{Q}}(\xi) + r + w$. This equation is understood in the sense of generalized functions analogously to (2.13).

Now we consider in more detail the properties of the function $\psi^{\mathbf{Q}}(\xi)$. Since X_t is a Lévy process under the measure $\mathbf{Q}$ then according to Lévy-Khintchine formula (2.2) we have

$$\psi^{\mathbf{Q}}(\xi) = \frac{1}{2}\sigma^2\xi^2 - i\mu\xi + \varphi(\xi) \tag{3.5}$$

where $\sigma \geq 0$, $\mu \in R$, and

$$\varphi(\xi) = -\int_{-\infty}^{\infty}(e^{iu\xi} - 1 - i\xi u I_{(-1,1)}(u))\Pi^{\mathbf{Q}}(du) \tag{3.6}$$

with the measure $\Pi^{\mathbf{Q}}$ satisfying the condition

$$\left|\int_{-\infty}^{\infty}\frac{u^2}{1+u^2}\Pi^{\mathbf{Q}}(du)\right| < \infty. \tag{3.7}$$

Lemma 3.1. (*See* [15] *for example.*) *For arbitrary* $\xi \in R$, $\quad \operatorname{Re}\psi^{\mathbf{Q}}(\xi) \geq 0$.

Proof. Since $\mathbf{Q}$ is a probability measure and $|e^{i\xi X_t}| = 1$ for $\xi \in R$, we have from (2.2) that

$$|E_{\mathbf{Q}}[e^{iX_1}]| = |e^{-\psi^{\mathbf{Q}}(\xi)}| \leq 1.$$

That is,

$$e^{-\operatorname{Re}\psi^{\mathbf{Q}}(\xi)} \leq 1$$

and $\operatorname{Re}\psi^{\mathbf{Q}}(\xi) \geq 0$. $\qquad\square$

Lemma 3.2. (*See* [28].) *The characteristic function* $\psi^{\mathbf{Q}}(\xi)$ *is continuous for arbitrary* $\xi \in R$, *and the function* $\varphi(\xi)$ *has at infinity the following asymptotic property:*

$$\lim_{\xi \to \infty}\frac{\varphi(\xi)}{\xi^2} = 0. \tag{3.8}$$

Proof. It is easy to see that the continuity of $\psi^{\mathbf{Q}}(\xi)$ follows from the representation (3.5)–(3.7).

Let $|\xi|$ be large. Then we have

$$|e^{iu\xi} - 1 - i\xi u| \leq \begin{cases} \text{const} \cdot |\xi u|^2, & \text{if} \quad |u| \leq |\xi|^{-1}, \\ \text{const} \cdot |\xi u|, & \text{if} \quad |\xi|^{-1} \leq |u| \leq 1. \end{cases}$$

Then the following inequalities hold,

$$\begin{aligned}
|\varphi(\xi)| \quad \leq \quad & \int_{|u|\geq 1}|e^{iu\xi} - 1|\Pi^{\mathbf{Q}}(du) + \text{const}\,|\xi|\int_{|\xi|^{-1/2}<u<1}|u|\Pi^{\mathbf{Q}}(du) \\
& + \text{const} \cdot \xi^2\int_{|u|\leq|\xi|^{-1/2}}u^2\Pi^{\mathbf{Q}}(du) \\
\leq \quad & 2\int_{|u|\geq 1}\Pi^{\mathbf{Q}}(du) + \text{const} \cdot |\xi|^{3/2}\int_{|\xi|^{-1/2}<|u|<1}u^2\Pi^{\mathbf{Q}}(du) \\
& + \text{const} \cdot \xi^2\int_{|u|\leq|\xi|^{-1/2}}u^2\Pi^{\mathbf{Q}}(du).
\end{aligned}$$

According to (3.7), if $|\xi| \to \infty$ then

$$\int_{|u|\geq 1} \Pi^{\mathbb{Q}}(du) \quad \text{is bounded,}$$

$$\int_{|\xi|^{-1/2}<u<1} u^2\Pi^{\mathbf{Q}}(du) \quad \text{is bounded, and}$$

$$\int_{|u|<|\xi|^{-1/2}} u^2\Pi^{\mathbf{Q}}(du) \quad \text{tends to zero.}$$

Thus we obtain (3.8). $\qquad\square$

We use the following restrictions on the function $\psi^{\mathbf{Q}}(\xi)$ which naturally follow from Lemmas 3.1 and 3.2. Namely, the function $\psi^{\mathbf{Q}}(\xi)$ must have the form (3.5)

$$\psi^{\mathbf{Q}}(\xi) = \frac{1}{2}\sigma^2\xi^2 - i\mu\xi + \varphi(\xi) \tag{3.9}$$

where $\sigma \geq 0$, $\mu \in R$, and $\varphi(\xi)$ is a continuous function of $\xi \in R$ and there exists a number $\nu \in (0,2)$ such that the function $\varphi(\xi)$ has the following behavior at infinity,

$$\varphi(\xi) \sim |\xi|^{\nu}. \tag{3.10}$$

(The notation $\theta(\xi) \sim \eta(\xi)$ means that the quotients

$$\frac{|\theta(\xi)|}{|\eta(\xi)|} \quad \text{and} \quad \frac{|\eta(\xi)|}{|\theta(\xi)|}$$

are bounded by some constant for all $|\xi|$ large enough.)

Thus we have the following cases of asymptotic behavior of the function $\psi^{\mathbf{Q}}(\xi)$ at infinity,

$$\sigma > 0, \quad \psi^{\mathbf{Q}}(\xi) \sim \frac{\sigma^2\xi^2}{2}, \tag{3.11}$$

$$\sigma = 0, \quad 1 \leq \nu < 2, \quad \psi^{\mathbf{Q}}(\xi) \sim |\xi|^{\nu}, \tag{3.12}$$

$$\sigma = 0, \quad \mu = 0, \quad 0 < \nu \leq 1, \quad \psi^{\mathbf{Q}}(\xi) \sim |\xi|^{\nu}, \tag{3.13}$$

$$\sigma = 0, \quad \mu \neq 0, \quad 0 < \nu < 1, \quad \psi^{\mathbf{Q}}(\xi) \sim \xi. \tag{3.14}$$

This work is devoted to the cases (3.12)–(3.13). The cases (3.11), (3.14) and some others will be considered elsewhere.

Now we consider some examples of function $\psi^{\mathbf{Q}}(\xi)$ (we take the examples 3.1–3.5 from the book [15, Chapter 3]).

Example 3.1 (Kobol Family). For Lévy processes from this family the characteristic exponent $\psi(\xi)$ can have the following forms,

i)
$$\psi(\xi) = -i\mu\xi + c_+\Gamma(-\nu)[\lambda_-^{\nu} - (\lambda_- - i\xi)^{\nu}]$$
$$+ c_-\Gamma(-\nu)[\lambda_+^{\nu} - (\lambda_+ + i\xi)^{\nu}] \tag{3.15}$$

where $\nu \in (0,1) \cup (1,2)$, $\mu \in R$, $c_\pm > 0$, $\lambda_\pm > 0$, and $\Gamma(u)$ is the Euler Gamma-function;

ii)
$$\psi(\xi) = -i\mu\xi + c_+[\ln(\lambda_- - i\xi) - \ln\lambda_-]$$
$$+ c_-[\ln(\lambda_+ + i\xi) - \ln\lambda_+] \tag{3.16}$$

where $\mu \in R$, $c_\pm > 0$, $\lambda_\pm > 0$;

iii)
$$\psi(\xi) = -i\mu\xi + c_+[\lambda_- \ln(\lambda_-) - (\lambda_- - i\xi)\ln(\lambda_- - i\xi)]$$
$$+ c_-[\lambda_+ \ln(\lambda_+) - (\lambda_+ + i\xi)\ln(\lambda_+ + i\xi)] \tag{3.17}$$

where $\mu \in R$, $c_\pm > 0$, $\lambda_\pm > 0$.

Example i) corresponds to cases (3.12)–(3.14). The cases ii) and iii) have the following (non-power) behavior at infinity,

$$\psi(\xi) + \mu\xi \sim \ln\xi \tag{3.18}$$

and

$$\psi(\xi) \sim \xi \ln\xi. \tag{3.19}$$

Example 3.2 (Normal Tempered Stable Lévy Processes). In this case the characteristic exponent is

$$\psi(\xi) = -i\mu\xi + \delta[(\alpha^2 - (\beta + i\xi)^2)^{\nu/2} - (\alpha^2 - \beta^2)^{\nu/2}] \tag{3.20}$$

where $\nu \in (0,2)$, $\mu \in R$, $\delta > 0$, $\beta \in R$, $\alpha > |\beta|$ (see (3.13)–(3.14)).

Example 3.3 (Normal Inverse Gaussian Processes). If we put in (3.20) $\nu = 1$ we obtain the characteristic exponent of a normal inverse Gaussian Process

$$\psi(\xi) = -i\mu\xi + \delta[(\alpha^2 - (\beta + i\xi)^2)^{1/2} - (\alpha^2 - \beta^2)^{1/2}] \tag{3.21}$$

(see (3.13)–(3.14)).

Example 3.4 (Variance Gamma Processes). The characteristic exponent for this case is

$$\psi(\xi) = -i\mu\xi + c[\ln(\alpha^2 - (\beta + i\xi)^2) - \ln(\alpha^2 - \beta^2)] \tag{3.22}$$

where $c > 0$, $\beta \in R$, $\alpha > |\beta| > 0$ (see (3.18)).

Example 3.5 (Generalized Hyperbolic Processes). For this case the characteristic function is

$$\exp(\psi(\xi)) = e^{i\mu\xi}\left(\frac{\alpha^2 - \beta^2}{\alpha^2 - (\beta + i\xi)^2}\right)^{\lambda/2}\frac{K_\lambda(\delta\sqrt{\alpha^2 - (\beta + i\xi)^2})}{K_\lambda(\delta\sqrt{\alpha^2 - \beta^2})} \tag{3.23}$$

where $\mu \in R$, $\beta \in R$, $\alpha > |\beta| > 0$, $\delta > 0$, $\lambda \in R$, and $K_\lambda(u)$ is the modified Bessel function of the third kind with index λ. An integral representation of $K_\lambda(u)$ is given by

$$K_\lambda(u) = \frac{1}{2}\int_0^\infty y^{\lambda-1}\exp[-0.5u(y + y^{-1})]dy.$$

Example 3.6 (Poisson Processes). For a Poisson process we have the following characteristic exponent (see, for example, [28])

$$\psi(\xi) = c(1 - e^{i\xi}). \tag{3.24}$$

It is easy to see that the Poisson process of the kind (3.24) has characteristic triplet $(0, c, c\Pi_1)$ where Π_1 is a discrete measure which is concentrated in the one point $u_0 = 1$ with weight equal to 1.

Let Π_d be discrete measure which is concentrated in the points $\{\mu_1, \mu_2, \ldots, \mu_n\}$ with weights $\{d_1, d_2, \ldots, d_n\}$ respectively. Suppose that the Lévy process X_t has the characteristic triplet (σ, μ, Π_d). According to formula (2.2) the characteristic exponent of X_t has the following form,

$$\psi(\xi) = \frac{1}{2}\sigma^2\xi^2 - i\left(\mu - \sum_{|u_j|<1} d_j u_j\right)\xi - \sum_{j=1}^{n}(e^{iu_j\xi} - 1)d_j. \tag{3.25}$$

It should be noted that (ACP)-condition does not hold in this case. In spite of that we can use the system (2.5)–(2.7) for a finding of option price (see Remark 2.1 d) of [15], p. 64).

Example 3.7 (Rational Characteristic Exponent). Let the measure Π_{r_1} be given by the following formula,

$$\Pi_{r_1}(dx) = \lambda_+ c_+ e^{\lambda_+ x}\chi_{(-\infty,0)}(x)dx + \lambda_- c_- e^{-\lambda_- x}\chi_{(0,\infty)}dx \tag{3.26}$$

where $c_\pm > 0$, $\lambda_\pm > 0$, and $\chi_{(-\infty,0)}(x)$, $\chi_{(0,\infty)}(x)$ are characteristic functions of the semi-axes.

Consider the Lévy process X_t with the triplet (σ, μ, Π_{r_1}). The characteristic exponent in this case is

$$\psi_{r_1}(\xi) = \frac{\sigma^2}{2}\xi^2 - i\gamma'\xi + \frac{ic_+\xi}{\lambda_+ + i\xi} - \frac{ic_-\xi}{\lambda_- - i\xi} \tag{3.27}$$

where

$$\gamma' = \gamma - \left(c_+\lambda_+ \int_{-1}^{0} ue^{\lambda_+ u}du + c_-\lambda_- \int_{0}^{1} ue^{-\lambda_- u}du\right)$$

$$= \gamma - c_+\frac{(1+\lambda_+)e^{-\lambda_+} - 1}{\lambda_+} + c_-\frac{(1+\lambda_-)e^{-\lambda_-} - 1}{\lambda_-}.$$

It should be noted that replacement of the factors $e^{\pm\lambda\pm x}$ by factors of the form $\sum_{j=1}^{m} d_j^{\pm}e^{\pm\lambda_j, \pm x}$ in (3.26) provides more general rational characteristic exponent.

4. Reducing to modified Wiener-Hopf equation. Necessary information from Toeplitz operator theory

4.1. Modified Wiener Hopf equation

Our basic equation (3.4) is defined on the interval $(0, a)$. Extend it to the whole real axis R, that is, rewrite this equation in the form

$$v_0(x, w) + F^{-1}(\psi^{\mathbf{Q}}(\xi) + r + w)Fv(x, w) = g(x) \tag{4.1}$$

where $v_0(x, w) \in H^{s_2}((a, \infty) \cup (-\infty, 0))$.

If $s_2 \in (-1/2, 1/2)$, then (see [38], [40])

$$H^{s_2}((a, \infty) \cup (-\infty, 0)) = H^{s_2}(a, \infty) \oplus H^{s_2}(-\infty, 0) \tag{4.2}$$

Thus in this case equation (4.1) is

$$v_1(x, w) + F^{-1}(\psi^{\mathbf{Q}}(\xi) + r + w)Fv(x, w) + v_2(x, w) = g(x) \tag{4.3}$$

where

$$
\begin{aligned}
g(x) &\in H^{s_2}(0, a) &\tag{4.4}\\
v_1(x, w) &\in H^{s_2}(a, \infty) &\tag{4.5}\\
v_2(x, w) &\in H^{s_2}(-\infty, 0) &\tag{4.6}
\end{aligned}
$$

If $s_2 \notin (-1/2, 1/2)$, the decomposition (4.2) does not hold. However, in this case we suppose that case $v_0 = v_1 + v_2$ where v_1, v_2 satisfy (4.5), (4.6) like we have in the good case. This additional requirement holds if the function $g(x)$ is good enough, for example (as we will see below) when $g(x)$ satisfies (4.4)

Apply the Fourier transform to equation (4.3). Denote

$$
\begin{aligned}
\Phi_a^-(\xi, w) &:= (\mathcal{F}v)(\xi, w) & (\in L_2(\mathbb{R}, s_1)); &\tag{4.7}\\
e^{-ia\xi}\Phi^-(\xi, w) &:= (\mathcal{F}v_1)(\xi, w) & (\in L_2(\mathbb{R}, s_2)); &\tag{4.8}\\
\Phi^+(\xi, w) &:= (\mathcal{F}v_2)(\xi, w) & (\in L_2(\mathbb{R}, s_2)); &\tag{4.9}\\
\hat{g}(\xi) &= (\mathcal{F}g)(\xi) & (\in L_2(\mathbb{R}, s_2)); &\tag{4.10}
\end{aligned}
$$

where $L_2(\mathbb{R}, s)$ is Hilbert space with the norm

$$\|\Phi\|_{L_2^s} = \int_{-\infty}^{\infty} |\Phi(\xi)|^2(1 + \xi^2)^s d\xi.$$

(Note that the "+" sign in the notation $\Phi^+(\xi, w)$ means that this function is analytic in the upper half-plane. Analogously the "−" sign in $\Phi_a^-(\xi, w)$ and $\Phi^-(\xi, w)$ means that these functions are analytic in the lower half-plane.)

Thus we obtain the following boundary value problem

$$e^{-ia\xi}\Phi^-(\xi, w) + a(\xi, w)\Phi_a^-(\xi, w) + \Phi^+(\xi, w) = \hat{g}(\xi). \tag{4.11}$$

This problem is called ([39]) the modified Wiener-Hopf equation and its solution is a triple $(\Phi^-, \Phi_a^-, \Phi^+)$ of unknown functions. We emphasize that these unknown functions are not arbitrary functions from $L_2(\mathbb{R}, s_{1,2})$. Namely $\Phi_a^-(\xi, w)$ is the Fourier Transform of a function with support belonging to $(0, a)$, and $\Phi^{\mp}(\xi, w)$

118   S.M. Grudsky

are Fourier Transforms of functions with support in the semi-axes $(0, \infty)$ and $(-\infty, 0)$ respectively. The classes where the solution of (4.10) are looked for will be introduced in the Section 5.

4.2. Toeplitz operators

We need some results from Toeplitz operator theory. Introduce the so-called analytic projectors

$$P^+ := \mathcal{F}^{-1}\chi_{(0,\infty)}\mathcal{F} \quad \text{and} \quad P^- := \mathcal{F}^{-1}\chi_{(-\infty,0)}\mathcal{F}.$$

The projectors $P^\pm$ are bounded linear operators in the spaces $L_2(\mathbb{R}, s)$ for $s \in (-1/2, 1/2)$ ([40], [41]).

Denote

$$L_2^\pm(\mathbb{R}, s) = P^\pm(L_2(\mathbb{R}, s)).$$

It is easy to see (see also [40], [41]) that

$$P^{\pm 2} = P^\pm, \quad P^+P^- = P^-P^+ = 0, \quad P^+ + P^- = I$$

where 0 and I are the zero and identity operators.

Let further $L_\infty(\mathbb{R})$ be the space of all measurable essentially bounded functions on the real axis $\mathbb{R}$ with the norm

$$\|a\|_{L_\infty(\mathbb{R})} = \operatorname*{ess\,sup}_{x \in \mathbb{R}} |a(x)| < \infty.$$

The operator

$$T(a) := P^+aP^+ : L_2^+(\mathbb{R}, s) \to L_2^+(\mathbb{R}, s)$$

is called a Toeplitz operator with symbol $a(x)$. If $a \in L_\infty(\mathbb{R})$ then $T(a)$ is a bounded operator on $L_2^+(\mathbb{R}, s)$ (for $s \in (-1/2, 1/2)$) and the conjugate operator $T^*(a) = T(\bar{a})$ also is bounded on the same spaces.

Definition 4.1. *The operator A acting on Hilbert space is called normally solvable if the subspace $\operatorname{im} A$ is closed, i.e., $\operatorname{im} A = \overline{\operatorname{im} A}$.*

We will use the following well-known fact from functional analysis.

Lemma 4.1. *If the operator A is normally solvable, then the Hilbert space H may be represented as the following direct sum,*

$$\operatorname{im} A \oplus \ker A^* = H.$$

Definition 4.2. *An operator A acting in Hilbert space H is called left-(right)-invertible if there exists an operator A_l^{-1} (A_r^{-1}) bounded on H such that $A_l^{-1} A = I$ $(AA_r^{-1} = I)$. A is called an invertible operator if there exists an operator A^{-1} bounded on H such that $AA^{-1} = A^{-1}A = I$.*

It should be noted that a one-side invertible operator is normally solvable. Moreover if A is left-invertible, then $\ker A = \{0\}$; if A is right-invertible, then $\operatorname{im} A = H$.

Introduce the following well-known subspace of $L_\infty(\mathbb{R})$ in the theory of Toeplitz operators. Namely, $H^\infty(\mathbb{R}) + C(\dot{\mathbb{R}})$ is the set of all essentially bounded functions $f(x)$ representable in the form $f(x) = h(x) + c(x)$ where $h(x)$ is bounded and analytic in the upper half-plane. That is, $h(x)$ belongs to the Hardy space $H^\infty(\mathbb{R})$, and $c(x)$ belongs to the space $C(\dot{\mathbb{R}})$, the set of all continuous on $\mathbb{R}$ functions such that $\lim\limits_{x \to +\infty} c(x) = \lim\limits_{x \to -\infty} c(x) = c_0 \in \mathbb{C}$.

We formulate the following well-known facts from the theory of Toeplitz operators connected with class $H^\infty(\mathbb{R}) + C(\dot{\mathbb{R}})$ ([40], [41]).

Lemma 4.2. *Let the symbol* $a(x) \in H^\infty(\mathbb{R}) + C(\dot{\mathbb{R}})$ ($\in \overline{H^\infty(\mathbb{R}) + C(\dot{\mathbb{R}})}$) *and* $\operatorname*{ess\,inf}_{x \in \mathbb{R}} |a(x)| > 0$. *Then the operator* $T(a)$ *is left-invertible (right-invertible) in the space* $L_2^+(\mathbb{R}, s)$, *for* $|s| < 1/2$.

Lemma 4.3 (Sarason Lemma ([42])). *Let the function* $c(x)$ *be continuous on* $\mathbb{R}$ *and let there exist* $\lim\limits_{x \to +\infty} c(x) = c_+$ *and* $\lim\limits_{x \to -\infty} c(x) = c_-$ *(in general* $c_+ \neq c_-$*), and let* $\lambda > 0$ *(*$\lambda < 0$*). Then we have*

$$e^{i\lambda x} c(x) \in H^\infty(\mathbb{R}) + C(\dot{\mathbb{R}}) \quad (\in \overline{H^\infty(\mathbb{R}) + C(\dot{\mathbb{R}})}).$$

4.3. Sectoriality

Our work is based essentially on the concept of sectoriality. Lemma 3.1 makes this possible. This subsection is devoted to necessary information from the theory of sectorial operators.

Definition 4.3. *A linear bounded operator acting on a Hilbert space* H *is called a sectorial operator if*

$$\inf_{\|x\|_H = 1} (Ax, x) := \varepsilon > 0 \tag{4.12}$$

where (Ax, y) *denotes the scalar product in* H.

If $a(x) \in L_\infty(\mathbb{R})$ then the operator of multiplication by the function $a(x)$ in the space $L_2(\mathbb{R}, s)$ is sectorial if and only if

$$\operatorname*{ess\,inf}_{x \in \mathbb{R}} \operatorname{Re} a(x) = \varepsilon > 0. \tag{4.13}$$

Definition 4.4. *We will call a function* $a(x) \in L_\infty(\mathbb{R})$ *sectorial if exists a number* $\theta \in (-\pi, \pi)$ *such that for the function* $a_\theta(x) := e^{i\theta} a(x)$ *the condition* (4.11) *holds.*

We formulate the famous result of Brown and Halmos (see, for example, [25, Theorem 2.2]).

Theorem 4.1. *Let* A *be a sectorial operator on a Hilbert space* H. *Then the operator* A *is invertible and the following estimate holds for the norm of the inverse operator,*

$$\|A^{-1}\|_H \leq 2\varepsilon^{-1}$$

where ε *is the value from* (4.12).

Let now G be a subspace of the Hilbert space $L_2^+(\mathbb{R}, s)$ and let $\mathcal{P}_G$ be the orthoprojector onto the space G. This means that an arbitrary function $f(x) \in L_2^+(\mathbb{R}, s)$ can be represented uniquely in the form

$$f(x) = g_1(x) + g_2(x) \tag{4.14}$$

where $g_1(x) \in G$, $g_2(x) \in G^\perp$, and $G^\perp$ denotes the orthogonal complement of the space G in $L_2^+(\mathbb{R}, s)$. Thus the following equation holds,

$$P^+ = \mathcal{P}_G^\perp + \mathcal{P}_G$$

where $\mathcal{P}_G^\perp$ is the orthoprojector onto $G^\perp$. Consider the operator

$$D = \mathcal{P}_G^\perp + P^+ a \mathcal{P}_G : L_2^+(\mathbb{R}, s) \to L_2^+(\mathbb{R}, s) \tag{4.15}$$

where the function a belongs to $L_\infty(\mathbb{R})$.

Theorem 4.2. *Let function $a(\in L_\infty(\mathbb{R}))$ be sectorial. Then the operator D (4.15) is invertible and for the solution x of the equation*

$$Dx = f, \quad f \in L_2^+(\mathbb{R}, s), \tag{4.16}$$

there holds the following estimate,

$$\|x_1\|_{L_2(\mathbb{R}, s)} \le 2\varepsilon^{-1} \|f_1\|_{L_2(\mathbb{R}, s)} \tag{4.17}$$

where $x_1 = \mathcal{P}_G x$, $f_1 = \mathcal{P}_G f$, and ε is the value from (4.11).

Proof. Consider the operator

$$D_1 := \mathcal{P}_G a \mathcal{P}_G : G \to G.$$

and

$$D_{1,\theta} = \mathcal{P}_G a_\theta \mathcal{P}_G : G \to G$$

where the function $a_\theta(x)(= e^{i\theta} a(x))$ and the number θ are from Definition 4.4.
We show that $D_{1,\theta}$ operator is sectorial. Let $x_1 \in G$ then

$$(\mathcal{P}_G a_\theta x_1, x_1)_{L_2(\mathbb{R})} = (a_\theta x_1, x_1)_{L_2(\mathbb{R})}.$$

Then

$$\mathrm{Re}(\mathcal{P}_G a_\theta x_1, x_1) = \mathrm{Re}\left(\int_{-\infty}^{\infty} a_\theta(t) |x_1(t)|^2 dt \right)$$

$$= \int_{-\infty}^{\infty} (\mathrm{Re}\, a_\theta(t)) |x_1(t)|^2 dt \ge \varepsilon(x_1, x_1).$$

Thus the operator $D_{1,\theta}$ is invertible and according to Theorem 4.1 we have

$$\|D_{1,\theta}^{-1}\|_{L_2(\mathbb{R}, s)} \le 2\varepsilon^{-1}$$

where ε is the value from (4.11).
Since $D_{1,\theta} = e^{i\theta} D_1$, then

$$\|D_1^{-1}\|_{L_2(\mathbb{R}, s)} < 2\varepsilon^{-1} \tag{4.18}$$

Now rewrite the equation (4.16) in the form

$$x_2 + P^+ a x_1 = f_1 + f_2 \tag{4.19}$$

where $x_2 = \mathcal{P}_G^{\perp} x$ and $f_2 = \mathcal{P}_G^{\perp} f$. Applying the projector $\mathcal{P}_G$ to the last equality we get

$$D_1 x_1 = f_1. \tag{4.20}$$

Since D_1 is invertible, the equation (4.20) has a unique solution

$$x_1 = D_1^{-1} f_1$$

and according to (4.18) we have (4.17). Further applying the projector $\mathcal{P}^{\perp}$ to (4.19) we get

$$x_2 + \mathcal{P}_G^{\perp} a x_1 = f_2.$$

Therefore $x_2 = f_2 - \mathcal{P}_G^{\perp} a D_1^{-1} f_1$. Thus for arbitrary $f \in L_2^+(\mathbb{R})$ the equation (4.16) has a unique solution in the form

$$x = (D_1^{-1} \mathcal{P}_G + \mathcal{P}_G^{\perp} - \mathcal{P}_G^{\perp} a D_1^{-1} \mathcal{P}_G) f$$

and consequently the operator D is invertible. $\qquad\square$

5. Unique solvability of the modified Wiener-Hopf equation in the space $L_2(\mathbb{R}, s)$

This section is central in this work. In order to obtain the theorem of solvability we apply the Matrix Riemann Boundary Problem approach (originally worked out in [24]–[27]) for some diffraction problems. It should be noted that this approach suits perfectly for barrier option problems.

Assume that function $\psi^{\mathbf{Q}}(\xi)$ satisfies the following conditions (see (3.9), (3.10))

$$\psi^{\mathbf{Q}}(\xi) = \frac{1}{2}\sigma^2 \xi^2 - i\mu\xi + \varphi(\xi). \tag{5.1}$$

We suppose (see (3.12), (3.13)) that

$$\sigma = 0, \tag{5.2}$$

that there exists such $\nu \in (0, 2)$ that the function

$$c(\xi) := \frac{\varphi(\xi)}{(1 + \xi^2)^{\nu/2}} \in L_\infty(\mathbb{R}), \tag{5.3}$$

for some $M > 0$ satisfies

$$\inf_{|\xi| \geq M} \operatorname{Re} c(\xi) = \varepsilon_1 > 0, \tag{5.4}$$

and that

$$\begin{cases} \text{if} \quad \mu \neq 0 \quad \text{then} \quad 1 < \nu < 2, \\ \text{if} \quad \mu = 0 \quad \text{then} \quad 0 < \nu < 2. \end{cases} \tag{5.5}$$

Finally we assume that

$$r + \operatorname{Re} w \geq \varepsilon_2 > 0. \tag{5.6}$$

It should be noted that the interest rate of the bond r is positive and the complex number w lies on the contour L of the inverse Laplace transform. Very often $L = \{z \in \mathbb{C} : \mathrm{Re}\, z = -\delta\}$ where $\delta(\geq 0)$ is as small as we wish. Thus the condition (5.6) is natural.

Introduce the function

$$c(\xi, w) := \frac{(\psi^{\mathbf{Q}}(\xi) + r + w)}{(1 + \xi^2)^{\nu/2}}. \tag{5.7}$$

Lemma 5.1. *Let the conditions* (5.1)–(5.6) *hold. Then the function* $c(\cdot, \xi)$ *is sectorial, and if the value* ε_2 *in* (5.6) *is independent of* w *then there exists a number* ε *independent of* w *such that*

$$\inf_{\xi \in \mathbb{R}} \mathrm{Re}\, c(\xi, w) \geq \varepsilon > 0. \tag{5.8}$$

Proof. According to Lemma 3.1 and conditions (5.2), (5.6) we have for $\xi \in \mathbb{R}$

$$\mathrm{Re}\, c(\xi, w) = \mathrm{Re}\left(\frac{\varphi(\xi)}{(1 + \xi^2)^{\nu/2}}\right) + \frac{r + \mathrm{Re}\, w}{(1 + \xi^2)^{\nu/2}} > 0.$$

According to (5.6),

$$\inf_{|\xi| \leq M} \mathrm{Re}\, c(\xi, w) \geq \frac{\varepsilon_2}{(1 + M^2)^{\nu/2}}.$$

Further according to (5.4) we get

$$\inf_{|\xi| \geq M} \mathrm{Re}\, c(\xi, w) \geq \varepsilon_1.$$

Now set

$$\varepsilon = \min\left(\frac{\varepsilon_2}{(1 + M^2)^{\nu/2}}, \varepsilon_1\right). \tag{5.9}$$

Then we obtain (5.8).

Finally with the help of (5.3) and (5.5) we see that $c(\xi, w) \in L_\infty(\mathbb{R})$. Thus $c(\xi, w)$ is sectorial. $\qquad\square$

We see that according to Lemma 5.1 if the condition (5.6) holds then function $c(\xi, w)$ is sectorial with $\theta = 0$. It should be noted that $c(\xi, w)$ could be sectorial even when the condition (5.6) does not hold. In particular we need the following result.

Lemma 5.2. *Let the conditions* (5.1)–(5.5) *hold and for* $w(\neq 0)$ *suppose that*

$$|\arg w| \leq \frac{\pi}{2} + \theta_0, \quad \theta_0 > 0, \tag{5.10}$$

Then there exists a number θ_0 *(small enough) such that the function* $c(w)$ *is sectorial with the same* ε *(see Definition 4.4) for all* w *satisfying the condition* (5.10).

Proof. According to Lemma 5.1 we have that the statement is true for the region $|\arg w| \leq \frac{\pi}{2}$.

The function $c(\xi, 0)$ is sectorial with parameter $\theta = 0$. This means that there exists a number $\theta_0 > 0$ such that the set $J_0 := \{z \in |z = c(\xi, 0), \xi \in R\}$ lies strictly within the region $|\arg z| < \frac{\pi}{2} - \theta_0$.

Consider the case that

$$\frac{\pi}{2} < \arg w \le \frac{\pi}{2} + \theta_0$$

Let $\varepsilon_1(>0)$ be the distance between J_0 and the line

$$R_{-\frac{\pi}{2}+\theta_0} = \left\{ z \in \mathbb{C} \mid z = re^{i(-\frac{\pi}{2}+\theta_0)}, r \in \mathbb{R} \right\}$$

Then the distance between the set

$$J_w := \{ z \in R \mid z = c(\xi, w), \xi \in R \}$$

and the semiplane

$$-\frac{\pi}{2} + \varphi_0 \le \arg z \le \frac{\pi}{2} + \varphi_0$$

is no smaller then ε_1, since

$$c(\xi, w) = c(\xi, 0) + \frac{|w|e^{i\arg w}}{(1+\xi^2)^{v/2}}$$

Thus the function $c(\xi, w)$ is sectorial with parameters $\theta = -\theta_0$ and $\varepsilon = \varepsilon_1$.

The case $-\frac{\pi}{2} - \varphi_0 \le \arg w < -\frac{\pi}{2}$ is considered analogously. $\qquad\square$

It should be noted that the hypothesis of Lemmas 5.1–5.2 hold for Lévy processes of the Kobol family in the case (3.15), for normal tempered stable Lévy processes (3.20) and for normal inverse Gaussian processes (3.21).

Now consider equation (4.11). It is convenient for us to make a change of variable $(\xi \mapsto -\xi)$ and denote

$$\widetilde{\Phi}^\pm(\xi, w) = \Phi^\mp(-\xi, w), \quad \widetilde{\Phi}_a^+(\xi, w) = \Phi_a^-(-\xi, w).$$

Then we can rewrite (4.10) in the form

$$e^{ia\xi}\widetilde{\Phi}^+(\xi, w) + (1+\xi^2)^{v/2}\widetilde{c}(\xi, w)\widetilde{\Phi}_a^+(\xi, w) + \widetilde{\Phi}^-(\xi, w) = \hat{g}(-\xi), \qquad (5.11)$$

where conditions (5.1)–(5.6) are satisfied and $\widetilde{c}(\xi, w) = c(-\xi, w)$ (see 5.7).

Furthermore we assume that conditions (4.6)–(4.9) hold for $s_1 = v/2 + s$, $s_2 = -v/2 + s$ where $|s| < \frac{1}{2}$. That is,

$$\widetilde{\Phi}^\pm(\xi, w) \in L_2^\pm(\mathbb{R}, -v/2 + s); \qquad (5.12)$$

$$\widetilde{\Phi}_a^+(\xi, w) \in L_2^+(\mathbb{R}, v/2 + s). \qquad (5.13)$$

Consider the so-called Wiener-Hopf factorization of the function $\gamma(\xi) := (1+\xi^2)^{v/2}$,

$$\gamma(\xi) = (1 + i\xi)^{v/2}(1 - i\xi)^{v/2} := \gamma_-(\xi)\gamma_+(\xi).$$

The cuts of the functions $\gamma_\pm(\xi) := (1 \mp i\xi)^{v/2}$ pass along the rays $\Gamma_\pm = \{z \in \mathbb{C} : z = \mp is, s \in [1, \infty)\}$ respectively. Thus the function $\gamma_+(\xi)$ is analytic in the upper half-plane and $\gamma_-(\xi)$ is analytic in the lower half-plane.

Divide all terms of (5.11) by $\gamma_-(\xi)$ and write

$$\Psi_a^+(\xi, w) \ := \ \gamma_+(\xi)\widetilde{\Phi}_a^+(\xi, w); \tag{5.14}$$

$$\Psi^\pm(\xi, w) \ := \ \frac{\widetilde{\Phi}^\pm(\xi, w)}{\gamma_\pm(\xi)}. \tag{5.15}$$

Then we obtain

$$e^{ia\xi}u(\xi)\Psi^+(\xi, w) + \widetilde{c}(\xi, w)\Psi_a^+(\xi, w) + \Psi^-(\xi, w) = \frac{\hat{g}(-\xi)}{\gamma_-(\xi)} \tag{5.16}$$

where

$$u(\xi) := \frac{\gamma_+(\xi)}{\gamma_-(\xi)} = \left(\frac{1 - i\xi}{1 + i\xi}\right)^{\nu/2}. \tag{5.17}$$

It is easy to see (see, for example, [40]) that

$$\Psi_a^+(\xi) \ \in \ L_2^+(\mathbb{R}, s); \tag{5.18}$$

$$\Psi^\pm(\xi) \ \in \ L_2^\pm(\mathbb{R}, s). \tag{5.19}$$

It should be noted that the functions $\widetilde{\Phi}_a^+(\xi, w)$ and $\Psi_a^+(\xi, w)$ belong to narrower classes of functions than $L_2^+(\mathbb{R}, \nu/2 + s)$ and $L_2^+(\mathbb{R}, s)$ respectively.

Namely, the following statement holds.

Lemma 5.3.

i) *The class of functions where the unknown function $\widetilde{\Phi}_a^+(\xi, w)$ is looked for coincides with the set of "+"-components of the solutions of the boundary value problem*

$$e^{-ia\xi}\Phi_a^+(\xi) = \Phi_a^-(\xi), \quad a > 0, \tag{5.20}$$

where $\Phi_a^\pm(\xi) \in L_2^\pm(\mathbb{R}, \nu/2 + s)$.

ii) *The class of functions where the unknown function $\Psi_a^+(\xi, w)$ is looked for coincides with the set of "+"-components of solutions of the boundary value problem*

$$e^{-ia\xi}\overline{u}(\xi)\Psi_a^+(\xi) = \Psi_a^-(\xi), \quad a > 0, \tag{5.21}$$

where $\Psi_a^\pm(\xi) \in L_2^\pm(\mathbb{R}, s), |s| < 1/2$.

Proof. The statement i) is well known ([43]–[44]).

We pass to the proof of ii). Multiply both sides of (5.20) by $\gamma_-(\xi)$ and denote

$$\Psi_a^+(\xi) := \gamma_+(\xi)\Phi_a^+(\xi) \quad \text{and} \quad \Psi_a^-(\xi) = \gamma_-(\xi)\Phi_a^-(\xi).$$

Then we obtain (5.21). $\qquad \square$

The problems (5.20) and (5.21) are called Riemann Boundary Value Problems with coefficients $e_a(\xi) := e^{-ia\xi}$ and $\overline{u_a}(\xi) := e^{-ia\xi}\overline{u}(\xi)$ respectively. Moreover the set of all functions $\Phi_a^+(\xi)$ ($\in L_2^+(\mathbb{R}, \nu/2 + s)$) satisfying problem (5.20) coincides with the kernel of the Toeplitz operator T_{e_a} in the space $L_2^+(\mathbb{R}, \nu/2 + s)$. Analogously the set of all functions $\Psi_a^+(\xi)$ satisfying the problem (5.21) coincides with the subspace $\ker T_{\overline{u}_a}\big|_{L_2^+(\mathbb{R}, s)}$.

Thus the components of the solution of the problem (5.11) are looked for in the following spaces

$$\widetilde{\Phi}_a^+(\xi, w) \;\in\; \ker T_{e_a}\Big|_{L_2^+(\mathbb{R}, \nu/2+s)}; \tag{5.22}$$

$$\widetilde{\Phi}^{\pm}(\xi, w) \;\in\; L_2^{\pm}(\mathbb{R}, -\nu/2+s). \tag{5.23}$$

Analogously the components of solution of the problem (5.16) are looked for in the spaces

$$\Psi_a^+(\xi, w) \;\in\; \ker T_{\overline{u}_a}\Big|_{L_2^+(\mathbb{R},s)}; \tag{5.24}$$

$$\Psi^{\pm}(\xi, w) \;\in\; L_2^{\pm}(\mathbb{R}, s). \tag{5.25}$$

Apply the projector P^+ to all terms of equation (5.16). Then we have

$$(T_{u_a}\Psi^+)(\xi, w) + P^+(\widetilde{c}(\xi, w)\Psi_a^+(\xi, w) = f^+(\xi) \tag{5.26}$$

where T_{u_a} is the Toeplitz operator with symbol

$$u_a(\xi) := e^{ia\xi}u(\xi), \tag{5.27}$$

and

$$f^+(\xi) = P^+(\hat{g}(-\xi)/\gamma_-(\xi)), \tag{5.28}$$

$$\Psi_a^+(\xi, w) \in \ker T_{\overline{u}_a}\Big|_{L_2^+(\mathbb{R},s)}, \tag{5.29}$$

$$\Psi^+(\xi, w) \in L_2^+(\mathbb{R}, s). \tag{5.30}$$

It is easy to observe that the problem (5.11), (5.22), (5.23) has a solution if and only if the problem (5.26)–(5.30) has a solution as well. Further the components of the solution of the first problem relate to the components of solution of the second problem by means of formulae (5.14)–(5.15).

Consider the function $u(\xi)$. It is easy to see that $u(\xi)$ is continuous on $\mathbb{R}$ and $\lim_{\xi \to \pm\infty} u(\xi) = e^{\mp i\pi\nu/2}$. Thus according to Lemma 4.3 $u_a(\xi) \in H^\infty(\mathbb{R}) + C(\dot{\mathbb{R}})$. Consequently according to Lemma 4.2 the Toeplitz operator T_{u_a} is left-invertible and according to Lemma 4.1 we have $\operatorname{im} T_{u_a} \oplus \ker T_{\overline{u}_a} = L_2^+(\mathbb{R}, s)$ since $T_{u_a}^* = T_{\overline{u}_a}$. Associate with this decomposition the pair of orthogonal projectors $\mathcal{P}_{u_a}^\perp$ and $\mathcal{P}_{u_a}$ $(\mathcal{P}_{u_a}(L_2^+(\mathbb{R}, s)) = \ker T_{\overline{u}_a}, \; \mathcal{P}_{u_a}^\perp(L_2^+(\mathbb{R}, s)) = \operatorname{im} T_{u_a})$ and consider the operator

$$D_{u_a} := \mathcal{P}_{u_a}^\perp + P^+\widetilde{c}(\xi, w)\mathcal{P}_{u_a} : L_2^+(\mathbb{R}, s) \to L_2^+(\mathbb{R}, s). \tag{5.31}$$

Associate with the operator (5.31) the following operator equation

$$(D_{u_a}Y^+)(\xi) = f^+(\xi), \quad Y^+(\xi) \in L_2^+(\mathbb{R}, s) \tag{5.32}$$

where $f^+(\xi)$ is defined by (5.28).

Lemma 5.4. *The problem (5.26)–(5.30) has a solution if and only if the equation (5.32) has a solution as well. Moreover if $Y^+(\xi)$ satisfies (5.32) then the following* *functions*

$$\Psi_a^+(\xi) \;\; = \;\; (\mathcal{P}_{u_a} Y^+)(\xi); \tag{5.33}$$

$$\Psi^+(\xi) \;\; = \;\; T_{u_a}^{-1}(\mathcal{P}_{u_a}^{\perp} Y^+)(\xi) \tag{5.34}$$

are a solution of 5.26. Here $T_{u_a}^{-1}$ is a left inverse of operator T_{u_a}.

Proof. Let $Y^+(\xi)$ be a solution of (5.32). Taking into account that for function $f(\xi)$ belonging to $\operatorname{im} T_{u_a}$, $(T_{u_a} T_{u_a}^{-1} f)(\xi) = f(\xi)$ and substituting (5.33)–(5.34) to equation (5.26) we obtain

$$T_{u_a} T_{u_a}^{-1}(\mathcal{P}_{u_a}^{\perp} Y^+)(\xi) + P^+(\tilde{c}(\xi, w)(\mathcal{P}_{u_a} Y^+)(\xi))$$
$$= (\mathcal{P}_{u_a}^{\perp} Y^+)(\xi) + P^+(\tilde{c}(\xi, w)(\mathcal{P}_{u_a} Y^+)(\xi)) = (D_{u_a} Y^+)(\xi) = f^+(\xi).$$

Conversely, let $(\Psi_a^+(\xi), \Psi^+(\xi))$ be a solution of (5.26). Then it is easy to check that the function

$$Y^+(\xi) := (T_{u_a} \Psi^+)(\xi, w) + \Psi_a^+(\xi)$$

is a solution of (5.32). $\qquad\square$

Theorem 5.1. *Let the function $c(\xi, w)$ (5.7) satisfy conditions (5.1)–(5.5) and w belong the region (5.10). Then the following statements are true:*

i) *The operator D_{u_a} (5.31) is invertible and for the solution of (5.32) the following estimate holds*

$$\|\mathcal{P}_{u_a} Y^+\|_{L_2(\mathbb{R}, s)} \leq 2\varepsilon^{-1} \|\mathcal{P}_{u_a} f^+\|_{L_2(\mathbb{R}, s)}.$$

where ε does not dependent of w.

ii) *The problem (5.26)–(5.30) has the unique solution*

$$\Psi_a^+(\xi, w) = (\mathcal{P}_{u_a} D_{u_a}^{-1} f^+)(\xi), \quad \Psi^+(\xi, w) = (T_{u_a}^{-1} \mathcal{P}_{u_a}^{\perp} D_{u_a}^{-1} f^+)(\xi). \tag{5.35}$$

iii) *The problem (5.11), (5.22), (5.23) has the unique solution*

$$\tilde{\Phi}_a^+(\xi, w) \;\; = \;\; \frac{1}{\gamma^+(\xi)} \left(\mathcal{P}_{u_a} D_{u_a}^{-1} P^+ \frac{\hat{g}(-\xi)}{\gamma_-(\xi)} \right); \tag{5.36}$$

$$\tilde{\Phi}^+(\xi, w) \;\; = \;\; \gamma^+(\xi) \left(T_{u_a}^{-1} \mathcal{P}_{u_a}^{\perp} D_{u_a}^{-1} P^+ \frac{\hat{g}(-\xi)}{\gamma_-(\xi)} \right); \tag{5.37}$$

$$\tilde{\Phi}^-(\xi, w) \;\; = \;\; \gamma_-(\xi) \Psi^-(\xi, w); \tag{5.38}$$

where the function $\Psi^-(\xi, w)$ can be found from the relation (5.16).

Proof. The statement i) follows directly from Theorem 4.2 and Lemma 5.2.

ii) This statement follows from i) and Lemma 5.4 since the function $Y^+(\xi) := (D_{u_a}^{-1} f^+)(\xi)$ is the unique solution of equation (5.32).

iii) It is easy to see that problems (5.11), (5.22), (5.23) and (5.26)–(5.30) have solutions simultaneously and they are connected according to formulae (5.14)–(5.15). Moreover the triple $(\Psi^+, \Psi_a^+, \Psi^-)$ satisfies the equation (5.16) if and only if the pair (Ψ^+, Ψ_a^+) satisfies the equation (5.26) and

$$\Psi^-(\xi, w) = \left(P^- \frac{\hat{g}(-\xi)}{\gamma_-(\xi)} \right) - P^-(u_a(\xi) \Psi^+(\xi, w)) - P^-(\tilde{c}(\xi, w) \Psi_a^+(\xi, w)). \qquad\square$$

6. Unique solvability of the problem (2.10)–(2.12) and the price of the double barrier option

We shall look for solutions of the problem (2.10)–(2.12) in the following functional space:

$$u(x,\tau) \in C^0([0,\infty), H^{\frac{\nu}{2}+s}(0,a)), \quad |s| < 1/2.$$

This means that for each fixed $\tau \le 0$ $u(\cdot,\tau) \in H^{\frac{\nu}{2}+s}(0,a)$, and the function $F(\tau) := \|u(\cdot,\tau)\|_{H^{\frac{\nu}{2}+s}}$ is continuous on $[0,\infty)$ with $\lim_{\tau\to\infty} F(\tau) = 0$. Applying by Laplace transform (3.1) on the function $u(x,\tau)$ we have (at least for w with $\mathrm{Re}\,w > 0$) that

$$v(\cdot,w) \in H^{\frac{\nu}{2}+s}(0,a)$$

Further we have for the function

$$\widetilde{\Phi}_a^+(\xi,w) = (Fv)(-\xi,w)$$

the problem (5.11), (5.22), (5.23). This problem has a unique solution of the form (5.36) and this solution has $L_2(\mathbb{R}, \frac{\nu}{2} + s)$-norm bounded uniformly by w belonging to the region (5.10).

Thus applying the inverse Fourier Transform to the function $\Phi_a^+(-\xi, \dot{w})$ and then applying the inverse Laplace transform we obtain that the problem (2.10)–(2.12) has the solution of the following form

$$u(x,\tau) = \frac{1}{(2\pi)^2 i} \int_{L_{\theta_0}} \int_{-\infty}^{\infty} \widetilde{\Phi}_a^+(-\xi, w) e^{i\xi x + \tau w} d\xi dw \tag{6.1}$$

Here $\widetilde{\Phi}_a^+(-\xi, w)$ is given by (5.36) and the contour L_{θ_0} is the boundary of the sector $K_{\theta_0} := \{z \in \mathbb{C} \,|\, |\arg z| \le \frac{\pi}{2} + \theta_0\}$ for $\theta_0 > 0$ small enough.

Theorem 6.1. *Let $\nu \in (0,2)$, let the function $g(x) \in H^{-\frac{\nu}{2}+s}(0,a)$, for some $s \in (-1/2, 1/2)$ and let the be characteristic exponent under a EMM Q, the function $\psi^Q(\xi)$ (3.5), such that the symbol $c(\xi,w)$ (given by formula (5.7)) satisfies the conditions (5.2)–(5.5).*

Then the problem (2.10)–(2.12) has a unique solution in the space

$$C^0([0,\infty), H^{\frac{\nu}{2}+s)}(0,a))$$

and this solution has the form (6.1).

This theorem follows from Theorem 5.1 and the fact that the function $e^{\tau w}$ decreases to zero as $e^{\tau \mathrm{Re} w}$. In fact, for w belonging to L_{θ_0} $\mathrm{Re}\,w < 0$ and $\mathrm{Re}\,w \to -\infty$ if w passes along L_{θ_0}.

Now we are ready consider problem of finding the option price $U(x,t)$ (2.4). According to Theorem 2.13 of [15], $U(x,t)$ is a bounded solution of the problem (2.5)–(2.7).

Theorem 6.2. *Let $g(x) \in L_\infty(0,a)$ and let the process X_t satisfy the (ACP)-condition. Then the problem (2.10)–(2.12) (and problem (2.5)–(2.7)) has no more than one solution.*

128   S.M. Grudsky

Suppose we have two bounded solutions $u_{1,2}(x,\tau)$ of the problem (2.10)–(2.12). Then the function $u_0(x,\tau) := u_2(x,\tau) - u_1(x,\tau)$ satisfies the following problem:

$$\frac{\partial u_0(x,\tau)}{\partial \tau} + (r - L_x^Q)u_0\,(x,\tau) = 0 \quad (x,\tau) \in (0,a) \times (0,\infty) \tag{6.2}$$

$$u_0(x,0) = 0 \qquad x \in (0,a) \tag{6.3}$$

$$u(x,\tau) = 0 \qquad x \in R\backslash(0,a), \tau \in (0,\infty) \tag{6.4}$$

Applying to (6.2)–(6.4) the Laplace transform we obtain for the function

$$v_0(v,w) := (Lu_0)(x,w)$$

the following problem,

$$(-L_x^Q + r + w)v_0(x,w) = 0 \qquad x \in (0,a) \tag{6.5}$$

$$v_0(x,w) = 0, \qquad x \in R\backslash(0,a) \tag{6.6}$$

The function $v_0(x,w)$ is bounded at least for all w with $\mathrm{Re}\,w > 0$. The problem (6.5)–(6.6) is understood in sense of generalized functions:

$$(v(x), P_{[0,a]}(F^{-1}(\psi^Q(-\xi) + r + \overline{w})Fv_0) = 0 \tag{6.7}$$

where $v(x)$ is arbitrary function of $S(R)$ such that supp $v(x) \subset (0,a)$. Let $\{v_n\} \in S(R)$ be a sequence of functions with supp $v_n(x) \in (0,a)$ and such that $v_n(x) \to v_0(x,w)$ in the weak sense. Then we have from (6.7) that

$$(v_n(x), F^{-1}(\psi^Q(-\xi) + r + \overline{w})Fv_0) = 0$$

or equivalently

$$(Fv_n(x), (\psi^Q(-\xi) + r + \overline{w})Fv_0) = 0. \tag{6.8}$$

Introduce the sequence of numbers

$$\ell_n := (Fv_n, (\psi^Q(-\xi) + r + \overline{w})Fv_n)$$

According to (6.8),

$$\lim_{n\to\infty} \ell_n = 0.$$

But on the other hand according to Lemma 3.1

$$\mathrm{Re}\,\ell_n = \int_{-\infty}^{\infty} \mathrm{Re}(\psi^Q(-\xi) + r + \overline{w})|(Fv_n)(\xi)|^2 ds \geq$$

$$\geq r\int_{-\infty}^{\infty} |(Fv_n)(\xi))^2 d\xi = r\int_{-\infty}^{\infty} |v_n(\xi)|^2 d\xi.$$

That is, for n larger enough

$$\mathrm{Re}\,\ell_n \geq \frac{r}{2}\|v(\cdot,w)\|_{L_2(\mathbb{R})}^2;$$

That is $v_0(\xi,w) \equiv 0$ for all w with $\mathrm{Re}\,w > 0$. Thus $u_0(x,\tau) \equiv 0$ and the theorem is proved.

We wish to obtain a bounded solution of the problem (2.5)–(2.7) or equivalently the problem (2.10)–(2.12).

For this we impose an additional condition. Namely, let for some $s \in (-1/2, 1/2)$

$$\frac{\nu}{2} + s > \frac{1}{2} \tag{6.9}$$

It is well known that in this case

$$H^{\frac{\nu}{2}+s}(0,a) \subset C[0,a] \tag{6.10}$$

where $C[0,a]$ is the space of continuous functions on the segment $[0,a]$ and for the function $f(x) \in H^{\frac{\nu}{2}+s}(0,a)$ the following inequality holds,

$$\sup_{x \in [0,a]} |f(x)| \leq M \|f\|_{H^{\frac{\nu}{2}+s}} \tag{6.11}$$

with $M > 0$ constant.

Theorem 6.3. *Let all conditions of Theorem 6.1 and inequality (6.9) hold. Then the solution of the problem (2.10)–(2.12) is bounded.*

Proof. According Theorem 6.1 the problem (2.10)–(2.10) has a unique solution in the space $C^0([0,\infty), H^{\frac{\nu}{2}+s}(0,a))$ having the form (6.1). In virtue of (6.10) and (6.11) this solution is a bounded function in x uniformly in $t \in [0,\infty)$. $\qquad \square$

Finally suppose that $g(x)$ is a piecewise smooth function on the segment $[0,a]$. It is easy to see that in this case $g(x) \in H^\mu(0,a)$ for any $\mu < \frac{1}{2}$. For arbitrary $\frac{\nu}{2} \in (0,1)$ we always can choose $s \in [0,1/2)$ such that condition (6.9) holds and moreover we have

$$\mu = -\frac{\nu}{2} + s < \frac{1}{2}.$$

Thus in this case according to Theorem 6.3 the problems (2.10)–(2.12) and (2.5)–(2.7) have bounded solutions. Since the Theorem 6.2 implies that this solution is unique, it has the form (6.1) and coincides with (2.4).

It should be noted that the condition for the function $g(x)$ to be piecewise smooth holds very often in option theory.

7. Stability of the solution with respect to small perturbations of the characteristic function

Rewrite the equation (5.26) in the form

$$P^+(u_a(\xi)\Psi^+(\xi, w)) + P^+(\widetilde{c}(\xi, w)\Psi_a^+(\xi, w)) = f^+(\xi). \tag{7.1}$$

Apply the projector P^+ to the equation (5.21)

$$P^+ \overline{u}_a \Psi_a^+(\xi, w) = 0. \tag{7.2}$$

Rewrite (7.1)–(7.2) as a matrix equation

$$(T_{B_a}\vec{\Psi})(\xi) = \vec{F}^+(\xi) \tag{7.3}$$

where the vector functions

$$\vec{\Psi}(\xi) := \left(\begin{array}{c} \Psi^+(\xi) \\ \Psi_a^+(\xi) \end{array} \right) \in L_2^{2+}(\mathbb{R}, s), \tag{7.4}$$

$$\vec{F}^+(\xi) := \left(\begin{array}{c} f^+(\xi) \\ 0 \end{array} \right) \in L_2^{2+}(\mathbb{R}, s), \tag{7.5}$$

and the matrix Toeplitz operator is defined in the usual way,

$$T_{B_a} := \mathbb{P}^+ B_a \Big|_{L_2^{2+}(\mathbb{R}, s)} \tag{7.6}$$

with the matrix symbol

$$B_a(\xi) = \left(\begin{array}{cc} u_a(\xi) & c(\xi, w) \\ 0 & u_a(\xi) \end{array} \right). \tag{7.7}$$

Here the vector analytic projector

$$\mathbb{P}^+ : L_2^2(\mathbb{R}, s) \to L_2^{2+}(\mathbb{R}, s)$$

is defined component-wise,

$$\mathbb{P}^+ := \left(\begin{array}{c} P^+ \\ P^+ \end{array} \right).$$

It is obvious that the problems (5.26), (5.29), (5.30) and (7.3)–(7.5) are equivalent. Moreover the following result follows from Lemma 5.4.

It should be noted that the norm in space $L^2(\mathbb{R}, s)$ is define by usual way

$$\|f_1, f_2\|_{L_2^2(\mathbb{R}, sw)} = (\|f_1\|_{L_2(\mathbb{R}, s)}^2 + \|f_2\|_{L_2(\mathbb{R}, s)}^2)^{1/2}$$

Lemma 7.1. *The matrix Toeplitz operator* T_{B_a} *(7.6) is invertible in the space* $L_2^{2+}(\mathbb{R}, s), |s| < 1/2$, *if and only if the operator* D_{u_a} *(5.31) is invertible in the space* $L_2^+(\mathbb{R}, s)$.

Thus the invertibility of the operator T_{B_a} follows from Theorem 5.1, i).

Theorem 7.1. *Let the function* $c(\xi, w)$ *(5.7) satisfy conditions (5.3)–(5.6). Then the operator* T_{B_a} *is invertible and*

$$\|T_{B_a}^{-1}\|_{L_2^2(\mathbb{R})} \leq M\varepsilon^{-1}$$

where ε *is given by (5.8), (5.9), and* $M > 0$ *is constant.*

Thus we can write the solution of the option problem with the help of the operator $T_{B_a}^{-1}$. Indeed, under the hypotheses of Theorem 7.1 the solution of equation (7.3) has the form

$$\left(\begin{array}{c} \Psi^+(\xi) \\ \Psi_a^+(\xi) \end{array} \right) = T_{B_a}^{-1} \left(\begin{array}{c} f^+ \\ 0 \end{array} \right).$$

So the formula (6.1) can be rewritten in the form

$$\mathcal{U}(x, t) = \frac{1}{(2\pi)^2 i} \int_{R_\sigma} \int_{-\infty}^{\infty} T_{B_a}^{-1} \left(\begin{array}{c} f^+(\xi) \\ 0 \end{array} \right) \Big|_2 e^{(T-t)w - i\xi x} d\xi dw \tag{7.8}$$

where $\vec{F}(\xi)\big|_2$ denotes the second component of the vector function $\vec{F}(\xi)$. Thus practical (approximate) solution of the equation (7.3) is an important problem. The following reasoning can be considered as a basis for some algorithms of approximate solution.

With equation (7.3) consider

$$T_{B_a^*}\vec{\Psi} = \vec{F}_0^* \tag{7.9}$$

where the approximate symbol of the Toeplitz operator $B_a^*(\xi)$ and right-hand member $\vec{F}_0^*$ have the forms

$$B_a^*(\xi) = \begin{pmatrix} u_a^*(\xi) & c^*(\xi,w) \\ 0 & u_a^*(\xi) \end{pmatrix}; \quad \vec{F}_0^*(\xi) = \begin{pmatrix} f^*(\xi); \\ 0 \end{pmatrix}$$

with the components satisfying the following conditions

$$\sup_{\xi \in \mathbb{R}} |u_a(\xi) - u_a^*(\xi)| \le \delta_0, \tag{7.10}$$

$$\sup_{\xi \in \mathbb{R}} |\widetilde{c}(\xi,w) - c^*(\xi,w)| \le \delta_0, \tag{7.11}$$

and

$$\|f^+ - f^{*+}\|_{L_2(\mathbb{R},s)} \le \delta_1 \tag{7.12}$$

where the numbers $\delta_0, \delta_1 > 0$ are sufficiently small.

The following theorem is a standard fact from the theory of Toeplitz operators ([40]–[41]).

Theorem 7.2. *Let the function $c(\xi,w)$ (5.7) satisfy conditions (5.1)–(5.6). Then for δ_0 small enough the operator $T_{B_a^*}$ is invertible, equation (7.9) has an unique solution $\vec{\Psi}^*(\xi)$ and the following estimate holds,*

$$\|\vec{\Psi} - \vec{\Psi}^*\|_{L_2^2(\mathbb{R},s)} \le M_0\delta + M_1\delta_1. \tag{7.13}$$

In particular,

$$\|\Psi_a^+ - \Psi_a^{*+}\|_{L_2(\mathbb{R},s)} \le M_0\delta + M_1\delta_1$$

where $\vec{\Psi} = \begin{pmatrix} \Psi^+ \\ \Psi_a^+ \end{pmatrix}$ and $\vec{\Psi}^* = \begin{pmatrix} \Psi^{*+} \\ \Psi_a^{*+} \end{pmatrix}$ are the solutions of equations (7.3) and (7.9) respectively, and $M_0, M_1 > 0$ are independent of δ_0, δ_1.

Proof. According to Theorem 7.1, the operator T_{B_a} is invertible. Therefore if δ_0 is small enough, then the operator $T_{B_a^*}$ is invertible also and

$$\|T_{B_a}^{-1} - T_{B_a^*}^{-1}\|_{L_2^2(\mathbb{R},s)} \le C\delta_0$$

where $C > 0$ is independent of δ_0. Thus equation (7.9) has the unique solution $\vec{\Psi}^* = T_{B_a^*}\vec{F}^*$ and we have the following inequalities,

$$\begin{aligned}
\|\vec{\Psi} - \vec{\Psi}^*\|_{L_2^2(\mathbb{R},s)} &= \|T_{B_a}^{-1}\vec{F} - T_{B_a^*}^{-1}\vec{F}^*\|_{L_2^2(\mathbb{R},s)} \\
&= \|(T_{B_a}^{-1} - T_{B_a^*}^{-1})\vec{F} + T_{B_a^*}^{-1}(\vec{F} - \vec{F}^*)\|_{L_2^2(\mathbb{R},s)} \\
&\le (C\|\vec{F}\|_{L_2^2(\mathbb{R},s)})\delta_0 + (\|T_{B_a^*}^{-1}\|_{L_2^2(\mathbb{R},s)})\delta_1.
\end{aligned}$$

Denote $M_0 := C\|\vec{F}\|_{L^2_2(\mathbb{R},s)}$ and $M_1 = 2\|T_{B_a}^{-1}\|_{L^2_2(\mathbb{R},s)}$. Then for δ_0, δ_1 small enough we have the evaluation (7.13). $\qquad\qquad\square$

Thus the approximate solution of our option problem can be written in the form (see (7.8))

$$\mathcal{U}^*(x,t) = \frac{1}{(2\pi)^2 i} \int_{R_\sigma} \int_{-\infty}^{\infty} T_{B_a^*}^{-1} \left(\begin{array}{c} f^{*+}(\xi) \\ 0 \end{array} \right)\Big|_2 e^{(T-t)w - i\xi x} d\xi\, dw. \qquad (7.14)$$

This formula can serve as the basis for an algorithm for the approximate solution of the double barrier option problem. We will present this algorithm in future work.

8. Conclusion

In this article we treat some power cases of characteristic functions (see (3.12)–(3.13)). These cases involve wide classes of Lévy processes which are used in option theory. However, there exist many other cases which could be considered with the help of the methods worked out in this article.

1. The case $\sigma > 0$ is important because it corresponds to the processes with non trivial Gaussian components. This case can be realized as the case $\nu < 2$ considered in these notes.
2. The case $\sigma = 0$, $\mu \neq 0$ and $0 < \nu < 1$ (see (3.14), (3.15), (3.16), (3.20), (3.21)).
3. Logarithmic cases (3.16) and (3.22) if $\mu = 0$.
4. Power logarithmic case (3.17).
5. Rational case (3.27). In this case not only the solvability theory can be worked out but one can obtain the solution in explicit form.
6. Periodic case. The Poisson process generates a periodic characteristic function (3.24). It is interesting to get explicit formulae and to analyze them in this case. (3.25) is very interesting also because here X_t is sum of a Gaussian process and a discrete-jumping process. In this area the theory of matrix Toeplitz operators with periodic and almost periodic symbols (worked out by Karlovich-Spitkovsky-Böttcher see [45]) could be applied.
7. General case. According to a famous result ([28, p. 13]) for an arbitrary triplet (a, γ, Π) with measure Π satisfying (2.3) there exists a Lévy process X_t with this characteristic triplet. The condition (2.3) is quite general . Thus there exist Lévy processes with characteristic function having discontinuities of the first type at infinity, semi almost periodic discontinuities and so on. It is very interesting to consider the double barrier option problem for the general case when characteristic function has the form (2.2)–(2.3).

Acknowledgment

The author wishes to thank the reviewer for very useful remarks.

References

[1] Kunimoto N., Ikeda M. Pricing options with curved boundaries. Math. Finance 2, 275–298, 1992.

[2] Geman H. and Yor M. Pricing and Hedging double-barrier options: A probabilistic approach. Math. Finance **6:4**, 365–378, 1996.

[3] Sidenius J. Double Barrier Options – Valuating by Path Counting. J. Computat. Finance, 1998, V1, No. 3.

[4] Pelsser A. Pricing double barrier options using Laplace transform. Finance Stochastics **4:1**, 95–104, 2000.

[5] Baldi P., Caramellino L., Iovino M.G. Pricing general Barrier Options using sharp larger deviations. Math. Finance 9:4, 291–322, 1999.

[6] Hui C.H. One-touch double barrier binary option values. Appl. Financial Econ. **6**, 343–346, 1996.

[7] Hui C.H. Time dependent barrier option values. J. Futures Markets, 17, 667–688, 1997.

[8] Wilmott P., Dewynne J., Howison S. Option pricing: Mathematical models and computations. Oxford: Oxford Financial Press, 1993.

[9] Raible R. Lévy processes in Finance: Theory, Numerics, and Empirical Facts. Dissertation. Mathematische Fakultät, Universität Freiburg im Breisgau, 2000.

[10] Barndorff-Nielsen O.E. Processes of normal inverse Gaussian Type. Finance and Stochastics **2**, 41–68, 1998.

[11] Madan D.B., Carr P. and Chang E.C. The variance Gamma process and option pricing. European Finance Review **2**, 79–105, 1998.

[12] Eberlein E. Application of generalized hyperbolic Lévy motions to Finance. In: Lévy processes: Theory and applications, O.E. Barndorff-Nielsen, T. Mikosh and S. Resnik (Eds.), Birkhäuser, 319–337, 2001.

[13] Bouchaud J.-P. and Potters M. Theory of financial risk. Cambridge University Press, Cambridge.

[14] Matacz A. Financial modelling and option theory with the truncated Lévy process. Intern. Journ. Theor. and Appl. Finance **3:1**, 143–160, 2000.

[15] Boyarchenko S.I., Levendorskiĭ Sergei. Non-Gaussian Merton-Black-Scholes theory. Advanced Series on Statistical Science and Applied Probability **9**, World-Scientific, Singapore, 2002.

[16] Boyarchenko S.I., Levendorskiĭ S.Z. On rational pricing of derivative securities for a family of non-Gaussian processes. Preprint 98/7, Institut für Mathematik, Universität Potsdam, Potsdam.

[17] Boyarchenko S.I., Levendorskiĭ S.Z. Option pricing and hedging under regular Lévy processes of exponential type. In: Trends in Mathematics. Mathematical Finance, M. Kohlman and S. Tang (Eds.), 121–130, 2001.

[18] Levendorskiĭ S.Z. and Zherder V.M. Fast option pricing under regular Lévy processes of exponential type. Submitted to Journal of Computational Finance, 2001.

[19] Boyarchenko S.I., Levendorskiĭ S.Z. Option pricing for truncated Lévy processes. Intern. Journ. Theor. and Appl. Finance **3:3**, 549–552, 2000.

[20] Boyarchenko S.I., Levendorskiĭ S.Z. Perpetual American options under Lévy processes. SIAM Journ. of Control and Optimization, Vol. 40 (2002) no. 6, pp. 1663–1696

[21] Mordecki E. Optimal stopping and perpetual options for Lévy processes. Talk presented at the 1 World Congress of the Bachelier Finance Society, June 2000.

[22] Boyarchenko S.I., and Levendorskiĭ S.Z. Barrier options and touch-and-out options under regular Lévy processes of exponential type. Annals of Applied Probability, Vol. 12 (2002), no. 4, pp. 1261–1298.

[23] Cont R., Voltchkova E. Integro-differential equations for Options prices in exponential Lévy models. Finance Stochastics. 9, 299–325 (2005).

[24] Grudsky S.M. Convolution equations on a finite interval with a small parameter multiplying the growing part of the symbol. Soviet Math. (Iz. Vuz) **34**, 7, 7–18, 1990.

[25] Böttcher A., Grudsky S.M. On the condition numbers of large semi-definite Toeplitz matrices. Linear Algebra and its Applications **279**, 1–3, 285–301, 1998.

[26] Grudsky S.M., Mikhalkovich S.S. Semisectoriality and condition numbers of convolution operators on the large finite intervals. Integro-differential operators, Proceedings of different universities, Rostov-on-Don, 5, 78–87, 2001.

[27] Grudsky S.M., Mikhalkovich S.S., and E. Ramíirez de Arellano. The Wiener-Hopf integral equation on a finite interval: asymptotic solution for large intervals with an application to acoustics. Proceedings of International Workshop on Linear Algebra, Numerical Functional Analysis and Wavelet Analysis, Allied Publisher Private Limite, India, 2003, 89–116.

[28] Bertoin J. Lévy processes. Cambridge University Press, Cambridge, 1996.

[29] Shiryaev A.N. Essentials of stochastic Finance. Facts, models, theory. World Scientific, Singapore Jersey London Hong Kong, 1999.

[30] Karatzas I. and Shreve S.E. Methods of mathematical Finance. Springer-Verlag, Berlin Heidelberg New York, 1998.

[31] Delbaen F. and Schachermayer W. A general version of the fundamental theorem of asset pricing. Math. Ann. **300**, 463–520, 1994.

[32] Eberlein E. and Jacod J. On the range of options prices. Finance and Stochastics **1**, 131–140, 1997.

[33] Fölemer H. and Schweizer M. Hedging of contingent claims under incomplete information. In: Applied Stochastic Analysis, M.H.A. Davies and R.J. Elliot (eds.), New York, Gordon and Bleach, 389–414, 1991.

[34] Keller U. Realistic modelling of financial derivatives. Dissertation. Mathematische Fakultät, Universität Freiburg im Breisgau, 1997.

[35] Kallsen J. Optimal portfolios for exponential Lévy processes. Mathematical Methods of Operations Research **51:3**, 357–374, 2000.

[36] Madan D.B. and Milne F. Option prising with VG martingale components. Mathem. Finance **1**, 39–55, 1991.

[37] Eberlein E., Keller U. and Prause K. New insights into smile, mispricing and value at risk: The hyperbolic model. Journ. of Business **71**, 371–406, 1998.

[38] Eskin G.I. Boundary problems for elliptic pseudo-differential equations. Nauka, Moscow, 1973 (Transl. of Mathematical Monographs, 52, Providence, Rhode Island: Amer. Math. Soc., 1980).

[39] Noble B. Methods based on the Wiener-Hopf technique for the solution of partial differential equations. International Series of Monographs on Pure and Applied Mathematics **7**, Pergamon Press, New York-London-Paris-Los Angeles, 1958.

[40] Gohberg I. and Krupnik N.Ya. Introduction to the Theory of One-dimensional Singular integral Operators. "Shtiintsa", Kishinev, 1973 (Transl. One-dimensional Linear Singular Integral Equations. Vol. I. Introduction, Vol II General Theory and Applications, Translated from the 1979 German Translation. Operator theory: Advances and Applications, 53 and 54. Birkhäuser Verlag, Basel, 1992).

[41] Böttcher A. and Silbermann B. Analysis of Toeplitz Operators. Springer-Verlag, Berlin, 1990.

[42] Sarason D. Toeplitz operators with semi-almost-periodic symbols. Duke Math. J. **44**, 2, 357–364, 1977.

[43] Akhiezer N.I. Lectures on Approximation Theory. Second, revised and enlarged edition, "Nauka", Moscow, 1965 (Transl. of first edition: Theory of Approximation, Frederick Ungar Publishing Co., New York, 1956).

[44] Dybin V.B., Grudsky S.M. Introduction to the theory of Toeplitz operators with infinite index. Birkhäuser Verlag, Basel-Boston-Berlin, Operator Theory: Advances and Applications, 2002.

[45] Böttcher A., Karlovich Yu., I Spitkovsky I.M. Convolution Operators and Factorization of Almost Periodic Matrix Functions. Operator Theory: Advances and Application. Vol. 131, Birkhäuser Verlag, 2002.

Sergei M. Grudsky
Departamento de Matemáticas
CINVESTAV del I.P.N.
México, D.F., México
e-mail: grudsky@math.cinvestav.mx

Operator Theory:
Advances and Applications, Vol. 170, 137–166
© 2006 Birkhäuser Verlag Basel/Switzerland

A Local-trajectory Method and Isomorphism Theorems for Nonlocal C^*-algebras

Yu.I. Karlovich

To Professor I.B. Simonenko on the occasion of his 70th birthday

Abstract. A nonlocal version of the Allan-Douglas local principle applicable to nonlocal C^*-algebras $\mathcal{B}$ associated with C^*-dynamical systems is elaborated. This local-trajectory method allows one to study the invertibility of elements $b \in \mathcal{B}$ in terms of invertibility of their local representatives. Isomorphism theorems for nonlocal C^*-algebras are established.

Mathematics Subject Classification (2000). Primary 46L45; Secondary 47A67, 47L65.

Keywords. C^*-algebra, C^*-dynamical system, crossed product, representation, state, amenable group, isomorphism theorem, invertibility, local-trajectory method, spectral measure.

1. Introduction

The local method elaborated by I.B. Simonenko [30] (also see [31]) and being a power tool for studying different classes of integral operators essentially changed the strategy of invertibility and Fredholm study of operator equations and exerted a big influence on further investigations. In particular, it allowed him to study operators of convolution type in cones, one-dimensional and multidimensional singular integral operators with discontinuous coefficients and related boundary value problems, to simplify constructing the symbol calculus for Banach algebras of one-dimensional singular integral operators with piecewise continuous coefficients, etc.

The subsequent versions of local principles by G.R. Allan [1], R.G. Douglas [12, Theorem 7.47], I. Gohberg and N. Krupnik [13, Chapter 5, Theorems 1.1 and 1.2] had an algebraic nature and a wider sphere of their applications. The most convenient of them is the Allan-Douglas local principle (see, e.g., [9, Theorem 1.34]) that supplies us by a canonical localization related to a central subalgebra $\mathcal{Z}$ of

Partially supported by PROMEP (México).

an initial algebra $\mathcal{A}$. The elements of the algebra $\mathcal{A}$ are of *local type* with respect to the central subalgebra $\mathcal{Z}$.

The present paper is devoted to an extension of the Allan-Douglas local principle to C^*-algebras associated with C^*-dynamical systems and, thus, generated by elements of *nonlocal type*. Namely, let $\mathcal{A}$ be a unital C^*-algebra, $\mathcal{Z}$ a central C^*-subalgebra of $\mathcal{A}$ with the same unit I, G an amenable [15] discrete group with unit e, $U : g \mapsto U_g$ a homomorphism of the group G onto a group $U_G = \{U_g : g \in G\}$ of unitary elements such that $U_{g_1 g_2} = U_{g_1} U_{g_2}$, $U_e = I$, and for every $g \in G$ the mappings $\alpha_g : a \mapsto U_g\, a\, U_g^*$ are *-automorphisms of the C^*-algebras $\mathcal{A}$ and $\mathcal{Z}$. Let $\mathcal{B} := C^*(\mathcal{A}, U_G)$ be the minimal C^*-algebra containing the unital C^*-algebra $\mathcal{A}$ and the group U_G. A local-trajectory method of studying invertibility of elements $b \in \mathcal{B}$ is elaborated and an isomorphism theorem for C^*-algebras of the form $\mathcal{B} := C^*(\mathcal{A}, U_G)$ is established. The work is based on a close relation between C^*-algebras associated with C^*-dynamical systems and the crossed products of C^*-algebras and groups of their automorphisms [26].

These results were partially published without proofs in [16] (also see [22]). On the other hand, the local-trajectory analogue of the Allan-Douglas local principle and the isomorphism theorem presented here give a powerful and convenient machinery for studying C^*-algebras of nonlocal type operators with discontinuous data, which has a lot of applications. In particular, it was applied to C^*-algebras of convolution type operators with discrete groups of shifts and oscillating coefficients [16], [17], [5], C^*-algebras of singular integral operators with amenable groups of shifts and piecewise quasi-continuous coefficients [7], the C^*-algebra of singular integral operators with semi-almost periodic matrix coefficients [8].

All these C^*-algebras possess composite families of non-homogeneous representations. Thus, in view of increasing variety of applications it appeared a necessity to expose such important techniques with details. The present paper gives corresponding results with the proofs contained earlier only in [18].

An isomorphism theorem and a local-trajectory method applicable to commutative C^*-algebras $\mathcal{A}$ extended by subexponential or admissible [2] groups G of unitary elements U_g ($g \in G$) generated the automorphisms $a \mapsto U_g a U_g^*$ of $\mathcal{A}$ were elaborated by A.B. Antonevich, V.V. Brenner, and A.V. Lebedev (see [2] and the references therein). These works revealed the importance of the topologically free action of the group G. The isomorphism theorem in the case of subexponential discrete groups G was proved making use of an estimate for the growth of the number of words of length n. A generalization of those results to arbitrary C^*-algebras $\mathcal{A}$ with non-trivial central subalgebras $\mathcal{Z}$ and arbitrary amenable discrete groups G, based on a related to pure states of $\mathcal{A}$ version of topologically free action of the group G on the maximal ideal space $M(\mathcal{Z})$ of $\mathcal{Z}$, was elaborated in [16], [18] (also see [8] where a version of the isomorphism theorem for a non-commutative C^*-algebra $\mathcal{A}$ and an Abelian group G was presented).

The same kind results for separable C^*-algebras $\mathcal{A}$ and amenable discrete groups G based on a weaker version of topologically free action of G on the space

$\operatorname{Prim} \mathcal{A}$ of primitive ideals of $\mathcal{A}$ were independently obtained in [23] (also see [3, Corollaries 12.9, 12.17, and Theorem 21.2] and the references therein). Note that isomorphism theorems and local-trajectory approaches were elaborated in [16], [18] and in [3, Theorems 16.21 and 21.3] also in the case of violation of topologically free action of the group G. These C^*-algebra methods are qualitatively different of those for studying Banach algebras of nonlocal bounded linear operators on Banach spaces (see, e.g., [22], [19], [20] and the references therein).

The paper is organized as follows. Section 2 contains definitions and necessary known results concerning amenable groups, crossed products of C^*-algebras and groups of their automorphisms, and regular representations of these products. Section 3 is devoted to isomorphism theorems. The proof of an isomorphism between the C^*-algebra $\mathcal{B} := C^*(\mathcal{A}, U_G)$ and the crossed product $\mathcal{A} \otimes_\alpha G$ of $\mathcal{A}$ and G plays the crucial role here. In Section 4 we present a local-trajectory analogue of the Allan-Douglas local principle which gives an invertibility criterion for nonlocal type elements in terms of invertibility of their local representatives associated to orbits of points in $M(\mathcal{Z})$. Here we also obtain several sufficient conditions guaranteeing the uniform boundedness of norms of inverses to these local representatives. Finally, Section 5 deals with the isomorphism theorem and trajectorial localization in case of violation of topologically free action of the group G, which leads to the appearance of massive sets of fixed points as well as non-homogeneous representations. Such situation is natural for C^*-algebras of nonlocal type operators with discontinuous data. The study is essentially based on applying spectral measures.

2. Crossed products and their regular representations

2.1. Definitions

Let $\mathcal{A}$ be a unital C^*-algebra, G a discrete group, and $\alpha : g \to \alpha_g$ a homomorphism of G into the group $\operatorname{Aut} \mathcal{A}$ of $*$-automorphisms of $\mathcal{A}$. The triple $\{\mathcal{A}, G, \alpha\}$ is called a C^*-*dynamical system*. To every C^*-dynamical system $\{\mathcal{A}, G, \alpha\}$ one can assign a canonical C^*-algebra $\mathcal{A} \otimes_\alpha G$. To define it, let us consider the involutive Banach algebra $l^1(G, \mathcal{A})$ of functions $x : G \to \mathcal{A}$ with at most countable sets of non-zero values and the norm $\|x\|_1 = \sum \|x(g)\|_{\mathcal{A}} < \infty$, and equip $l^1(G, \mathcal{A})$ with the linear space operations and with the following product and involution:

$$(xy)(g) = \sum_{h \in G} x(h)\alpha_h[y(h^{-1}g)], \quad x^*(g) = \alpha_g\big([x(g^{-1})]^*\big) \quad (g \in G;\ x, y \in l^1(G, \mathcal{A})).$$

Consider now the C^*-seminorm $\|x\|_0 := \sup_\pi \|\pi(x)\| \leq \|x\|_1$, where π runs through the set of all representations of $l^1(G, \mathcal{A})$ in Hilbert spaces, and show that $\|\cdot\|_0$ is actually a C^*-norm. To this end we take the universal representation π_u of $\mathcal{A}$ in a Hilbert space H_u and the representation $F : l^1(G, \mathcal{A}) \to \mathcal{B}\big(l^2(G, H_u)\big)$ defined by

$$(F(x)f)(g) = \sum_{h \in G} \pi_u(\alpha_g(x(h)))f(gh) \quad (g \in G,\ x \in l^1(G, \mathcal{A}),\ f \in l^2(G, H_u)).$$

Then $\|F(x)\| \leq \|x\|_0$ and $\|\pi_u(x(g))\| = \|x(g)\|_{\mathcal{A}}$ for all $g \in G$ (see, e.g., [24, Theorem 3.4.1]). Consider the imbedding operators $I_g : H_u \to l^2(G, H_u)$ and their left inverses $\Pi_g : l^2(G, H_u) \to H_u$ defined for $\varphi \in H_u$ and $f \in l^2(G, H_u)$ by

$$(I_g\varphi)(g) = \varphi, \quad (I_g\varphi)(h) = 0 \ \text{ for } \ h \neq g, \quad \Pi_g f = f(g) \ (g, h \in G). \tag{2.1}$$

Finally, since $\Pi_e F(x) I_g = \pi_u(x(g))$ for all $x \in l^1(G, \mathcal{A})$ and all $g \in G$, we obtain

$$\|x\|_1 \geq \|x\|_0 \geq \|F(x)\| \geq \|\Pi_e F(x) I_g\| = \|\pi_u(x(g))\| = \|x(g)\|_{\mathcal{A}}, \quad g \in G. \tag{2.2}$$

Hence, $x = 0$ if $\|x\|_0 = 0$, and therefore $\|\cdot\|_0$ is a C^*-norm.

The completion of $l^1(G, \mathcal{A})$ by the new norm $\|\cdot\|_0$ is called the *crossed product* of $\mathcal{A}$ and G, and is denoted by $\mathcal{A} \otimes_\alpha G$ (see [10, Definition 2.7.2]). Thus $\mathcal{A} \otimes_\alpha G$ is the enveloping C^*-algebra of the involutive Banach algebra $l^1(G, \mathcal{A})$. One may visualize the crossed product $\mathcal{A} \otimes_\alpha G$ as a skew tensor product between $C^*(G)$ and $\mathcal{A}$. In particular, $\mathbb{C} \otimes_\alpha G = C^*(G)$, the group C^*-algebra.

Let π be a representation of $\mathcal{A}$ in a Hilbert space H. The *left (right) regular representation* of $\mathcal{A} \otimes_\alpha G$ induced by π is the representation $\pi \times \lambda$ (respectively, $\pi \times \varrho$) in $l^2(G, H)$ defined on functions $x \in l^1(G, \mathcal{A})$ by the formulas

$$[(\pi \times \lambda)(x)f](g) = \sum_{s \in G} \pi\left(\alpha_{g^{-1}}[x(s)]\right) f(s^{-1}g), \quad f \in l^2(G, H), \tag{2.3}$$

$$\left([(\pi \times \varrho)(x)f](g) = \sum_{s \in G} \pi\left(\alpha_g[x(s)]\right) f(gs), \quad f \in l^2(G, H)\right) \tag{2.4}$$

and extended by continuity to the whole C^*-algebra $\mathcal{A} \otimes_\alpha G$ due to the inequality

$$\|(\pi \times \lambda)(x)\| \leq \|x\|_0 \quad (\|(\pi \times \varrho)(x)\| \leq \|x\|_0).$$

These two representations are unitarily equivalent. Indeed, define the isomorphism $f \mapsto f'$ of $l^2(G, H)$ onto itself by $f'(g) = f(g^{-1})$, $g \in G$. Then, by (2.3) and (2.4),

$$[(\pi \times \lambda)(x)f'](g) = \sum_{s \in G} \pi(\alpha_{g^{-1}}[x(s)])f(g^{-1}s) = [(\pi \times \varrho)(x)f]'(g), \quad g \in G.$$

Let $\pi_u \times \lambda$ be the left regular representation of the C^*-algebra $\mathcal{A} \otimes_\alpha G$ in the Hilbert space $l^2(G, H_u)$ induced by the universal representation π_u of $\mathcal{A}$. The C^*-algebra $(\pi_u \times \lambda)(\mathcal{A} \otimes_\alpha G)$ is called the *reduced crossed product* of $\mathcal{A}$ and G (see [26, 7.7.4]).

2.2. Amenable groups

Amenable groups constitute a natural maximal class of groups for which one can establish an isomorphism of two C^*-algebras associated with C^*-dynamical systems. According to [15, § 1.2], a discrete group G is called *amenable* if the C^*-algebra $l^\infty(G)$ of all bounded complex-valued functions on G with sup-norm has a *left invariant* (or *right invariant*) *mean*, that is, a state ρ satisfying the condition

$$\rho(f) = \rho({}_sf) \ (\text{respectively}, \ \rho(f) = \rho(f_s)) \ \text{for all } s \in G \text{ and all } f \in l^\infty(G),$$

where $({}_sf)(g) = f(s^{-1}g)$, $(f_s)(g) = f(gs)$, $g \in G$. By definition (see, e.g., [24, p. 89]), a state on a C^*-algebra A is a positive linear functional on A of norm 1.

For discrete groups the existence of left or right invariant mean is equivalent to the existence of two-sided invariant mean [15]. By [15, § 1.2], the set of amenable groups is closed with respect to the passage to subgroups and quotient groups.

A group G is called *subexponential* (see [2]), if for every finite set $K \subset G$, $\lim_{n \to \infty} |K^n|^{1/n} = 1$, where $|K^n|$ is the number of different words of length n constructed from elements $g \in K$. Besides finite groups, the set of subexponential groups contains all commutative groups, groups of polynomial growth, finitely generated groups of growth bigger than polynomial and smaller than exponential (see, e.g., [2]). According to [4], all subexponential groups are amenable.

By [21, p. 17], a group G is *solvable of class k* if the following subgroups exist: an Abelian normal subgroup $A_0 \subset G$, an Abelian normal subgroup A_1 of $\widetilde{G}_1 = G/A_0$, an Abelian normal subgroup A_2 of $\widetilde{G}_2 = \widetilde{G}_1/A_1$, and so on, where this chain breaks off at the kth step, that is, $\widetilde{G}_k = \{e\}$. By [15], all solvable groups are amenable. Since there exist solvable groups of exponential growth with a finite set of generators [14, p. 43], the set of amenable groups is strictly wider than the set of subexponential groups.

Thus, the class of amenable groups contains all finite groups, commutative groups, subexponential groups, and solvable groups. On the other hand, if a discrete group G contains the free discrete group F_2 with two generators, then G is not amenable [15].

2.3. Isomorphism of crossed products and their regular representations

Inequalities (2.2) obtained for $x \in l^1(G, \mathcal{A})$ imply that for every $g \in G$ the map $E_g : l^1(G, \mathcal{A}) \to \mathcal{A}$ given by $E_g x = x(g)$ extends to a bounded linear operator $E_g : \mathcal{A} \otimes_\alpha G \to \mathcal{A}$, and $\|E_g\| = 1$ in view of (2.2) and the relations $\|x_g\|_0 \leq \|x_g\|_1 = \|x_g(g^{-1})\|_{\mathcal{A}}$ where $x_g \in l^1(G, \mathcal{A})$ and $x_g(h) = 0$ for $h \neq g^{-1}$.

Let $S_{\mathcal{A}}$ denote the set of all states (positive linear functionals of norm 1) on $\mathcal{A}$. A state μ on $\mathcal{A}$ is said to be *pure* if it majorizes only the positive linear functionals on $\mathcal{A}$ which have the form $c\mu$ ($0 \leq c \leq 1$). Let $P_{\mathcal{A}}$ stand for the set of all pure states on $\mathcal{A}$. By [10, Theorem 2.3.15]), if $\mathcal{A}$ is a unital C^*-algebra, then $S_{\mathcal{A}}$ is a convex weak* compact set of the dual space $\mathcal{A}^*$, $P_{\mathcal{A}}$ is the set of all extreme points of $S_{\mathcal{A}}$, and $S_{\mathcal{A}}$ is the weak* closed convex hull of $P_{\mathcal{A}}$.

For $\mu \in S_{\mathcal{A}}$ and $x \in \mathcal{A} \otimes_\alpha G$, we put

$$(\delta_e \times \mu)(x) = \mu(E_e x) = \mu[x(e)]. \tag{2.5}$$

Obviously, $\delta_e \times \mu$ is a state of the C^*-algebra $\mathcal{A} \otimes_\alpha G$. Let $\pi_{\delta_e \times \mu}$ denote the GNS-representation associated to $\delta_e \times \mu$ (see [11, Proposition 2.4.4] or [24, Section 3.4]). By analogy with (2.2), for the left regular representation $\pi_u \times \lambda$, we obtain

$$\|x\|_0 \geq \|(\pi_u \times \lambda)(x)\| \geq \|\Pi_e (\pi_u \times \lambda)(x) I_e\| = \|\pi_u(x(e))\| = \|x(e)\|_{\mathcal{A}}, \quad x \in l^1(G, \mathcal{A}).$$

Therefore $\|x(e)\|_{\mathcal{A}} \leq \|(\pi_u \times \lambda)(x)\|$ for all $x \in \mathcal{A} \otimes_\alpha G$, whence due to (2.5),

$$\mathrm{Ker}\,(\pi_u \times \lambda) \subset \mathrm{Ker}\,(\delta_e \times \mu). \tag{2.6}$$

As $\mathrm{Ker}\,\pi_{\delta_e \times \mu}$ is the largest closed two-sided ideal of $\mathcal{A} \otimes_\alpha G$ which is contained in $\mathrm{Ker}\,(\delta_e \times \mu)$ (see, e.g., [11, Corollary 2.4.10]), relation (2.6) implies that

$$\mathrm{Ker}\,(\pi_u \times \lambda) \subset \mathrm{Ker}\,\pi_{\delta_e \times \mu} \subset \mathrm{Ker}\,(\delta_e \times \mu). \tag{2.7}$$

Lemma 2.1. *For the left regular representation $\pi_u \times \lambda$ of $\mathcal{A} \otimes_\alpha G$, we have*

$$\mathrm{Ker}\,(\pi_u \times \lambda) = \bigcap_{\mu \in P_\mathcal{A}} \mathrm{Ker}\,\pi_{\delta_e \times \mu}. \tag{2.8}$$

Proof. Since the universal representation π_u of the C^*-algebra $\mathcal{A}$ is of the form $\pi_u = \bigoplus_{\mu \in S_\mathcal{A}} \pi_\mu$, the corresponding left regular representation $\pi_u \times \lambda$ of the C^*-algebra $\mathcal{A} \otimes_\alpha G$ is of the form $\pi_u \times \lambda = \bigoplus_{\mu \in S_\mathcal{A}} (\pi_\mu \times \lambda)$. Hence

$$\mathrm{Ker}\,(\pi_u \times \lambda) = \bigcap_{\mu \in S_\mathcal{A}} \mathrm{Ker}\,(\pi_\mu \times \lambda). \tag{2.9}$$

Let us show that

$$\mathrm{Ker}\,(\pi_\mu \times \lambda) = \mathrm{Ker}\,\pi_{\delta_e \times \mu}. \tag{2.10}$$

Indeed, let ξ_μ be the cyclic vector of the representation π_μ of the C^*-algebra $\mathcal{A}$ in the Hilbert space H_μ defined by a state $\mu \in S_\mathcal{A}$. Then $I_e(\xi_\mu)$ is a cyclic vector for the representation $\pi_\mu \times \lambda$ of the C^*-algebra $\mathcal{A} \otimes_\alpha G$ in the Hilbert space $l^2(G, H_\mu)$. In view of (2.3) and [11, Proposition 2.4.4],

$$\big((\pi_\mu \times \lambda)(x) I_e(\xi_\mu),\, I_e(\xi_\mu)\big) = (\pi_\mu[x(e)]\xi_\mu,\, \xi_\mu) = \mu[x(e)].$$

On the other hand, by (2.5) and again by [11, Proposition 2.4.4],

$$\mu[x(e)] = (\delta_e \times \mu)(x) = \big(\pi_{\delta_e \times \mu}(x)\, \xi_{\delta_e \times \mu},\, \xi_{\delta_e \times \mu}\big), \tag{2.11}$$

where $\xi_{\delta_e \times \mu}$ is the cyclic vector for the representation $\pi_{\delta_e \times \mu}$ of the C^*-algebra $\mathcal{A} \otimes_\alpha G$ in the Hilbert space $H_{\delta_e \times \mu}$ determined by the state $\delta_e \times \mu \in S_{\mathcal{A} \otimes_\alpha G}$. Thus

$$\big((\pi_\mu \times \lambda)(x) I_e(\xi_\mu),\, I_e(\xi_\mu)\big) = \big(\pi_{\delta_e \times \mu}(x)\, \xi_{\delta_e \times \mu},\, \xi_{\delta_e \times \mu}\big),$$

and hence, by [24, Theorem 5.1.4], the representations $\pi_{\mu \times \lambda}$ and $\pi_{\delta_e \times \mu}$ are unitarily equivalent, which implies (2.10). From (2.9) and (2.10) it follows that

$$\mathrm{Ker}\,(\pi_u \times \lambda) = \bigcap_{\mu \in S_\mathcal{A}} \mathrm{Ker}\,\pi_{\delta_e \times \mu}. \tag{2.12}$$

It remains to prove that

$$\bigcap_{\mu \in S_\mathcal{A}} \mathrm{Ker}\,\pi_{\delta_e \times \mu} = \bigcap_{\mu \in P_\mathcal{A}} \mathrm{Ker}\,\pi_{\delta_e \times \mu}. \tag{2.13}$$

Let $x \in \bigcap_{\mu \in P_\mathcal{A}} \mathrm{Ker}\,\pi_{\delta_e \times \mu}$. By (2.11), $x(e) \in \mathrm{Ker}\,\mu$ for all $\mu \in P_\mathcal{A}$. Therefore, since $S_\mathcal{A}$ is the weak* closed convex hull of $P_\mathcal{A}$, we conclude that $x(e) \in \mathrm{Ker}\,\nu$ for all $\nu \in S_\mathcal{A}$. Again by (2.11), $x \in \mathrm{Ker}\,(\delta_e \times \nu)$ for all $\nu \in S_\mathcal{A}$. Thus, the closed two-sided ideal $\bigcap_{\mu \in P_\mathcal{A}} \mathrm{Ker}\,\pi_{\delta_e \times \mu}$ of the C^*-algebra $\mathcal{A} \otimes_\alpha G$ is contained in every set $\mathrm{Ker}\,(\delta_e \times \nu)$ $(\nu \in S_\mathcal{A})$. Hence, due to (2.7),

$$\bigcap_{\mu \in P_\mathcal{A}} \mathrm{Ker}\,\pi_{\delta_e \times \mu} \subset \mathrm{Ker}\,\pi_{\delta_e \times \nu} \subset \mathrm{Ker}\,(\delta_e \times \nu) \quad (\nu \in S_\mathcal{A}).$$

Thus,

$$\bigcap_{\mu \in P_\mathcal{A}} \mathrm{Ker}\,\pi_{\delta_e \times \mu} \subset \bigcap_{\nu \in S_\mathcal{A}} \mathrm{Ker}\,\pi_{\delta_e \times \nu},$$

which proves (2.13) because the inverse inclusion is obvious.

Finally, (2.12) and (2.13) immediately give (2.8). $\qquad\square$

If C^*-algebras $\mathcal{A}$ and $\mathcal{B}$ are (isometrically) *-isomorphic, we will write $\mathcal{A} \cong \mathcal{B}$.

Theorem 2.2. (cf. [26]) *The following assertions are equivalent:*

(i) *G is an amenable discrete group;*

(ii) $\mathrm{Ker}\,(\pi_u \times \lambda) = \{0\}$ *(equivalently, $\mathcal{A} \otimes_\alpha G \cong (\pi_u \times \lambda)(\mathcal{A} \otimes_\alpha G)$);*

(iii) $\bigcap_{\mu \in P_A} \mathrm{Ker}\,\pi_{\delta_e \times \mu} = \{0\};$

(iv) $\bigcap_{g \in G} \mathrm{Ker}\,E_g = \{0\}.$

Proof. (i)$\Leftrightarrow$(ii)$\Leftrightarrow$(iii). According to [26, Theorems 7.7.7 and 7.3.9], a discrete group G is amenable if and only if the crossed product $\mathcal{A} \otimes_\alpha G$ is *-isomorphic to the reduced crossed product $(\pi_u \times \lambda)(\mathcal{A} \otimes_\alpha G)$, which is equivalent to $\mathrm{Ker}\,(\pi_u \times \lambda) = \{0\}$. By Lemma 2.1, the latter is equivalent to (iii).

(ii)$\Leftrightarrow$(iv). Obviously, it is sufficient to prove that

$$\mathrm{Ker}\,(\pi_u \times \lambda) = \bigcap_{g \in G} \mathrm{Ker}\,E_g. \tag{2.14}$$

As $E_g x = \Pi_e (\pi_u \times \lambda)(x) I_{g^{-1}}$ for all $x \in \mathcal{A} \otimes_\alpha G$ and all $g \in G$ where Π_e and $I_{g^{-1}}$ are given by (2.1), we get $\mathrm{Ker}\,(\pi_u \times \lambda) \subset \mathrm{Ker}\,E_g$ and thus $\mathrm{Ker}\,(\pi_u \times \lambda) \subset \bigcap_{g \in G} \mathrm{Ker}\,E_g$. Conversely, let $x \in \bigcap_{g \in G} E_g$. Then all $x(g) = 0$ and, in view of the equalities

$$\Pi_g(\pi_u \times \lambda)(x)I_h = \pi_u\big(\alpha_{g^{-1}}(x(gh^{-1}))\big) \quad (g, h \in G),$$

we conclude that $\big((\pi_u \times \lambda)(x)\varphi, f\big) = 0$ for all vectors $\varphi, f \in l^2(G, H_u)$. Hence $\|(\pi_u \times \lambda)(x)\| = 0$, that is, $x \in \mathrm{Ker}\,(\pi_u \times \lambda)$. Thus, $\bigcap_{g \in G} \mathrm{Ker}\,E_g \subset \mathrm{Ker}\,(\pi_u \times \lambda)$, which completes the proof of (2.14). $\square$

3. Isomorphism theorems

3.1. Starting assumptions

Let $\mathcal{A}$ be a unital C^*-algebra, $\mathcal{Z}$ a central C^*-subalgebra of $\mathcal{A}$ with the same unit I, G a discrete group with unit e, $U : g \mapsto U_g$ a homomorphism of the group G onto a group $U_G = \{U_g : g \in G\}$ of unitary elements such that $U_{g_1 g_2} = U_{g_1} U_{g_2}$ and $U_e = I$. Assume that

(A1) *for every $g \in G$ the mappings $\alpha_g : a \mapsto U_g a U_g^*$ are *-automorphisms of the C^*-algebras $\mathcal{A}$ and $\mathcal{Z}$;*

(A2) *G is an amenable discrete group.*

Let $M := M(\mathcal{Z})$ be the maximal ideal space of the (commutative) C^*-algebra $\mathcal{Z}$. By the Gelfand-Naimark theorem [25, § 16], $\mathcal{Z} \cong C(M)$ where $C(M)$ is the C^*-algebra of all continuous complex-valued functions on M. Under assumption (A1), identifying the non-zero multiplicative linear functionals φ_m of the algebra $\mathcal{Z}$ and the maximal ideals $m = \mathrm{Ker}\,\varphi_m \in M$, we obtain the homomorphism $g \mapsto \beta_g(\cdot)$ of the group G into the homeomorphism group of M according to the rule

$$z(\beta_g(m)) = (\alpha_g(z))(m), \quad z \in \mathcal{Z}, \ m \in M, \ g \in G,$$

where $z(\cdot) \in C(M)$ is the Gelfand transform of the element $z \in \mathcal{Z}$.

Let $\mathcal{B} := C^*(\mathcal{A}, U_G)$ be the minimal C^*-algebra containing the unital C^*-algebra $\mathcal{A}$ and the group U_G. By virtue of (A1), $\mathcal{B}$ is the closure of the set $\mathcal{B}^0$

consisting of the elements $b = \sum a_g U_g$ where $a_g \in \mathcal{A}$ and g runs through finite subsets of G. Along with the C^*-algebra $\mathcal{B}$ we consider the crossed product $\mathcal{A} \otimes_\alpha G$.

The isomorphism of the C^*-algebras $\mathcal{B} = C^*(\mathcal{A}, U_G)$ and $\mathcal{A} \otimes_\alpha G$ is a key result in the theory of nonlocal C^*-algebras. To formulate it, in addition to (A1) and (A2), we need to introduce one more assumption.

Let $P_\mathcal{A}$ be the set of all pure states on the C^*-algebra $\mathcal{A}$ equipped with the induced weak* topology, and let J_m denote the closed two-sided ideal of $\mathcal{A}$ generated by the maximal ideal $m \in M$ of the central C^*-algebra $\mathcal{Z} \subset \mathcal{A}$. By [8, Lemma 4.1], if $\mu \in P_\mathcal{A}$, then $\operatorname{Ker} \mu \supset J_m$ where $m := \mathcal{Z} \cap \operatorname{Ker} \mu \in M$, and therefore

$$P_\mathcal{A} = \bigcup_{m \in M} \mathcal{P}_m, \quad \mathcal{P}_m := \mathcal{P}_m(\mathcal{A}) := \{\nu \in P_\mathcal{A} : \operatorname{Ker} \nu \supset J_m\}. \qquad (3.1)$$

Let the following version of *topologically free* action of the group G hold:

(A3) *for every finite set $G_0 \subset G$ and every nonempty open set $W \subset P_\mathcal{A}$ there exists a state $\nu \in W$ such that $\beta_g(m_\nu) \neq m_\nu$ for all $g \in G_0 \setminus \{e\}$, where $m_\nu := \mathcal{Z} \cap \operatorname{Ker} \nu \in M$.*

We say that the group G *acts freely* on M if the group $\{\beta_g : g \in G\}$ of homeomorphisms of M onto itself *acts freely* on M, that is, if $\beta_g(m) \neq m$ for all $g \in G \setminus \{e\}$ and all $m \in M$. Obviously, if the group G acts freely on M, then (A3) is fulfilled automatically.

If the C^*-algebra $\mathcal{A}$ is commutative, then the set $P_\mathcal{A}$ of all pure states of $\mathcal{A}$ coincides with the set of non-zero multiplicative linear functionals of $\mathcal{A}$ (see, e.g., [10, Corollary 2.3.21]). Therefore, choosing $\mathcal{Z} = \mathcal{A}$ and identifying the set of non-zero multiplicative linear functionals of $\mathcal{A}$ with the maximal ideal space $M(\mathcal{A})$ of $\mathcal{A}$, we can rewrite (A3) in the form

(A$_0$) *for every finite set $G_0 \subset G$ and every nonempty open set $W \subset M(\mathcal{A})$ there exists a point $m_0 \in W$ such that $\beta_g(m_0) \neq m_0$ for all $g \in G_0 \setminus \{e\}$.*

3.2. Isomorphism of the C^*-algebras $\mathcal{B}$ and $\mathcal{A} \otimes_\alpha G$

For every $g \in G$, we consider the mapping $\widetilde{E}_g : \mathcal{B}^0 \to \mathcal{A}$ defined on the elements $b = \sum_{h \in G_0} a_h U_h \in \mathcal{B}^0$ with $a_h \in \mathcal{A}$ by

$$\widetilde{E}_g \left(\sum_{h \in G_0} a_h U_h \right) = a_g \qquad (3.2)$$

where G_0 is a finite subset of G and $a_g = 0$ if $g \notin G_0$.

Lemma 3.1. [8, Lemma 4.3] *If* (A1) *and* (A3) *hold, then the mappings $\widetilde{E}_g : \mathcal{B}^0 \to \mathcal{A}$ $(g \in G)$ given by* (3.2) *extend to bounded linear operators $\widetilde{E}_g : \mathcal{B} \to \mathcal{A}$ of norm one.*

Theorem 3.2. *Under assumptions* (A1)–(A3)*, $\mathcal{B} \cong \mathcal{A} \otimes_\alpha G$.*

Proof. Consider the $*$-homomorphism F of $l^1(G, \mathcal{A})$ into $\mathcal{B}$ which is given by

$$Fx = \sum_{g \in G} x(g) U_g \quad (x \in l^1(G, \mathcal{A})). \qquad (3.3)$$

Since $F(l^1(G, \mathcal{A}))$ is a dense subset of the C^*-algebra $\mathcal{B}$ and since every C^*-algebra has a faithful representation in a Hilbert space H [24, Theorem 3.4.1], we conclude that $\|Fx\|_{\mathcal{B}} \leq \|x\|_0 = \sup_\pi \|\pi(x)\|$. Hence the mapping $F : l^1(G, \mathcal{A}) \to \mathcal{B}$ extends by continuity to a *-homomorphism $F : \mathcal{A} \otimes_\alpha G \to \mathcal{B}$ and, by [11, Corollary 1.8.3],

$$(\mathcal{A} \otimes_\alpha G)/\mathrm{Ker}\, F \cong \mathcal{B}. \tag{3.4}$$

In virtue of (3.4), it remains to show that $\mathrm{Ker}\, F = \{0\}$. Since assumptions (A1) and (A3) are fulfilled, the operator $\widetilde{E}_g$ is well defined according to Lemma 3.1, and $E_g = \widetilde{E}_g F$ for all $g \in G$. Hence $\delta_e \times \mu = \mu \widetilde{E}_e F$ for all $\mu \in P_{\mathcal{A}}$, and therefore $\mathrm{Ker}\, F \subset \mathrm{Ker}\, (\delta_e \times \mu)$. Then from [11, Corollary 2.4.10] it follows that

$$\mathrm{Ker}\, F \subset \mathrm{Ker}\, \pi_{\delta_e \times \mu} \subset \mathrm{Ker}\, (\delta_e \times \mu),$$

whence

$$\mathrm{Ker}\, F \subset \bigcap_{\mu \in P_{\mathcal{A}}} \mathrm{Ker}\, \pi_{\delta_e \times \mu}. \tag{3.5}$$

Finally, (3.5), (A2) and Theorem 2.2 imply that $\mathrm{Ker}\, F = \{0\}$. $\square$

3.3. Isomorphism of two nonlocal C^*-algebras

Along with the C^*-algebra $\mathcal{B} = C^*(\mathcal{A}, U_G)$ we consider the C^*-algebra $\mathcal{B}' = C^*(\mathcal{A}', U'_G)$ where $\mathcal{A}'$ is a unital C^*-algebra *-isomorphic to the C^*-algebra $\mathcal{A}$, $U' : g \mapsto U'_g$ is a homomorphism of the group G onto a group $U'_G = \{U'_g : g \in G\}$ of unitary elements such that $U'_{g_1 g_2} = U'_{g_1} U'_{g_2}$ and $U'_e = I'$ where I' is the unit of the C^*-algebra $\mathcal{A}'$. Finally, we assume that

(A4) *there is a C^*-algebra isomorphism $\varphi : \mathcal{A} \to \mathcal{A}'$ such that*

$$\varphi(U_g a U_g^*) = U'_g \varphi(a)(U'_g)^* \quad \text{for all } g \in G \text{ and all } a \in \mathcal{A}.$$

Using Theorem 3.2 one can prove the following isomorphism theorem.

Theorem 3.3. *If conditions* (A1)–(A4) *hold, then there is a* (*unique*) *C^*-algebra isomorphism $\Phi : \mathcal{B} \to \mathcal{B}'$ such that $\Phi|\mathcal{A} = \varphi$ and $\Phi(U_g) = U'_g$ for all $g \in G$.*

Proof. Let F' be the *-homomorphism of the involutive Banach algebra $l^1(G, \mathcal{A})$ into the C^*-algebra $\mathcal{B}'$ defined by $F'x = \sum \varphi(x(g))U'_g$ for all $x \in l^1(G, \mathcal{A})$. It extends by continuity to a C^*-algebra homomorphism $F' : \mathcal{A} \otimes_\alpha G \to \mathcal{B}'$. Consider the central subalgebra $\mathcal{Z}' = \varphi(\mathcal{Z})$ of the C^*-algebra $\mathcal{A}'$. In view of the isomorphism $\mathcal{Z}' \cong \mathcal{Z} \cong C(M)$, the maximal ideal space of $\mathcal{Z}'$ can be identified with M, and the Gelfand transform on $\mathcal{Z}'$ can be defined by $z'(m) = [\varphi^{-1}(z')](m)$ for all $z' \in \mathcal{Z}'$ and all $m \in M$. In virtue of (A4), the mappings $\alpha'_g : a' \mapsto U'_g a' U'_g$ $(g \in G)$ are *-automorphisms of the C^*-algebras $\mathcal{A}'$ and $\mathcal{Z}'$, which induce the previous action of G on M because

$$z'(\beta_g(m)) = [\varphi^{-1}(z')](\beta_g(m)) = [\alpha_g(\varphi^{-1}(z'))](m) = [\varphi^{-1}(\alpha'_g(z'))](m) = [\alpha'_g(z')](m).$$

Thus, (A1) holds for the C^*-algebra $\mathcal{A}'$. Since the mapping $\nu \mapsto \nu \circ \varphi^{-1}$ is an isometric isomorphism of $\mathcal{A}^*$ onto $(\mathcal{A}')^*$ and also a bijection of $P_{\mathcal{A}}$ onto $P_{\mathcal{A}'}$ and a bijection of $\mathcal{P}_m(\mathcal{A})$ given by (3.1) onto $\mathcal{P}_m(\mathcal{A}') := \{\nu' \in P_{\mathcal{A}'} : \mathrm{Ker}\, \nu' \supset \varphi(J_m)\}$, assumption (A3) is also fulfilled for the C^*-algebra $\mathcal{A}'$. In view of (A1) for $\mathcal{A}'$, the set $F'(\mathcal{A} \otimes_\alpha G)$ is dense in $\mathcal{B}'$. Analogously to Theorem 3.2 from (A2) it follows that $\mathrm{Ker}\, F' = \{0\}$, and hence F' is a C^*-algebra isomorphism of $\mathcal{A} \otimes_\alpha G$ onto $\mathcal{B}'$ in view of [11, Corollary 1.8.3]. As $F : \mathcal{A} \otimes_\alpha G \to \mathcal{B}$ also is a C^*-algebra isomorphism by

Theorem 3.2, we conclude that $\mathcal{B} \cong \mathcal{B}'$. The C^*-algebra isomorphism $\Phi = F' \circ F^{-1}$ of $\mathcal{B}$ onto $\mathcal{B}'$ acts on the dense subset $\mathcal{B}^0 \subset \mathcal{B}$ by the rule $\Phi\left(\sum a_g U_g\right) = \sum \varphi(a_g) U'_g$, where $a_g \in \mathcal{A}$, and is extended to all $\mathcal{B}$ by continuity. Obviously, Φ is a unique extension of φ to a $*$-isomorphism of $\mathcal{B}$ onto $\mathcal{B}'$. $\qquad\square$

Corollary 3.4. *If assumptions* (A1)–(A3) *are fulfilled, π is an isometric representation of the C^*-algebra $\mathcal{A}$ in a Hilbert space H, and Φ is the representation of the C^*-algebra $\mathcal{B} = C^*(\mathcal{A}, U_G)$ in the Hilbert space $l^2(G, H)$ such that*

$$(\Phi(a)f)(g) = \pi(\alpha_g(a))f(g), \quad (\Phi(U_h)f)(g) = f(gh) \quad \big(a \in \mathcal{A};\, h, g \in G;\, f \in l^2(G, H)\big),$$

then any $b \in \mathcal{B}$ is invertible if and only if so is the operator $\Phi(b) \in \mathcal{B}(l^2(G, H))$.

Indeed, since the C^*-algebras $\mathcal{A}$ and $\Phi(\mathcal{A})$ are $*$-isomorphic and since

$$\Phi(\alpha_g(a)) = \Phi(U_g)\Phi(a)\Phi(U_g^*) \quad \text{for all} \quad a \in \mathcal{A} \quad \text{and all} \quad g \in G,$$

from Theorem 3.3 it follows that the C^*-algebras $\mathcal{B}$ and $\Phi(\mathcal{B})$ also are $*$-isomorphic, which gives Corollary 3.4 due to the inverse closedness of C^*-algebras.

4. Trajectorial localization

4.1. Main results

Let the unital C^*-algebras $\mathcal{Z}$, $\mathcal{A}$, and $\mathcal{B} = C^*(\mathcal{A}, U_G)$ satisfy all the conditions of Subsection 3.1. In this section we establish an invertibility criterion for elements $b \in \mathcal{B}$ in terms of the invertibility of their local representatives associated with the G-orbits of points $m \in M$, where M is the compact space of maximal ideals of the central algebra $\mathcal{Z}$. As a result, we will get a nonlocal version of the Allan-Douglas local principle.

For every $m \in M$, let $G(m) := \{\beta_g(m) : g \in G\}$ be the G-orbit of the point m, let J_m be the closed two-sided ideal of the algebra $\mathcal{A}$ generated by the maximal ideal m of the algebra $\mathcal{Z}$, and let $\mathcal{H}_m$ be the Hilbert space of an isometric representation $\widetilde{\pi}_m : \mathcal{A}/J_m \to \mathcal{B}(\mathcal{H}_m)$. We also consider the canonical $*$-homomorphism $\varrho_m : \mathcal{A} \to \mathcal{A}/J_m$ and the representation

$$\pi'_m : \mathcal{A} \to \mathcal{B}(\mathcal{H}_m), \quad a \mapsto (\widetilde{\pi}_m \circ \varrho_m)(a).$$

Since $\alpha_g(J_{\beta_g(m)}) = J_m$ for all $g \in G$ and all $m \in M$ in view of (A1), we infer that all the mappings $\mathcal{P}_m \to \mathcal{P}_{\beta_g(m)}$, $\mu_m \mapsto \mu_{\beta_g(m)} := \mu_m \circ \alpha_g$ are bijections, and therefore the quotient algebras $\mathcal{A}/J_{\beta_g(m)}$ and $\mathcal{A}/J_m$ are $*$-isomorphic. Then the spaces $\mathcal{H}_{\beta_g(m)}$ can be chosen equal for all $g \in G$.

Given $X \subset M$, let $\Omega(X)$ be the set of G-orbits of all points $m \in X$, let $H_\omega = \mathcal{H}_m$ where $m = m_\omega$ is an arbitrary fixed point of an orbit $\omega \in \Omega$ and $\Omega = \Omega(M)$, and let $l^2(G, H_\omega)$ be the Hilbert space of all functions $f : G \mapsto H_\omega$ such that $f(g) \neq 0$ for at most countable set of points $g \in G$ and $\sum \|f(g)\|_{H_\omega}^2 < \infty$. For every $\omega \in \Omega$ we consider the representation $\pi_\omega : \mathcal{B} \to \mathcal{B}(l^2(G, H_\omega))$ defined by

$$[\pi_\omega(a)f](g) = \pi'_m(\alpha_g(a))f(g), \quad [\pi_\omega(U_h)f](g) = f(gh)$$

for all $a \in \mathcal{A}$, all $g, h \in G$, and all $f \in l^2(G, H_\omega)$.

Applying Theorem 3.3 we obtain the following invertibility criterion.

Theorem 4.1. *If assumptions* (A1)–(A3) *are fulfilled, then an element* $b \in \mathcal{B}$ *is invertible (left invertible, right invertible) in* $\mathcal{B}$ *if and only if for every orbit* $\omega \in \Omega$ *the operator* $\pi_\omega(b)$ *is invertible (left invertible, right invertible) on the space* $l^2(G, H_\omega)$ *and, in the case of infinite* Ω,

$$\sup\left\{\left\|(\pi_\omega(b))^{-1}\right\| : \omega \in \Omega\right\} < \infty$$

(respectively, there is a uniformly bounded family $\{\pi_\omega^l(b)\}_{\omega \in \Omega}$ *of left inverse operators with self-adjoint projections* $\pi_\omega(b)\pi_\omega^l(b)$, *or there is a uniformly bounded family* $\{\pi_\omega^r(b)\}_{\omega \in \Omega}$ *of right inverse operators with self-adjoint projections* $\pi_\omega^r(b)\pi_\omega(b))$.

Proof. According to [25, § 23, Corollaries 2 and 3], the left (right) invertibility of an element $b \in \mathcal{B}$ is equivalent to the two-sided invertibility of the element b^*b (respectively, bb^*) in the C^*-algebra $\mathcal{B}$. Moreover, these cases are reduced one to another by passing to adjoint operators.

If b^*b is invertible in $\mathcal{B}$, then the element $b^l := (b^*b)^{-1}b^*$ is a left inverse to b, and bb^l is a self-adjoint idempotent. On the other hand, if $b \in \mathcal{B}$ is left invertible in $\mathcal{B}$ and b^l is its left inverse such that bb^l is a self-adjoint idempotent, then

$$b^l(b^l)^*(b^*b) = b^l(bb^l)^*b = b^l(bb^l)b = I,$$

whence we infer that $(b^*b)^{-1} = b^l(b^l)^*$ in view of the invertibility of b^*b.

Analogously, the operator $\pi_\omega(b)$ is left invertible on the space $l^2(G, H_\omega)$ if and only if the operator $\pi_\omega(b^*)\pi_\omega(b)$ is two-sided invertible on the same space $l^2(G, H_\omega)$. In addition, from the relations

$$\pi_\omega^l(b) = (\pi_\omega(b^*)\pi_\omega(b))^{-1}\pi_\omega(b^*), \quad (\pi_\omega(b^*)\pi_\omega(b))^{-1} = \pi_\omega^l(b)(\pi_\omega^l(b))^*$$

it follows the equivalence of the uniform boundedness of the norms for the operators $\pi_\omega^l(b)$ and $(\pi_\omega(b^*)\pi_\omega(b))^{-1}$ if the element bb^l is self-adjoint.

As a result, the left (right) invertibility criterion for any $b \in \mathcal{B}$ is equivalent to the two-sided invertibility criterion for the element b^*b (respectively, bb^*). Thus, it only remains to prove the two-sided invertibility criterion for elements $b \in \mathcal{B}$.

Consider the representation $\pi = \bigoplus_{\omega \in \Omega} \pi_\omega$ of the C^*-algebra $\mathcal{A}$ in the Hilbert space $H = \bigoplus_{\omega \in \Omega} l^2(G, H_\omega)$. Let us show that π is an isomorphism of $\mathcal{A}$ onto $\pi(\mathcal{A})$. To this end it is sufficient to prove that $\|\pi(a)\| = \|a\|$ for all $a \in \mathcal{A}$.

By [11, Proposition 2.7.1], $\|a\| = \sup\left\{\sqrt{\mu(a^*a)} : \mu \in P_\mathcal{A}\right\}$ where supremum is attained (see [24, Theorem 5.1.11]). In virtue of the bijection $\nu \mapsto \nu \circ \varrho_m$ of the set $\widetilde{\mathcal{P}}_m$ of all pure states of the quotient algebra $\mathcal{A}/J_m$ onto $\mathcal{P}_m$ (see [11, Proposition 2.11.8(i)]) and of the bijection $\mu \mapsto \mu \circ \alpha_g$ of $\mathcal{P}_m$ onto $\mathcal{P}_{\beta_g(m)}$, we get

$$\left\|\pi'_m(\alpha_g(a))\right\| = \left\|\varrho_m(\alpha_g(a))\right\| = \sup\left\{\sqrt{(\nu \circ \varrho_m \circ \alpha_g)(a^*a)} : \nu \in \widetilde{\mathcal{P}}_m\right\}$$

$$= \sup\left\{\sqrt{\mu(a^*a)} : \mu \in \mathcal{P}_{\beta_g(m)}\right\}$$

for all $a \in \mathcal{A}$, all $g \in G$, and all $m \in \omega$. Hence, for a fixed $m \in \omega$, we obtain

$$\|\pi_\omega(a)\| = \sup\left\{\|\pi'_m(\alpha_g(a))\| : g \in G\right\} = \sup\left\{\sqrt{\mu(a^*a)} : \mu \in \bigcup_{g \in G} \mathcal{P}_{\beta_g(m)}\right\}.$$

Finally, taking into account (3.1) we conclude that

$$\|\pi(a)\| = \sup\left\{\|\pi_\omega(a)\| : \omega \in \Omega\right\} = \sup\left\{\sqrt{\mu(a^*a)} : \mu \in P_\mathcal{A}\right\} = \|a\|.$$

The isomorphism $\varphi = \pi$ of $\mathcal{A}$ onto $\pi(\mathcal{A})$ automatically satisfies (A4) because

$$\pi_\omega(U_g a U_g^*) = \pi_\omega(U_g)\pi_\omega(a)\pi_\omega(U_g^*) \quad \text{for all } a \in \mathcal{A}, \ g \in G, \text{ and } \omega \in \Omega.$$

Since assumptions (A1)–(A3) for the C^*-algebra $\mathcal{B}$ are also fulfilled, Theorem 3.3 implies that $\mathcal{B} \cong \pi(\mathcal{B})$. The C^*-algebra $\pi(\mathcal{B})$ consists of the operator functions

$$\pi(b) : \Omega \to \mathcal{B}\big(\bigoplus_{\omega \in \Omega} l^2(G, H_\omega)\big), \quad \omega \mapsto \pi_\omega(b)$$

equipped with the norm $\|\pi(b)\| = \sup\left\{\|\pi_\omega(b)\| : \omega \in \Omega\right\}$ for every $b \in \mathcal{B}$.

Therefore, taking into account the inverse closedness of C^*-algebras, we conclude that an element $b \in \mathcal{B}$ is invertible in the C^*-algebra $\mathcal{B}$ if and only if the operator function $\pi(b)$ is invertible in the C^*-algebra $\mathcal{B}\big(\bigoplus_{\omega \in \Omega} l^2(G, H_\omega)\big)$, that is, for every $\omega \in \Omega$, the operator $\pi_\omega(b)$ is invertible on the space $l^2(G, H_\omega)$ and in the case of infinite Ω the norms of inverse operators $(\pi_\omega(b))^{-1}$ are uniformly bounded with respect to $\omega \in \Omega$. $\qquad\square$

In the next theorem we formulate an additional (to (A1)–(A3)) condition allowing us to remove the condition of uniform boundedness of the norms for corresponding (two- or one-sided) inverse operators.

Theorem 4.2. *If assumptions* (A1)–(A3) *are satisfied and in the case of infinite* Ω, *for every irreducible representation* π *of the C^*-algebra $\mathcal{B}$ in a Hilbert space, there exists a G-orbit* $\omega \in \Omega$ *such that*

$$\operatorname{Ker}\pi_\omega \subset \operatorname{Ker}\pi, \tag{4.1}$$

then any element $b \in \mathcal{B}$ is invertible (left invertible, right invertible) on the space H if and only if for every orbit $\omega \in \Omega$ the operator $\pi_\omega(b)$ is invertible (left invertible, right invertible) on the space $l^2(G, H_\omega)$.

Proof. As in Theorem 4.1, it is sufficient to consider only the case of the two-sided invertibility of the elements $b \in \mathcal{B}$. According to the proof of Theorem 4.1 the C^*-algebras $\mathcal{B}$ and $\big(\bigoplus_{\omega \in \Omega} \pi_\omega\big)(\mathcal{B})$ are isometrically isomorphic. From here it follows that the invertibility of the element $b \in \mathcal{B}$ implies the invertibility of all the operators $\pi_\omega(b)$, $\omega \in \Omega$. In the case of finite set Ω the converse is also true.

Let now Ω be infinite and let π be a non-zero irreducible representation of the C^*-algebra $\mathcal{B}$ in a Hilbert space. Fix $\omega \in \Omega$ such that (4.1) holds. Hence there is a canonical *-homomorphism $\varphi_\omega : \mathcal{B}/\operatorname{Ker}\pi_\omega \to \mathcal{B}/\operatorname{Ker}\pi$. Since $\mathcal{B}/\operatorname{Ker}\pi_\omega \cong \pi_\omega(\mathcal{B})$ and $\mathcal{B}/\operatorname{Ker}\pi \cong \pi(\mathcal{B})$, the *-homomorphism φ_ω generates a *-homomorphism $\widetilde{\varphi}_\omega : \pi_\omega(\mathcal{B}) \to \pi(\mathcal{B})$. Therefore the invertibility of $\pi_\omega(b)$ for all $\omega \in \Omega$ implies the invertibility of $\pi(b)$ for all non-zero irreducible representations π of the C^*-algebra

$\mathcal{B}$ in Hilbert spaces, which in its turn implies the invertibility of b in $\mathcal{B}$. Indeed, by [11, Lemma 3.3.6], $\|b\| = \max_\pi \|\pi(b)\|$ where π runs through the set of all irreducible representations of $\mathcal{B}$ in Hilbert spaces, and therefore an element $b \in \mathcal{B}$ is invertible if and only if the operator $\pi(b)$ is invertible for every irreducible representation π (also see [27, Chapter 5, Proposition 1.10]). $\qquad\square$

Definition 4.3. A family of representations π_s $(s \in \mathcal{S})$ of a C^*-algebra $\mathcal{B}$ is said to be *sufficient* if for every element $b \in \mathcal{B}$, the invertibility of all operators $\pi_s(b)$ $(s \in \mathcal{S})$ implies the invertibility of b.

Thus, under the conditions of Theorem 4.2, the family of the representations π_ω $(\omega \in \Omega)$ is sufficient for the C^*-algebra $\mathcal{B}$ along with the family of all irreducible representations of $\mathcal{B}$ in Hilbert spaces.

4.2. Auxiliary results

Here we prove two auxiliary assertions on spectral measures, which will allow us to find sufficient conditions for the fulfillment of assumption (4.1) in Theorem 4.2.

Let M be a compact Hausdorff space and H a Hilbert space. By [25, p. 249], a *spectral measure* $P(\cdot)$ is a map from the σ-algebra of all Borel sets of M into the set of orthogonal projections in $\mathcal{B}(H)$ such that for every $\xi \in H$ the function $\Delta \to (P(\Delta)\xi, \xi)$ is the restriction to Borel sets of a measure on M defined by some integral on $C(M)$. Hence, $P(\emptyset) = 0$, $P(M) = I$, $P(\Delta_1 \cap \Delta_2) = P(\Delta_1)P(\Delta_2)$ for all Borel sets $\Delta_1, \Delta_2 \subset M$, and $P(\Delta_1 \cup \Delta_2) = P(\Delta_1) + P(\Delta_2)$ if $\Delta_1 \cap \Delta_2 = \emptyset$.

Theorem 4.4. [25, p. 249]. *Let π be a representation of a unital commutative C^*-algebra $\mathcal{Z}$ in a Hilbert space H, let M be the maximal ideal space of $\mathcal{Z}$ and let $z(\cdot) \in C(M)$ be the Gelfand transform of an element $z \in \mathcal{Z}$. Then there exists a unique spectral measure $P_\pi(\cdot)$ which commutes with all bounded linear operators belonging to the algebra $\pi(\mathcal{Z})$ and to its commutant, and such that*

$$\pi(z) = \int_M z(m)dP_\pi(m) \quad for\ all \quad z \in \mathcal{Z}.$$

By the analogy with [2, Lemma 5.2] one can prove the following.

Lemma 4.5. *Let H be a Hilbert space, X a compact Hausdorff topological space with a countable base, G a group of homeomorphisms of X onto itself, and let $P(\cdot) : X \to \mathcal{B}(H)$ be a spectral measure satisfying the condition: $P(\Delta) = I$ for any non-empty open G-invariant set $\Delta \subset X$, that is, $g(\Delta) = \Delta$ for all $g \in G$. Then there is a point $x \in X$ such that the G-orbit $G(x) = \{g(x) : g \in G\}$ is dense in X.*

Proof. Let W be the set of points in X such that their G-orbits are not dense in X. The closure $\overline{G(x)}$ of the orbit $G(x)$ is a G-invariant closed subset of X. For every point $x \in W$, the set $X \backslash \overline{G(x)}$ is open, non-empty, and it contains a non-empty open set U_j from the countable base of open sets in X. Then the set $V_j := \bigcup_{g \in G} g(U_j)$ is G-invariant, open, non-empty, and it is contained in $X \setminus \overline{G(x)}$. Hence, by the condition of the lemma, $P(V_j) = I$. As a result, for every point $x \in W$, there exists a set U_j in the countable base of open sets in X such that $x \in X \setminus V_j$ where $V_j = \bigcup_{g \in G} g(U_j)$, and $P(X \setminus V_j) = 0$. Consequently, $W \subset \bigcup_j (X \setminus V_j)$, where j runs through an at most countable set. For the Borel set $Y := \bigcup_j (X \setminus V_j)$ we get $0 \leq P(Y) \leq \sum_j P(X \setminus V_j) = 0$. Then $P(X \setminus Y) = I$, whence $X \setminus Y \neq \emptyset$. Therefore in view of the inclusion $W \subset Y$ there exists a point $x_0 \in X \setminus W$. For this point, by the definition of the set W, $\overline{G(x_0)} = X$. $\square$

Let $\mathfrak{R}(M)$ be the σ-algebra of all Borel subsets in M, and

$$\mathfrak{R}_G(M) := \{\Delta \in \mathfrak{R}(M) : \beta_g(\Delta) = \Delta \text{ for all } g \in G\}.$$

Let π be a representation of the C^*-algebra $\mathcal{B} = C^*(\mathcal{A}, U_G)$ in a Hilbert space H and let $P_\pi(\cdot)$ be a spectral measure determined by a unital central C^*-subalgebra $\mathcal{Z} \subset \mathcal{A}$ with the unit I of $\mathcal{B}$ and by the representation π according to Theorem 4.4.

Since $az = za$ for all $a \in \mathcal{A}$ and all $z \in \mathcal{Z}$, we get

$$\pi(a)P_\pi(\Delta) = P_\pi(\Delta)\pi(a) \text{ for all } \Delta \in \mathfrak{R}(M) \text{ and all } a \in \mathcal{A}. \tag{4.2}$$

Lemma 4.6. *If* (A1) *is fulfilled and π is a representation of the C^*-algebra $\mathcal{B} = C^*(\mathcal{A}, U_G)$ in a Hilbert space H, then*

$$\pi(U_g)P_\pi(\Delta)\pi(U_g^{-1}) = P_\pi(\Delta) \text{ for all } \Delta \in \mathfrak{R}_G(M) \text{ and all } g \in G. \tag{4.3}$$

Proof. Fix $\Delta \in \mathfrak{R}_G(M)$ and $g \in G$. Decompose the restriction $\pi : \mathcal{Z} \to \mathcal{B}(H)$ of the representation $\pi : \mathcal{B} \to \mathcal{B}(H)$ into the direct sum of cyclic representations of $\mathcal{Z}$ in pairwise orthogonal subspaces H_α of the Hilbert space H. Since $H = \bigoplus_\alpha H_\alpha$, it follows that every vector $\xi \in H$ is represented in the form $\xi = \sum_\alpha \xi_\alpha$, where $\xi_\alpha \in H_\alpha$, $\xi_\alpha \neq 0$ only for an at most countable set of indices α (own for every ξ), and $\|\xi\|_H^2 = \sum_\alpha \|\xi_\alpha\|_{H_\alpha}^2 < \infty$. As the representations $\pi_\alpha = \pi|_{H_\alpha}$ of $\mathcal{Z}$ in the spaces H_α are cyclic, there are isometric isomorphisms $\psi_\alpha : H_\alpha \to L_2(M, \mu_\alpha)$ such that the operators $\pi(z) \in \pi(\mathcal{Z})$ and the spectral measures $P_\pi(\Delta)$ have the form

$$\pi(z) = \bigoplus_\alpha \psi_\alpha^{-1} z(\cdot)\psi_\alpha, \quad P_\pi(\Delta) = \bigoplus_\alpha \psi_\alpha^{-1} \chi_\Delta \psi_\alpha,$$

where χ_Δ is the characteristic function of the Borel set $\Delta \subset M$ and μ_α is the measure on M induced by the integral $I_\alpha(z(\cdot)) := (\pi(z)\xi_\alpha^0, \xi_\alpha^0)$ where $\xi_\alpha^0 \in H_\alpha$ is a cyclic vector (see [25, p. 248]). Since all Borel sets are summable, that is, $\mu_\alpha(\Delta) = I_\alpha(\chi_\Delta)$ for every $\Delta \in \mathfrak{R}(M)$, by Proposition IX in [25, p. 127]), there exist open sets $u_{n,\alpha} \supset \Delta$ and closed sets $v_{n,\alpha} \subset \Delta$ such that $\mu_\alpha(u_{n,\alpha} \setminus v_{n,\alpha}) < n^{-1}$.

Let H' be a subspace H of the form $\bigoplus_{\alpha \in \Omega} H_\alpha$ where Ω is a finite set. Then for the open sets $u_n = \bigcap_{\alpha \in \Omega} u_{n,\alpha}$ and the closed sets $v_n = \bigcup_{\alpha \in \Omega} v_{n,\alpha}$ we deduce that $\mu_\alpha(u_n \setminus v_n) < n^{-1}$ ($\alpha \in \Omega$). Since β_g is a homeomorphism of M onto itself, since $\beta_g(\Delta) = \Delta$, and since every point $t \in \Delta$ is a closed set in M, we conclude that for the open neighborhood u_n of the set Δ there is an open set $u_n' \supset \Delta$ such that $\Delta \subset (u_n' \cup \beta_g^{-1}(u_n')) \subset u_n$. Analogously, passing to the complement, we can find closed subsets $v_n' \subset \Delta$ such that $v_n \subset (v_n' \cap \beta_g^{-1}(v_n')) \subset \Delta$. Then for all $\alpha \in \Omega$,

$$\mu_\alpha(u_n' \setminus v_n') < n^{-1}, \quad \mu_\alpha(u_n' \cup \beta_g^{-1}(u_n')) - \mu_\alpha(v_n' \cap \beta_g^{-1}(v_n')) < n^{-1}. \tag{4.4}$$

By the Urysohn lemma, there are functions $\eta_n(\cdot) \in C(M)$ such that $0 \le \eta_n(m) \le 1$ for all $m \in M$, $\eta_n(m) = 1$ if $m \in v'_n$, and $\eta_n(m) = 0$ if $m \in M \setminus u'_n$. Let η_n, $\eta_n(g)$ be elements in $\mathcal{Z}$ associated to the functions $\eta_n(\cdot)$, $(\eta_n \circ \beta_g)(\cdot) \in C(M)$. Then

$$\pi(\eta_n) = \bigoplus_\alpha \psi_\alpha^{-1} \eta_n(\cdot)\psi_\alpha, \quad \pi(\eta_n(g)) = \bigoplus_\alpha \psi_\alpha^{-1}(\eta_n \circ \beta_g)(\cdot)\psi_\alpha \quad (\in \pi(\mathcal{Z})).$$

Therefore for every $\xi = \sum_{\alpha \in \Omega} \xi_\alpha \in H'$, setting $\xi_\alpha(\cdot) = \psi_\alpha \xi_\alpha$, we get

$$\left\| (P_\pi(\Delta) - \pi(\eta_n))\xi \right\|_H^2 = \sum_{\alpha \in \Omega} \left\| (P_\pi(\Delta) - \pi(\eta_n))\xi \right\|_{H_\alpha}^2$$

$$= \sum_{\alpha \in \Omega} \int_M |\chi_\Delta(m) - \eta_n(m)|^2 |\xi_\alpha(m)|^2 d\mu_\alpha(m) \le \sum_{\alpha \in \Omega} \int_{u'_n \setminus v'_n} |\xi_\alpha(m)|^2 d\mu_\alpha(m),$$

whence it follows due to (4.4) that $P_\pi(\Delta) = \operatorname*{s-lim}_{n\to\infty} \pi(\eta_n)$ on the space H'. Thus,

$$\pi(U_g)P_\pi(\Delta)\pi(U_g^{-1}) = \operatorname*{s-lim}_{n\to\infty} \pi(U_g)\,\pi(\eta_n)\pi(U_g^{-1}) = \operatorname*{s-lim}_{n\to\infty} \pi(\eta_n(g)) \quad \text{on} \quad H'.$$

Moreover, $\operatorname*{s-lim}_{n\to\infty}[\pi(\eta_n(g)) - \pi(\eta_n)] = 0$ on H' according to (4.4) because

$$\eta_n(\beta_g(m)) - \eta_n(m) = 0 \quad \text{for} \quad m \in M \setminus \left([u'_n \cup \beta_g^{-1}(u'_n)] \setminus [v'_n \cap \beta_g^{-1}(v'_n)] \right).$$

Consequently, $\pi(U_g)P_\pi(\Delta)\pi(U_g^{-1}) = P_\pi(\Delta)$ on H'. Since the set of vectors $\xi = \sum \xi_\alpha$ with finite numbers of non-zero entries $\xi_\alpha \in H_\alpha$ is dense in H, the latter equality holds on the whole H. $\qquad\square$

Equalities (4.2) and (4.3) immediately imply the following.

Corollary 4.7. *If $\Delta \in \mathfrak{R}_G(M)$ and $b \in \mathcal{B}$, then $\pi(b)P_\pi(\Delta) = P_\pi(\Delta)\pi(b)$ for every representation π of the C^*-algebra $\mathcal{B} = C^*(\mathcal{A}, U_G)$ in a Hilbert space.*

4.3. Sufficient families of representations

In this subsection we will find sufficient conditions for the fulfillment of (4.1). Let $\overline{\omega}$ be the closure of an orbit $\omega \in \Omega$, and let ω' be the set of all limit points of ω.

Theorem 4.8. *Suppose (A1)–(A3) are satisfied, and either*

(A5) *the C^*-algebra $\mathcal{Z}$ is separable, and $\bigcap_{m \in \omega} J_m = \bigcap_{m \in \overline{\omega}} J_m$ for every G-orbit $\omega \in \Omega$ such that $\overline{\omega} = \omega'$; or*

(A6) *for every point $m \in M$, every net $\{m_s\} \subset M$ which tends to m and does not contain a subnet consisting of points of one and the same G-orbit, and every element $a \in \bigcap_{g \in G} J_{\beta_g(m)}$, there exists a subnet $m_v \to m$ such that $\lim_v \|\pi_{\omega_v}(a)\| = 0$, where ω_v are the G-orbits of the points m_v.*

Then for every irreducible representation π of the C^*-algebra $\mathcal{B}$ there exists a G-orbit $\omega \in \Omega$ satisfying (4.1).

Proof. We divide the proof into several steps.

Step 1. Let (A1)–(A3) hold, and let π be an irreducible representation of the C^*-algebra $\mathcal{B}$ in a Hilbert space H_π. If π is the zero irreducible representation of dimension one, then the assertion is obviously valid.

Now let π be a non-zero irreducible representation, and F be the C^*-algebra homomorphism of $\mathcal{A} \otimes_\alpha G$ into $\mathcal{B}$ which was defined in (3.3). In view of (A1)–(A3) and by the proof of Theorem 3.2, F is an isomorphism of $\mathcal{A} \otimes_\alpha G$ onto $\mathcal{B}$. Then the non-zero irreducible representation π of the C^*-algebra $\mathcal{B}$ in H_π generates the non-zero irreducible representation $\widetilde{\pi} = \pi F$ of the C^*-algebra $\mathcal{A} \otimes_\alpha G$ in H_π. Since $\operatorname{Ker} \pi_\omega \subset \operatorname{Ker} \pi$ if and only if

$$F^{-1}\operatorname{Ker} \pi_\omega \subset \operatorname{Ker} \widetilde{\pi}, \tag{4.5}$$

it remains to prove the existence of a G-orbit $\omega \in \Omega$ satisfying (4.5).

Step 2. Fix a point $m \in M$ and its G-orbit $\omega = G(m)$. Show that

$$F^{-1}\operatorname{Ker} \pi_\omega = \Big\{ x \in \mathcal{A} \otimes_\alpha G : \ x(g) = E_g x \in \bigcap_{h \in G} J_{\beta_h(m)} \ \text{ for all } \ g \in G \Big\}. \tag{4.6}$$

Indeed, consider the imbedding operators $I_g : H_\omega \to l^2(G, H_\omega)$ and their left inverse operators $\Pi_g : l^2(G, H_\omega) \to H_\omega$ given by (2.1). Then for $x \in \mathcal{A} \otimes_\alpha G$,

$$\pi'_m(\alpha_h[x(g)]) = \Pi_h \pi_\omega(Fx) I_{hg} \ \text{ for all } \ g, h \in G. \tag{4.7}$$

If $\pi_\omega(Fx) = 0$, then due to (4.7), $\pi'_m(\alpha_h[x(g)]) = 0$. From here, in view of the equality $\pi'_m = \widetilde{\pi}_m \circ \varrho_m$ where $\widetilde{\pi}_m$ is the isometric representation of the quotient algebra $\mathcal{A}/J_m$ in H_ω and ϱ_m is the natural homomorphism $\mathcal{A} \to \mathcal{A}/J_m$, we get $\varrho_m(\alpha_h[x(g)]) = 0$, that is, $\alpha_h[x(g)] \in J_m$. Hence $x(g) \in \bigcap_{h \in G} J_{\beta_h(m)}$ for all $g \in G$.

Conversely, if $x \in \mathcal{A} \otimes_\alpha G$ and $x(g) \in \bigcap_{h \in G} J_{\beta_h(m)}$ for all $g \in G$, then $\pi'_m(\alpha_h[x(g)]) = 0$ for all $g, h \in G$, whence by (4.7), $\Pi_h \pi_\omega(Fx) I_g = 0$ for all $g, h \in G$. Consequently,

$$\big\| \pi_\omega(Fx) \big\| = \sup \Big\{ \big| (\pi_\omega(Fx)\varphi, \psi) \big| : \ \varphi, \psi \in l^2(G, H_\omega), \ \|\varphi\| = \|\psi\| = 1 \Big\} = 0,$$

and thus $\pi_\omega(Fx) = 0$, which completes the proof of (4.6).

Step 3. Consider the $*$-automorphisms

$$\gamma_g = \pi \circ \alpha_g : \ \pi(a) \mapsto \pi(U_g)\pi(a)\pi(U_g^*) \quad (g \in G)$$

of the C^*-algebra $\pi(\mathcal{A})$, the homomorphism $\gamma : g \mapsto \gamma_g$ of the group G into the group $\operatorname{Aut} \pi(\mathcal{A})$, and $*$-homomorphism Φ_π of the involutive Banach algebra $l^1(G, \mathcal{A}) \subset \mathcal{A} \otimes_\alpha G$ onto $l^1(G, \pi(\mathcal{A})) \subset \pi(\mathcal{A}) \otimes_\gamma G$ given by $(\Phi_\pi x)(g) = \pi[x(g)]$ for $g \in G$. If $\widehat{\pi}$ is a representation of $l^1(G, \pi(\mathcal{A}))$ in a Hilbert space, then $\widehat{\pi}\Phi_\pi$ is a representation of $l^1(G, \mathcal{A})$ in the same Hilbert space, whence we infer that

$$\|\Phi_\pi x\|_{\pi(\mathcal{A}) \otimes_\gamma G} = \sup_{\widehat{\pi}} \|\widehat{\pi}\Phi_\pi x\| \leq \|x\|_{\mathcal{A} \otimes_\alpha G} \quad \text{for all } \ x \in l^1(G, \mathcal{A}).$$

Consequently, Φ_π extends by continuity to a $*$-homomorphism of the C^*-algebra $\mathcal{A} \otimes_\alpha G$ onto the C^*-algebra $\pi(\mathcal{A}) \otimes_\gamma G$.

We also consider the representation $F_\pi : \pi(\mathcal{A}) \otimes_\gamma G \to \mathcal{B}(H)$ defined on $l^1(G, \pi(\mathcal{A}))$ by $F_\pi x = \sum_{g \in G} x(g)\pi(U_g)$ and extended by continuity to the whole

C^*-algebra $\pi(\mathcal{A}) \otimes_\gamma G$. From the commutativity of the diagram

$$
\begin{array}{ccc}
\mathcal{A} \otimes_\alpha G & \xrightarrow{\ \Phi_\pi\ } & \pi(\mathcal{A}) \otimes_\gamma G \\
F \downarrow & & F_\pi \downarrow \\
\mathcal{B} & \xrightarrow{\ \pi\ } & \pi(\mathcal{B})
\end{array}
$$

which is readily checked for the set $l^1(G, \mathcal{A})$ dense in $\mathcal{A} \otimes_\alpha G$, it follows that

$$\widetilde{\pi} = \pi F = F_\pi \Phi_\pi. \tag{4.8}$$

In view of (4.8) the relation (4.5) is fulfilled if

$$F^{-1} \operatorname{Ker} \pi_\omega \subset \operatorname{Ker} \Phi_\pi \quad \text{for some} \ \omega \in \Omega. \tag{4.9}$$

Since for $x \in \mathcal{A} \otimes_\alpha G$, $\Phi_\pi x \in \pi(\mathcal{A}) \otimes_\gamma G$, we conclude by Theorem 2.2(iv) that $\Phi_\pi x = 0$ if and only if $\pi[x(g)] = 0$ for all $g \in G$. Therefore for the fulfillment of (4.9) it remains to prove due to (4.6) that there exists a point $m \in M$ such that

$$\bigcap_{h \in G} J_{\beta_h(m)} \subset \operatorname{Ker} \pi. \tag{4.10}$$

Step 4. Let (A5) hold together with (A1)–(A3). Let $P_\pi(\cdot) : M \to \mathcal{B}(H_\pi)$ be the spectral measure associated with the unital central C^*-subalgebra $\mathcal{Z} \subset \mathcal{A}$ and with the representation π of the C^*-algebra $\mathcal{B}$ in a Hilbert space H_π.

By assumption (A5), the C^*-algebra $\mathcal{Z}$ is separable. Hence (see, e.g., [11, Lemma 3.3.3, Proposition 3.3.4]), the maximal ideal space M of the algebra $\mathcal{Z}$ admits a countable base of open sets. Consider the collection $\mathcal{Y}$ of all G-invariant closed sets $Y \subset M$ such that $P_\pi(Y) = I$. The set $\mathcal{Y}$ is partially ordered by the inclusion. Then every its linearly ordered subset T has a lower bound coinciding with the intersection Y_T of all elements in T. Indeed, as Y_T is a G-invariant closed subset of M which is contained in every element of T, it remains to show that $P_\pi(Y_T) = I$. To this end we consider the open set $M \setminus Y_T = \bigcup_{Y \in T}(M \setminus Y)$. In virtue of the countability of the topology base $\mathcal{U}$ on M, every open set $M \setminus Y$ is a union of an at most countable collection of open sets $u \in \mathcal{U}$. For these $u \subset M \setminus Y$ in view of $P_\pi(M \setminus Y) = 0$ we get $P_\pi(u) = 0$. As a result, $M \setminus Y_T$ also is a union of an at most countable collection $\mathcal{U}_T$ of sets $u \in \mathcal{U}$ such that $P_\pi(u) = 0$. But then

$$P_\pi(M \setminus Y_T) \leq \sum_{u \in \mathcal{U}_T} P_\pi(u) = 0.$$

Hence $P_\pi(M \setminus Y_T) = 0$, $P_\pi(Y_T) = I$, and thus $Y_T (\in \mathcal{Y})$ is a lower bound of the set T. Then by Zorn's lemma (see, e.g., [28, p. 3]), every linearly ordered (by the inclusion) set $T \subset \mathcal{Y}$ possesses a lower bound that also is a minimal element of $\mathcal{Y}$.

Now let X be a minimal by the inclusion G-invariant closed subset of M such that $P_\pi(X) = I$. Then the set X is a topological space with the topology induced from M and having a countable base. In view of the minimality of X, for every non-empty G-invariant open (in the topology induced from M) set $\Delta \subset X$ we get $P_\pi(\Delta) \neq 0$ because $P_\pi(X \setminus \Delta) \neq I$ and $P_\pi(\Delta) = I - P_\pi(X \setminus \Delta)$. Since $\Delta \in \mathfrak{R}_G(M)$ and the representation π is irreducible, we derive from Corollary 4.7

and [11, Proposition 2.3.1(i)] that either the projection $P_\pi(\Delta) = 0$, or $P_\pi(\Delta) = I$. But $P_\pi(\Delta) \neq 0$ and therefore $P_\pi(\Delta) = I$.

Since X is a closed subset of M and $P_\pi(X) = I$, the restriction of the spectral measure $P_\pi(\cdot) : M \to \mathcal{B}(H_\pi)$ to the set X is a spectral measure $P_\pi(\cdot) : X \to \mathcal{B}(H_\pi)$. Then by Lemma 4.5 there is a point $m \in X$ such that $\overline{G(m)} = X$. Here we need take into account that a net $\{m_s\}$ of points $m_s \in X$ converges in the topological space M and in the topological space X with the topology induced from M only simultaneously. Consequently, for every set $\Delta \subset X$ its closures in M and X coincide. The set of limit points for $\Delta \subset X$ also does not depend of which topology we take: the initial topology of M or the induced (from M) topology of X. Therefore in what follows these notions for sets $\Delta \subset X$ we may consider with respect to the initial topology of M.

For the orbit $\omega = G(m)$, we have the two cases: (a) $\overline{\omega} \neq \omega' \subset \overline{\omega}$; (b) $\omega' = \overline{\omega}$.

Step 5. Show that in the case (a) the set $\overline{\omega} \setminus \omega'$ is at most countable. Indeed, the set $\overline{\omega} \setminus \omega'$ consists of the isolated points of the orbit ω. For every point $\tau \in \overline{\omega} \setminus \omega'$, we choose its open neighborhood u_τ which does not contain other points of ω. Further for every u_τ choose an element v_τ in the base of open subsets of M such that $\tau \in v_\tau \subset u_\tau$. Then v_τ $(\tau \in \overline{\omega} \setminus \omega')$ are pairwise different elements in the base of open subsets of M, and in view of the countability of this base, the set $\overline{\omega} \setminus \omega'$ is at most countable.

Since $\omega' \subset \overline{\omega}$ and $\omega' \neq \overline{\omega}$, ω' is a G-invariant closed subset of M properly contained in $\overline{\omega}$. But by Step 4, $\overline{\omega} = X$ is a minimal (by the inclusion) G-invariant closed subset of M such that $P_\pi(X) = I$. Hence $P_\pi(\omega') \neq I$, and then $P_\pi(\omega') = 0$ in view of the irreducibility of the representation π.

As a result, in the case (a) $P_\pi(\overline{\omega} \setminus \omega') = I$ for the at most countable set $\overline{\omega} \setminus \omega'$.

By [29] (also see [6, Proposition 8.6]) each element $a \in \bigcap_{\tau \in \omega} J_\tau$ can be written in the form $a = a_\tau c_\tau$ for all $\tau \in \omega$, where $a_\tau \in \mathcal{A}$, $c_\tau \in \mathcal{Z}$ and the Gelfand transform $c_\tau(\cdot)$ vanishes at the point τ. Let χ_τ be the characteristic function of the one-point set $\{\tau\}$. Then

$$P_\pi(\{\tau\})\pi(c_\tau) = \int_M \chi_\tau(m) dP_\pi(m) \int_M c_\tau(m)\, dP_\pi(m) = \int_M \chi_\tau(m) c_\tau(m)\, dP_\pi(m) = 0,$$

and hence (4.10) in the case (a) follows from the equalities

$$
\begin{aligned}
\pi(a) &= P_\pi(\overline{\omega} \setminus \omega')\pi(a) = \sum_{\tau \in \overline{\omega} \setminus \omega'} P_\pi(\{\tau\})\pi(a) \\
&= \sum_{\tau \in \overline{\omega} \setminus \omega'} P_\pi(\{\tau\})\pi(a_\tau)\pi(c_\tau) = \sum_{\tau \in \overline{\omega} \setminus \omega'} \pi(a_\tau)P_\pi(\{\tau\})\pi(c_\tau) = 0.
\end{aligned}
$$

Step 6. By assumption (A5), in the case (b) we have $\bigcap_{\tau \in \omega} J_\tau = \bigcap_{\tau \in \overline{\omega}} J_\tau$. Analogously to [6, Proposition 8.6] one can prove that every element $a \in \bigcap_{\tau \in \overline{\omega}} J_\tau$ can be represented in the form $a = a_{\overline{\omega}} c_{\overline{\omega}}$ where $a_{\overline{\omega}} \in \mathcal{A}$, $c_{\overline{\omega}} \in Z$ and the Gelfand transform $c_{\overline{\omega}}(m) = 0$ for all $m \in \overline{\omega}$. Then denoting by $\chi_{\overline{\omega}}$ the characteristic function of the set $\overline{\omega}$ and using the equality $\chi_{\overline{\omega}}(m)c_{\overline{\omega}}(m) = 0$ for $m \in M$, we get

$$P_{\pi}(\overline{\omega})\pi(c_{\overline{\omega}}) = \int_M \chi_{\overline{\omega}}(m)dP_{\pi}(m) \int_M c_{\overline{\omega}}(m)dP_{\pi}(m) = \int_M \chi_{\overline{\omega}}(m)c_{\overline{\omega}}(m)dP_{\pi}(m) = 0.$$

From here, taking into account the equality $P_{\pi}(\overline{\omega}) = I$, we obtain

$$\pi(a) = P_{\pi}(\overline{\omega})\pi(a) = P_{\pi}(\overline{\omega})\pi(a_{\overline{\omega}})\pi(c_{\overline{\omega}}) = \pi(a_{\overline{\omega}})P_{\pi}(\overline{\omega})\pi(c_{\overline{\omega}}) = 0,$$

which proves (4.10) in case (b).

Thus (4.10) has been proved under the conditions (A1)–(A3) and (A5).

Step 7. Let assumption (A6) hold along with (A1)–(A3). Consider the state p of the C^*-algebra $\mathcal{B}$, which is defined by $p(\cdot) = (\pi(\cdot)\xi, \xi)$ where $\xi \in H_{\pi}$ and $\|\xi\| = 1$. Since π is a non-zero irreducible representation of the C^*-algebra $\mathcal{B}$ in a Hilbert space H_{π}, by [11, Proposition 2.3.1(iii)] every non-zero vector $\xi \in H_{\pi}$ is cyclic. Let π_p be the representation of the C^*-algebra $\mathcal{B}$ in the Hilbert space H_p with the cyclic vector ξ_p which is defined by the state p. Then by [11, Proposition 2.4.4(iv)], $p(\cdot) = (\pi_p(\cdot)\xi_p, \xi_p)$. Since $(\pi(b)\xi, \xi) = (\pi_p(b)\xi_p, \xi_p)$ for all $b \in \mathcal{B}$, the representations π and π_p are unitarily equivalent by [11, Proposition 2.4.1(ii)]. Then $\mathrm{Ker}\,\pi = \mathrm{Ker}\,\pi_p$, and since π is an irreducible representation, the representation π_p is also irreducible, whence by [24, Theorem 5.1.6] p is a pure state of the C^*-algebra $\mathcal{B}$.

Consider the smallest closed two-sided ideal $\mathcal{J}_{G(m)}$ of the C^*-algebra $\mathcal{B}$, containing the closed two-sided ideal $\bigcap_{h \in G} J_{\beta_h(m)}$ of the C^*-algebra $\mathcal{A}$. It is easily seen that the set of operators of the form $\sum_{g \in G_0} a_g U_g$, where G_0 is an arbitrary finite subset of the group G and all $a_g \in \bigcap_{h \in G} J_{\beta_h(m)}$, is dense in $\mathcal{J}_{G(m)}$.

By [11, Corollary 2.4.10], $\mathrm{Ker}\,\pi_p$ is the largest closed two-sided ideal of the C^*-algebra $\mathcal{B}$, which is contained in $\mathrm{Ker}\,p$. Hence the closed two-sided ideal $\mathcal{J}_{G(m)}$ is contained in $\mathrm{Ker}\,\pi_p$ and $\mathrm{Ker}\,p$ only simultaneously. Thus it remains to prove that for the pure state $p \in P_{\mathcal{B}}$ there is a point $m \in M$ such that

$$\mathcal{J}_{G(m)} \subset \mathrm{Ker}\,p, \tag{4.11}$$

because this implies (4.10) in view of $\bigcap_{h \in G} J_{\beta_h(m)} \subset \mathcal{J}_{G(m)} \subset \mathrm{Ker}\,\pi_p = \mathrm{Ker}\,\pi$.

Consider the pure state pF of the C^*-algebra $\mathcal{A} \otimes_{\alpha} G$, where F is the isomorphism of $\mathcal{A} \otimes_{\alpha} G$ onto $\mathcal{B}$ given by (3.3). Since by Theorem 2.2(iii),

$$\bigcap_{\mu \in P_{\mathcal{A}}} \mathrm{Ker}\,\pi_{\delta_e \times \mu} = \{0\},$$

from [11, Proposition 3.4.2(ii)] it follows that the pure state pF of the C^*-algebra $\mathcal{A} \otimes_{\alpha} G$ is a weak* limit of the net of the states $\eta_s = (\pi_{\delta_e \times \mu_s}(\cdot)\xi_s, \xi_s)$ where $\mu_s \in P_{m_s}$ ($m_s \in M$), $\xi_s \in H_{\delta_e \times \mu_s}$, and $\|\xi_s\| = 1$.

Consider the net $\{m_s\} \subset M$ corresponding to the net $\{\eta_s\}$.

Step 8. Assume that the net $\{m_s\}$ contains a subnet $\{m_{\gamma}\}$ which is entirely contained in some orbit ω. We show that for this orbit

$$\mathrm{Ker}\,\pi_{\omega} \subset \mathrm{Ker}\,p. \tag{4.12}$$

Since, by (4.6), $\mathcal{J}_{G(m)} \subset \mathrm{Ker}\,\pi_{\omega}$, (4.12) will imply the desired inclusion (4.11).

156   Yu.I. Karlovich

Fix a point $m \in \omega$, let $\widetilde{E}_e : \mathcal{B} \to \mathcal{A}$ be the C^*-algebra homomorphism defined by Lemma 3.1 and consider the bounded operator

$$\pi'_m \widetilde{E}_e : \mathcal{B} \to \mathcal{B}(H_\omega), \quad b \mapsto \pi'_m(\widetilde{E}_e b).$$

Obviously, $\widetilde{E}_g = E_g F^{-1}$, where F is the isomorphism of $\mathcal{A} \otimes_\alpha G$ onto $\mathcal{B}$ given by (3.3), and E_g is defined in Subsection 2.3. If $b \in \mathcal{B}$ and $\pi_\omega(b) = 0$, then, by (4.6),

$$\widetilde{E}_e b = E_e F^{-1} b \in \bigcap_{h \in G} J_{\beta_h(m)} \subset \mathcal{J}_{G(m)}$$

and thus, $\pi'_m(\widetilde{E}_e b) = 0$. Hence $\operatorname{Ker} \pi_\omega \subset \operatorname{Ker}(\pi'_m \widetilde{E}_e)$ and therefore

$$\widehat{E}_e : \pi_\omega(\mathcal{B}) \to \mathcal{B}(H_\omega), \quad \pi_\omega(b) \mapsto \pi'_m(\widetilde{E}_e b)$$

is a well-defined bounded operator. Since $\widehat{E}_e(\pi_\omega(I))$ is the identity operator on H_ω, we have $\|\widehat{E}_e\| = 1$. As $\delta_e \times \mu_m = \nu_m \widetilde{\pi}_m^{-1} \widehat{E}_e \pi_\omega F$ where ν_m is the pure state of the C^*-algebra $\mathcal{A}/J_m$ induced by a state $\mu_m \in \mathcal{P}_m$, we get $\operatorname{Ker}(\delta_e \times \mu_m) \supset F^{-1} \operatorname{Ker} \pi_\omega$. Then

$$\operatorname{Ker} \pi_{\delta_e \times \mu_m} \supset F^{-1} \operatorname{Ker} \pi_\omega. \tag{4.13}$$

Further, in virtue of the bijection $\mu_{\beta_g(m)} = \mu_m \circ \alpha_g$ between pure states $\mu_m \in \mathcal{P}_m$ and $\mu_{\beta_g(m)} \in \mathcal{P}_{\beta_g(m)}$, we have for $x \in \mathcal{A} \otimes_\alpha G$ and $\tau = \beta_g(m)$,

$$(\pi_{\delta_e \times \mu_\tau}(x)\xi_\tau, \, \xi_\tau) = (\delta_e \times \mu_\tau)(x) = \mu_\tau(x(e)) = \mu_m(\alpha_g(x(e)))$$

$$= (\delta_e \times \mu_m)(I_g x I_g^*) = (\pi_{\delta_e \times \mu_m}(I_g x I_g^*)\xi_m, \, \xi_m) = (\pi_{\delta_e \times \mu_m}(x)\xi, \, \xi),$$

where $I_g(g) = I$, $I_g(h) = 0$ for $h \neq g$, $\xi = \pi_{\delta_e \times \mu_m}(I_g^*)\xi_m$, and ξ_τ, ξ_m are cyclic vectors defined by the states $\delta_e \times \mu_\tau$, $\delta_e \times \mu_m$. Since ξ also is a cyclic vector for $\pi_{\delta_e \times \mu_t}$ (in view of the invertibility of I_g^* in $\mathcal{A} \otimes_\alpha G$), the representations $\pi_{\delta_e \times \mu_\tau}$ and $\pi_{\delta_e \times \mu_t}$ are unitarily equivalent by [11, Proposition 2.4.1(ii)]. Consequently,

$$\operatorname{Ker} \pi_{\delta_e \times \mu_\tau} = \operatorname{Ker} \pi_{\delta_e \times \mu_m} \quad \text{for all} \ \ m \in M \ \ \text{and all} \ \ \tau \in G(m). \tag{4.14}$$

It follows from (4.13) and (4.14) that

$$\operatorname{Ker} \pi_{\delta_e \times \mu_\tau} \supset F^{-1} \operatorname{Ker} \pi_\omega \quad \text{for all} \ \ \tau \in \omega. \tag{4.15}$$

Since $pF = \lim_\gamma (\pi_{\delta_e \times \mu_\gamma}(\cdot)\xi_\gamma, \, \xi_\gamma)$ where $\mu_\gamma \in \mathcal{P}_{m_\gamma}$, all m_γ belong to the orbit ω, $\xi_\gamma \in H_{\delta_e \times \mu_\gamma}$, and $\|\xi_\gamma\| = 1$, the relation (4.15) implies (4.12).

Step 9. Now let a net $\{m_s\} \subset M$ does not contain a subnet which is entirely contained in some orbit ω. As M is a compact, the Bolzano-Weierstrass theorem (see, e.g., [28, p. 98]) implies that the net $\{m_s\}$ contains a subnet convergent to a point $m \in M$. Without loss of generality we assume that the initial net $m_s \to m$.

By the continuity of the state p, for proving (4.11) it is sufficient to show that $p(b_m) = 0$ for every b_m in the (dense in $\mathcal{J}_{G(m)}$) set

$$\mathcal{J}_{G(m)}^0 := \left\{ b_m = \sum_{g \in G_0} a_g U_g : a_g \in \bigcap_{h \in G} J_{\beta_h(m)} \right\}$$

where G_0 runs through the finite subsets of G. Fix $b_m \in \mathcal{J}_{G(m)}^0$. According to (A6), the net $m_s \to m$ contains a subnet $m_v \to m$ such that $\lim_v \|\pi_{\omega_v}(a_g)\| = 0$

for all $g \in G_0$ where $\omega_v = G(m_v)$. Then in view of the finiteness of G_0, from the estimate $\|\pi_{\omega_v}(b_m)\| \leq \sum_{g \in G_0} \|\pi_{\omega_v}(a_g)\|$ it follows that

$$\lim_v \|\pi_{\omega_v}(b_m)\| = 0. \qquad (4.16)$$

Consider the net $\eta_v = (\pi_{\delta_e \times \mu_v}(\cdot)\xi_v, \, \xi_v)$ corresponding to the net m_v, where $\mu_v \in \mathcal{P}_{m_v}$, $\xi_v \in H_{\delta_e \times \mu_v}$, and $\|\xi_v\| = 1$. By (4.15), $\mathrm{Ker}\,(\pi_{\delta_e \times \mu_v} F^{-1}) \supset \mathrm{Ker}\,\pi_{\omega_v}$, and then

$$|\eta_v(F^{-1}b_m)| \leq \|\pi_{\delta_e \times \mu_v}(F^{-1}b_m)\| \leq \|\pi_{\omega_v}(b_m)\|,$$

which in view of (4.16) implies the relation $p(b_m) = \lim_v \eta_v(F^{-1}b_m) = 0$ and hence gives (4.11).

Thus, (4.11) has been proved under the conditions (A1)–(A3) and (A6) too, which completes the proof of the theorem. $\qquad\square$

Theorems 4.2 and 4.8 immediately imply the following.

Theorem 4.9. *If conditions* (A1)–(A3) *hold and in the case of infinite* Ω *assumption* (A5) *or* (A6) *is satisfied, then an element* $b \in \mathcal{B}$ *is invertible (left invertible, right invertible) in* $\mathcal{B}$ *if and only if for every orbit* $\omega \in \Omega$ *the operator* $\pi_\omega(b)$ *is invertible (left invertible, right invertible) on the space* $l^2(G, H_\omega)$.

Corollary 4.10. *If assumptions* (A1)–(A3) *are satisfied, the* C^*-*algebra* $\mathcal{Z}$ *is separable, and* $\omega' \neq \overline{\omega}$ *for all* $\omega \in \Omega$, *then an element* $b \in \mathcal{B}$ *is invertible (left invertible, right invertible) in* $\mathcal{B}$ *if and only if for every orbit* $\omega \in \Omega$ *the operator* $\pi_\omega(b)$ *is invertible (left invertible, right invertible) on the space* $l^2(G, H_\omega)$.

4.4. Minimization of sufficient families of representations

Let us minimize a sufficient family of representations for $\mathcal{B}$. For $Y \subset M$, put

$$\mathcal{P}_Y := \bigcup_{m \in Y} \mathcal{P}_m(\mathcal{A}), \quad J_Y := \bigcap_{\mu \in \mathcal{P}_Y} \mathrm{Ker}\,\pi_{\delta_e \times \mu},$$

where $\mathcal{P}_M = P_{\mathcal{A}}$ in virtue of (3.1). We replace (A3) by the stronger assumption

(A3′)   *there is a closed set* $M_0 \subset M$ *such that for every finite set* $G_0 \subset G$ *and for every nonempty open set* $W \subset P_{\mathcal{A}}$ *there exists a state* $\nu \in W$ *such that* $\beta_g(m_\nu) \neq m_\nu$ *for all* $g \in G_0 \setminus \{e\}$, *where the point* $m_\nu = Z \cap \mathrm{Ker}\,\nu$ *belongs to the* G-*orbit* $G(M_0)$ *of the set* M_0.

In the case $M_0 \neq M$, making use of (A3′), we strengthen assertion (iii) in Theorem 2.2, and also sharpen Theorems 4.1, 4.2, 4.8, 4.9, and Corollary 4.10.

By the definition of the set M_0, every pure state $\tau \in P_{\mathcal{A}}$ is contained in the weak* closure of the set $\mathcal{P}_{G(M_0)}$. Then the corresponding state $\delta_e \times \tau$ of the C^*-algebra $\mathcal{A} \otimes_\alpha G$ belongs to the weak* closure of the state set $\{\delta_e \times \mu : \mu \in \mathcal{P}_{G(M_0)}\}$. Hence there exists a net $\mu_s \in \mathcal{P}_{G(M_0)}$ such that $\delta_e \times \tau = \lim_s(\delta_e \times \mu_s)$. Thus

$$\bigcap_{\mu \in \mathcal{P}_{G(M_0)}} \mathrm{Ker}\,\pi_{\delta_e \times \mu} \subset \bigcap_{\mu \in \mathcal{P}_{G(M_0)}} \mathrm{Ker}\,(\delta_e \times \mu) \subset \mathrm{Ker}\,(\delta_e \times \tau). \qquad (4.17)$$

Since $\mathrm{Ker}\,\pi_{\delta_e \times \tau}$ is the largest two-sided ideal in $\mathrm{Ker}\,(\delta_e \times \tau)$, relations (4.17) imply

$$\bigcap_{\mu \in \mathcal{P}_{G(M_0)}} \mathrm{Ker}\,\pi_{\delta_e \times \mu} \subset \mathrm{Ker}\,\pi_{\delta_e \times \tau} \subset \mathrm{Ker}\,(\delta_e \times \tau),$$

whence

$$\bigcap_{\mu \in \mathcal{P}_{G(M_0)}} \operatorname{Ker} \pi_{\delta_e \times \mu} \subset \bigcap_{\tau \in P_{\mathcal{A}}} \operatorname{Ker} \pi_{\delta_e \times \tau}.$$

From here it follows in view of the obvious converse inclusion that

$$\bigcap_{\tau \in P_{\mathcal{A}}} \operatorname{Ker} \pi_{\delta_e \times \tau} = \bigcap_{\mu \in \mathcal{P}_{G(M_0)}} \operatorname{Ker} \pi_{\delta_e \times \mu}. \tag{4.18}$$

Finally, we infer from (4.18) and (4.14) that

$$\bigcap_{\tau \in P_{\mathcal{A}}} \operatorname{Ker} \pi_{\delta_e \times \tau} = \bigcap_{\mu \in M_0} \operatorname{Ker} \pi_{\delta_e \times \mu}. \tag{4.19}$$

Thus Theorem 2.2(iii) and (4.19) give the following.

Lemma 4.11. *Under assumptions* (A1), (A2) *and* (A3′),

$$J_{M_0} := \bigcap_{\mu \in M_0} \operatorname{Ker} \pi_{\delta_e \times \mu} = \{0\}.$$

Theorem 4.12. *If* (A1), (A2) *and* (A3′) *are satisfied, Theorem 4.1 remains valid under the replacement of* Ω *by* $\Omega_0 := \Omega(M_0)$.

Proof. As in Theorem 4.1, it is sufficient to consider only the case of the two-sided invertibility of an operator $b \in \mathcal{B}$.

Consider the representation $\pi = \bigoplus_{\omega \in \Omega_0} \pi_\omega$ of the C^*-algebra $\mathcal{B}$ in the Hilbert space $\bigoplus_{\omega \in \Omega_0} l^2(G, H_\omega)$. Show that π is an isomorphism of $\mathcal{A}$ onto $\pi(\mathcal{A})$. To this end it is sufficient to prove that $\|\pi(a)\| = \|a\|$ for all $a \in \mathcal{A}$.

By the definition of the set M_0, any pure state $\mu \in P_{\mathcal{A}}$ is contained in the weak* closure of the set $\mathcal{P}_{G(M_0)}$. Then it follows from [11, Proposition 2.7.1] that

$$\|a\| = \sup \left\{ \sqrt{\mu(a^*a)} : \mu \in P_{\mathcal{A}} \right\} = \sup \left\{ \sqrt{\mu(a^*a)} : \mu \in \mathcal{P}_{G(M_0)} \right\}.$$

Repeating the corresponding part of the proof of Theorem 4.1 we obtain

$$\|\pi(a)\| = \sup \left\{ \|\pi_\omega(a)\| : \omega \in \Omega_0 \right\} = \sup \left\{ \sqrt{\mu(a^*a)} : \mu \in \mathcal{P}_{G(M_0)} \right\} = \|a\|.$$

Since (A1)–(A4) hold, we get $\mathcal{B} \cong \pi(\mathcal{B})$. Hence an element $b \in \mathcal{B}$ is invertible in $\mathcal{B}$ if and only if for every orbit $\omega \in \Omega_0$ the operator $\pi_\omega(b)$ is invertible on the space $l^2(G, H_\omega)$, and $\sup\{\|(\pi_\omega(b))^{-1}\| : \omega \in \Omega_0\} < \infty$ in the case of infinite Ω_0. $\square$

Remark 4.13. Theorem 4.2 and its proof remain true under the replacements of (A3) by (A3′) and Ω by Ω_0.

Let (A6′) mean (A6) with M replaced by M_0.

Theorem 4.14. *If* (A1), (A2), (A3′) *and* (A6′) *hold, then for any irreducible representation* π *of the* C^*-*algebra* $\mathcal{B}$ *there exists a* G-*orbit* $\omega \in \Omega_0$ $(= \Omega(M_0))$ *for which* (4.1) *is satisfied.*

Proof. For the validity of Theorem 4.14, according to Steps 1–3 and 7 of the proof of Theorem 4.8, it is sufficient to show that for any pure state $p \in P_{\mathcal{B}}$ there exists a point $t \in M_0$ for which (4.11) holds.

Further we proceed to Steps 8–9 of the proof of Theorem 4.8 with the following modifications. In virtue of Lemma 4.11 and [11, Proposition 3.4.2(ii)], the pure state pF of the C^*-algebra $\mathcal{A} \otimes_\alpha G$ is the weak* limit of a state net $\eta_s = (\pi_{\delta_e \times \mu_s}(\cdot)\xi_s, \xi_s)$, where $\mu_s \in \mathcal{P}_{m_s}$, $m_s \in M_0$, and $\xi_s \in H_{\delta_e \times \mu_s}$, $\|\xi_s\| = 1$.

In view of the closedness of M_0 in the compact topological space M, the set M_0 is a compact subspace of M. Then either the corresponding net $\{m_s\} \subset M_0$ contains a subnet which is entirely contained in some orbit $\omega \in \Omega_0$, or we can extract a subnet of the net $\{m_s\} \subset M_0$, which converges to a point $m \in M_0$.

Repeating the remaining part of arguments of Steps 8–9 in the proof of Theorem 4.8, we complete the proof of Theorem 4.14. $\qquad\square$

Theorem 4.14 and Remark 4.13 imply the following.

Theorem 4.15. *If* (A1), (A2), (A3$'$) *hold and, in case* $\operatorname{card}\Omega_0 = \infty$, *assumption* (A6$'$) *is also satisfied, then an element* $b \in \mathcal{B}$ *is invertible (left invertible, right invertible) in the* C^*-*algebra* $\mathcal{B}$ *if and only if for every orbit* $\omega \in \Omega_0$, *the operator* $\pi_\omega(b)$ *is invertible (left invertible, right invertible) on the space* $l^2(G, H_\omega)$.

5. The isomorphism theorem and trajectorial localization again

In this section we establish analogues of the isomorphism theorem and the trajectorial localization method in case (A3) is violated. These analogues serve studying the invertibility of functional operators with shifts having massive sets of fixed points, and also studying the Fredholmness of convolution type operators (in particular, singular integral operators) with shifts and discontinuous coefficients and presymbols (see, e.g., [16], [17]).

Since every C^*-algebra has a faithful (injective) representation on a Hilbert space (see, e.g., [24, Theorem 3.4.1]), we assume in this section without loss of generality that $\mathcal{B} = C^*(\mathcal{A}, U_G)$ is a C^*-subalgebra of the C^*-algebra $\mathcal{B}(H)$ of all bounded linear operators acting on a Hilbert space H.

5.1. Auxiliary results

Consider the σ-algebra $\mathfrak{R}(M)$ of all Borel sets of the compact M of maximal ideals of a central subalgebra $\mathcal{Z} \subset \mathcal{A}$, and also the spectral measure $P(\cdot) : \mathfrak{R}(M) \to \mathcal{B}(H)$ corresponding to the algebra $\mathcal{Z}$ and its identical representation in the Hilbert space H. Then for all $z \in \mathcal{Z}$, all $\Delta \in \mathfrak{R}(M)$, and all $\xi \in H$,

$$z = \int_M z(m)dP(m), \quad P(\Delta) = \int_\Delta dP(m), \quad zP(\Delta) = P(\Delta)z. \qquad (5.1)$$

Given a set $\Delta \subset M$, let $\partial\Delta$ denote the boundary of Δ, $\operatorname{Int}\Delta = \Delta \setminus \partial\Delta$ the interior of Δ, Δ' the set of all cluster points of Δ, $\overline{\Delta} = \Delta \cup \Delta'$ the closure of Δ, and $\Delta^c := M \setminus \Delta$ the complement of Δ. Let now $\Delta \in \mathfrak{R}(M)$ and let $\widetilde{\Delta}$ denote the set of all points $m \in M$ such that $P(u_m \cap \Delta) \neq 0$ for every open neighborhood $u_m \subset M$ of m. Obviously, $\widetilde{\Delta}$ is a closed subset of $\overline{\Delta}$.

Lemma 5.1. *If* $\Delta \in \mathfrak{R}(M)$ *and* $P(\Delta) \neq 0$, *then* $\widetilde{\Delta} \neq \emptyset$ *and* $\mathcal{Z}_\Delta := P(\Delta)\mathcal{Z} \cong C(\widetilde{\Delta})$.

Proof. Let $\Delta \in \mathfrak{R}(M)$ and $P(\Delta) \neq 0$. If $\widetilde{\Delta} = \emptyset$, then choosing a finite covering of the compact $\overline{\Delta}$ by open neighborhoods $u_k = u_{m_k}$ $(k = 1, 2, \ldots, n)$ satisfying $P(u_k \cap \Delta) = 0$, we arrive at the contradiction

$$0 \leq P(\Delta) \leq \sum\nolimits_{k=1}^{n} P(u_k \cap \Delta) = 0.$$

With every element $P(\Delta)z \in \mathcal{Z}_\Delta$ we associate the function $z(\cdot)|_{\widetilde{\Delta}} \in C(\widetilde{\Delta})$, where $z(\cdot)$ is the Gelfand transform of $z \in \mathcal{Z}$. This mapping is well defined. Indeed, let u_t be an arbitrary open neighborhood of a point $t \in \widetilde{\Delta}$, and let $\xi \in H$. Then the equality $P(\Delta)z = 0$ implies in view of the relations $P(u_t \cap \Delta) \neq 0$, (5.1), and

$$\left\| P(\Delta)z\xi \right\|^2 = \left\| \int_\Delta z(m)dP(m)\xi \right\|^2 = \int_\Delta |z(m)|^2 d\big(P(m)\xi, \xi\big)$$

$$\geq \int_{u_t \cap \Delta} |z(m)|^2 d\big(P(m)\xi, \xi\big) \geq \left(\min \left\{ |z(m)| : m \in \overline{u_t \cap \Delta} \right\} \left\| P(u_t \cap \Delta)\xi \right\| \right)^2$$

that $\min \left\{ |z(m)| : m \in \overline{u_t \cap \Delta} \right\} = 0$ for every point $t \in \widetilde{\Delta}$ and every neighborhood u_t. Then $z(t) = 0$ for all $t \in \widetilde{\Delta}$. Thus the map $\psi : P(\Delta)z \mapsto z(\cdot)|_{\widetilde{\Delta}}$ is an algebraic *-homomorphism of the C^*-algebra $\mathcal{Z}_\Delta$ onto $C(\widetilde{\Delta})$.

Prove the injectivity of ψ. Let $z \in \mathcal{Z}$ and $z(t) = 0$ for all $t \in \widetilde{\Delta}$. Let $M_\varepsilon = \{t \in M : |z(t)| \geq \varepsilon\}$. Then $\overline{\Delta} \cap M_\varepsilon$ is a compact subset of $M \setminus \widetilde{\Delta}$, whence its covering by open neighborhoods u_t that satisfying $P(u_t \cap \Delta) = 0$ contains a finite subcovering u_{t_k} $(k = 1, 2, \ldots, n)$. Set $u = \bigcup_{k=1}^{n} u_{t_k}$. Then $P(u \cap \Delta) = 0$. Since $P(\Delta \setminus u) = P(\Delta) \neq 0$ and hence $\|P(\Delta \setminus u)\| = 1$, we obtain

$$\|P(\Delta)z\| = \|P(\Delta \setminus u)z\| = \left\| \int_{\Delta \setminus u} z(m)dP(m) \right\|$$

$$\leq \max \left\{ |z(m)| : m \in \overline{\Delta \setminus u} \right\} \|P(\Delta \setminus u)\| \leq \varepsilon.$$

Consequently, $\|P(\Delta)z\| = 0$ because ε is arbitrary, whence $\operatorname{Ker}\psi = \{0\}$. Then, by [11, Corollary 1.8.3], ψ is an isometric *-isomorphism of $\mathcal{Z}_\Delta$ onto $C(\widetilde{\Delta})$. $\qquad \square$

By Lemma 5.1, the maximal ideal space $M(\mathcal{Z}_\Delta)$ of the C^*-algebra $\mathcal{Z}_\Delta$ can be identified with $\widetilde{\Delta}$.

Lemma 5.2. *If $\Delta \in \mathfrak{R}(M)$ and $\operatorname{Int}\Delta \neq \emptyset$, then*

(i) $P(\Delta) \neq 0$,

(ii) $\overline{\operatorname{Int}\Delta} \subset \widetilde{\Delta} \subset \overline{\Delta}$,

(iii) $\widetilde{\Delta} = \overline{\operatorname{Int}\Delta}$ *in case* $P(\Delta \setminus \overline{\operatorname{Int}\Delta}) = 0$.

Proof. Let Δ be a Borel subset of M with the non-empty interior.

(i) If $P(\Delta) = 0$, then $P(\Delta^c) = I$ for $\Delta^c = M \setminus \Delta$. Therefore, by Lemma 5.1, $P(\Delta^c)\mathcal{Z} \cong C(\widetilde{\Delta^c})$, whence $C(M) \cong \mathcal{Z} = P(\Delta^c)\mathcal{Z} \cong C(\widetilde{\Delta^c})$. But this is impossible because $\widetilde{\Delta^c} \subset \overline{\Delta^c}$ and $\overline{\Delta^c} \neq M$ in virtue of the relations $M \setminus \overline{\Delta^c} = \operatorname{Int}\Delta \neq \emptyset$. Consequently, $P(\Delta) \neq 0$.

(ii) Obviously, $\operatorname{Int}(u_t \cap \Delta) \neq \emptyset$ for every open neighborhood u_t of every point $t \in \overline{\operatorname{Int}\Delta}$. Then, by (i), $P(u_t \cap \Delta) \neq 0$, whence $t \in \widetilde{\Delta}$ and thus $\overline{\operatorname{Int}\Delta} \subset \widetilde{\Delta}$. On the other hand $\widetilde{\Delta} \subset \bar{\Delta}$, and (ii) has been proved.

(iii) Let $\operatorname{Int}\Delta \neq \emptyset$ and $P(\Delta \setminus \overline{\operatorname{Int}\Delta}) = 0$. If $t \in \bar{\Delta} \setminus \overline{\operatorname{Int}\Delta}$, there is an open neighborhood $u_t \subset M$ of t such that $u_t \cap \overline{\operatorname{Int}\Delta} = \emptyset$. Then $u_t \cap \Delta = u_t \cap (\Delta \setminus \overline{\operatorname{Int}\Delta})$, whence

$$0 \leq P(u_t \cap \Delta) = P(u_t \cap (\Delta \setminus \overline{\operatorname{Int}\Delta})) \leq P(\Delta \setminus \overline{\operatorname{Int}\Delta}) = 0.$$

Thus $P(u_t \cap \Delta) = 0$, which means that $\widetilde{\Delta} \cap (\bar{\Delta} \setminus \overline{\operatorname{Int}\Delta}) = \emptyset$. Then using (ii) we get

$$\widetilde{\Delta} = \widetilde{\Delta} \cap \bar{\Delta} = (\widetilde{\Delta} \cap \overline{\operatorname{Int}\Delta}) \cup (\widetilde{\Delta} \cap (\bar{\Delta} \setminus \overline{\operatorname{Int}\Delta})) = \widetilde{\Delta} \cap \overline{\operatorname{Int}\Delta} = \overline{\operatorname{Int}\Delta},$$

which completes the proof. $\qquad\square$

Corollary 5.3. *If Δ is a non-empty open subset of M, then* $\overline{\operatorname{Int}\Delta} = \widetilde{\Delta} = \bar{\Delta}$.

5.2. An analogue of the isomorphism theorem: (A3) is violated

By Corollary 4.7, $P(\Delta)bP(\Delta) = P(\Delta)b = bP(\Delta)$ for every $b \in \mathcal{B} = C^*(\mathcal{A}, U_G)$ if $\Delta \in \mathfrak{R}_G(M)$. Let $\Delta \in \mathfrak{R}_G(M)$, $P(\Delta) \neq 0$, and $G_\Delta := \{g \in G : P(\Delta)U_g = P(\Delta)\}$. Obviously, G_Δ is a normal subgroup of G. Consider the unitary representation

$$U_\Delta : G/G_\Delta \to \mathcal{B}(P(\Delta)H), \quad h \mapsto U_{h,\Delta}$$

where $U_{h,\Delta} := P(\Delta)U_g$ for every g belonging to a coset $h \in G/G_\Delta$. In virtue of (A1) and Corollary 4.7, for every $h \in G/G_\Delta$, the mapping $a \mapsto U_{h,\Delta}aU_{h,\Delta}^*$ is a $*$-automorphism of the C^*-algebra $\mathcal{Z}_\Delta$, where $\mathcal{Z}_\Delta \cong C(\widetilde{\Delta})$ by Lemma 5.1. Therefore, the quotient group G/G_Δ admits a homomorphic realization as a group of homeomorphisms of the compact $\widetilde{\Delta} \subset M$ onto itself by the rule:

$$h(x) = g(x) \quad \text{for all } x \in \widetilde{\Delta}, \ g \in h, \ h \in G/G_\Delta.$$

In G/G_Δ we extract the normal subgroup

$$\widetilde{G}_\Delta := \{h \in G/G_\Delta : h(x) = x \text{ for all } x \in \widetilde{\Delta}\}$$

and denote by $\mathcal{A}_\Delta$ the C^*-algebra $P(\Delta)C^*(\mathcal{A}, U_{G_\Delta^0})$ generated by the operators $P(\Delta)a$ $(a \in \mathcal{A})$ and $P(\Delta)U_g$ $(g \in G_\Delta^0)$ where

$$G_\Delta^0 := \{g \in G : g(x) = x \text{ for all } x \in \widetilde{\Delta}\}.$$

Obviously, $G_\Delta \subset G_\Delta^0$ and $\widetilde{G}_\Delta = G_\Delta^0/G_\Delta$. From Lemma 4.6 and the equalities

$$P(\Delta)U_g z U_g^* = P(\Delta)z \quad (z \in \mathcal{Z}, \ g \in G_\Delta^0)$$

it follows that $\mathcal{Z}_\Delta$ is a central subalgebra of the C^*-algebra $\mathcal{A}_\Delta$.

Given groups G_1, G_0 and homomorphism $\psi : G_1 \to \operatorname{Aut} G_0$, the *semidirect product* $G_1 \rtimes G_0$ of the groups G_1 and G_0 is defined as the Cartesian product $G_1 \times G_0$ with the group operation $(g_1, g_0)(g_1', g_0') = (g_1 g_1', [\psi(g_1')^{-1} g_0]g_0')$, where $g_1, g_1' \in G_1$, $g_0, g_0' \in G_0$ [21, p. 20]. Then $G \rtimes G_0$ also is a group. Assume that

(A7) *the set M can be partitioned into an at most countable collection $\mathcal{R}$ of subsets $\Delta \in \mathfrak{R}_G(M)$ such that for every $\Delta \in \mathcal{R}$,*

(i) $P(\Delta) \neq 0$ and the quotient group G/G_Δ is represented as a semidirect product $\widehat{G}_\Delta \rtimes \widetilde{G}_\Delta$ of its subgroups $\widehat{G}_\Delta$ and $\widetilde{G}_\Delta$;

(ii) if $\widehat{G}_\Delta \neq \{e\}$, then (A3) is satisfied under the replacements of $\mathcal{A}, \mathcal{Z}, G, M$ by $\mathcal{A}_\Delta, \mathcal{Z}_\Delta, \widehat{G}_\Delta, \widehat{\Delta}$, respectively.

If (A7) holds and $\Delta \in \mathcal{R}$, then every group $\widehat{G}_\Delta$ is isomorphic to the quotient group $(G/G_\Delta)/\widetilde{G}_\Delta$, and in virtue of (A2) the groups G/G_Δ, $\widehat{G}_\Delta$ and $\widetilde{G}_\Delta$ are amenable (see, e.g., [15, § 1.2]). From condition (i) in (A7) it follows that for every $h \in \widehat{G}_\Delta$ and every $s \in \widetilde{G}_\Delta$ there exists a $q \in \widetilde{G}_\Delta$ such that $U_{h,\Delta} U_{s,\Delta} U_{h,\Delta}^* = U_{q,\Delta}$. Then according to (A1) every element $h \in \widehat{G}_\Delta$ generates the *-automorphisms $\alpha_{h,\Delta} : a \mapsto U_{h,\Delta} a U_{h,\Delta}^*$ of the C^*-algebra $\mathcal{A}_\Delta$ and its central subalgebra $\mathcal{Z}_\Delta$. Therefore under assumptions (A1), (A2), (A7), the C^*-algebras $\mathcal{B}_\Delta := P(\Delta)\mathcal{B}$ possess all the properties (A1)–(A3). Consequently, from Theorem 3.2 it follows that the C^*-algebras $\mathcal{B}_\Delta = P(\Delta)C^*(\mathcal{A}, U_G)$ are *-isomorphic to the C^*-algebras $\mathcal{A}_\Delta \otimes_{\alpha_\Delta} \widehat{G}_\Delta$ where $\alpha_\Delta : h \mapsto \alpha_{h,\Delta}$ is a homomorphism of the group $\widehat{G}_\Delta$ into the group of *-automorphisms of the C^*-algebra $\mathcal{A}_\Delta$.

Again along with the C^*-algebra $\mathcal{B} = C^*(\mathcal{A}, U_G)$ we consider the C^*-algebra $\mathcal{B}' = C^*(\mathcal{A}', U_G')$. Let φ be an isomorphism of the C^*-algebra $\mathcal{A}$ onto the C^*-algebra $\mathcal{A}'$, which satisfies (A4), and let $P(\Delta)$ and $P'(\Delta)$ be the spectral measures for the central algebras $\mathcal{Z} \subset \mathcal{A}$ and $\mathcal{Z}' = \varphi(\mathcal{Z}) \subset \mathcal{A}'$. By (A4), the automorphisms

$$\alpha_g : a \mapsto U_g a U_g^* \quad \text{and} \quad \alpha_g' : \varphi(a) \mapsto U_g' \varphi(a)(U_g')^*$$

induce the same action of the group G on M. Then (see Corollary 4.7)

$$b' P'(\Delta) = P'(\Delta)b' \quad \text{for all } \Delta \in \mathfrak{R}_G(M) \text{ and all } b' \in \mathcal{B}'.$$

Theorem 5.4. *If assumptions* (A1), (A2), (A4), (A7) *are satisfied and for every* $\Delta \in \mathcal{R}$ *the isomorphism* $\varphi : \mathcal{A} \to \mathcal{A}'$ *induces an isomorphism* φ_Δ *of the C^*-algebra* $\mathcal{A}_\Delta$ *onto the C^*-algebra* $\mathcal{A}_\Delta' := P'(\Delta)C^*(\mathcal{A}', U_{G_\Delta^0}')$ *that satisfies the conditions*

$$\varphi_\Delta(P(\Delta)a) = P'(\Delta)\varphi(a), \quad \varphi_\Delta(U_{h,\Delta}U_{s,\Delta}U_{h,\Delta}^*) = U_{h,\Delta}'U_{s,\Delta}'(U_{h,\Delta}')^* \qquad (5.2)$$

for all $a \in \mathcal{A}$, *all* $h \in \widehat{G}_\Delta$, *and all* $s \in \widetilde{G}_\Delta$, *then* $\mathcal{B} \cong \mathcal{B}'$.

Proof. Clearly, if $\widehat{G}_\Delta = \{e\}$, then φ_Δ is an isomorphism of $\mathcal{B}_\Delta$ onto $\mathcal{B}_\Delta' = P'(\Delta)\mathcal{B}'$.

Let $\widehat{G}_\Delta \neq \{e\}$. For every $a \in \mathcal{A}$, every $h \in \widehat{G}_\Delta$ and every $g \in h$, we infer from Corollary 4.7, from the first equality in (5.2), and from (A4) that

$$\varphi_\Delta\big(U_{h,\Delta}P(\Delta)aU_{h,\Delta}^*\big) = \varphi_\Delta\big(P(\Delta)(U_g a U_g^*)\big) = P'(\Delta)\varphi(U_g a U_g^*)$$

$$= P'(\Delta)U_g' \varphi(a)(U_g')^* = U_{h,\Delta}'P'(\Delta)\varphi(a)(U_{h,\Delta}')^* = U_{h,\Delta}'\varphi_\Delta(P(\Delta)a)(U_{h,\Delta}')^*.$$

Further, the second equality in (5.2) implies that

$$\varphi_\Delta(U_{h,\Delta}U_{s,\Delta}U_{h,\Delta}^*) = U_{h,\Delta}'\varphi_\Delta(U_{s,\Delta})(U_{h,\Delta}')^* \quad \text{for all } h \in \widehat{G}_\Delta \text{ and all } s \in \widetilde{G}_\Delta.$$

Thus, for every $\Delta \in \mathcal{R}$ the isomorphism $\varphi_\Delta : \mathcal{A}_\Delta \to \mathcal{A}_\Delta'$ satisfies (A4) of the form

$(A4)_\Delta \quad \varphi_\Delta(U_{h,\Delta} a U_{h,\Delta}^*) = U_{h,\Delta}'\varphi_\Delta(a)(U_{h,\Delta}')^*$ for all $a \in \mathcal{A}_\Delta$ and all $h \in \widehat{G}_\Delta$.

Since assumptions (A1)–(A3) for the C^*-algebras $\mathcal{B}_\Delta$ also hold, we infer from Theorem 3.3 that each mapping φ_Δ uniquely extends to a *-isomorphism Φ_Δ of the C^*-algebra $\mathcal{B}_\Delta$ onto $\mathcal{B}'_\Delta$ if to put $\Phi_\Delta(U_{h,\Delta}) = U'_{h,\Delta}$ for every $h \in \widehat{G}_\Delta$.

With every operator $b = \sum a_g U_g$ $(a_g \in \mathcal{A})$ from the non-closed subalgebra $\mathcal{B}^0 \subset \mathcal{B}$ we associate the operator $\Phi(b) = \sum \varphi(a_g) U'_g \in \mathcal{B}'$. By Corollary 4.7, the mapping $b \mapsto \bigoplus_{\Delta \in \mathcal{R}} P(\Delta)b$ is an isometrically isomorphic imbedding of $\mathcal{B}$ into $\bigoplus_{\Delta \in \mathcal{R}} \mathcal{B}_\Delta$. On the other hand, $\Phi_\Delta(P(\Delta)b) = P'(\Delta)\Phi(b)$ for all $b \in \mathcal{B}^0$ and all $\Delta \in \mathcal{R}$. Then for all $b \in \mathcal{B}^0$ we obtain

$$\|b\| = \sup_{\Delta \in \mathcal{R}} \|P(\Delta)b\| = \sup_{\Delta \in \mathcal{R}} \|\Phi_\Delta(P(\Delta)b)\| = \sup_{\Delta \in \mathcal{R}} \|P'(\Delta)\Phi(b)\| = \|\Phi(b)\|.$$

Hence, Φ is an isometric *-isomorphism of $\mathcal{B}^0$ onto $(\mathcal{B}')^0$, which extends by continuity to a *-isomorphism of $\mathcal{B}$ onto $\mathcal{B}'$. $\square$

5.3. Trajectorial localization: (A3) is violated

Let $\Omega_\Delta := \Omega(\widetilde{\Delta})$ be the set of $\widehat{G}_\Delta$-orbits of points $m \in \widetilde{\Delta}$. Note that each $\widehat{G}_\Delta$-orbit $\widehat{G}_\Delta(m)$ coincides with the G-orbit $G(m)$ for $m \in \widetilde{\Delta}$ if $\Delta \in \mathfrak{R}_G(M)$. For every orbit $\omega \in \Omega_\Delta$, by analogy with the space $l^2(G, H_\omega)$ and the representation π_ω, we construct the space $l^2(\widehat{G}_\Delta, H_{\omega,\Delta})$ and the representation $\pi_{\omega,\Delta}$ of the algebra $\mathcal{B}_\Delta$ in $l^2(\widehat{G}_\Delta, H_{\omega,\Delta})$, replacing $\mathcal{A}, G, U, M$ by $\mathcal{A}_\Delta, \widehat{G}_\Delta, U_\Delta, \widetilde{\Delta}$. Namely, let $m = m_\omega$ be an arbitrary point of an orbit $\omega \in \Omega_\Delta$, let $H_{\omega,\Delta} := \mathcal{H}_{m,\Delta}$ be a Hilbert space of an isometric representation $\widetilde{\pi}_{m,\Delta}$ of the quotient algebra $\mathcal{A}_\Delta/J_{m,\Delta}$ where $J_{m,\Delta}$ is the closed two-sided ideal of the algebra $\mathcal{A}_\Delta$ generated by the maximal ideal $m \in \widetilde{\Delta}$ of the central subalgebra $\mathcal{Z}_\Delta$, let $\varrho_{m,\Delta}$ be the natural homomorphism $\mathcal{A}_\Delta \to \mathcal{A}_\Delta/J_{m,\Delta}$, and let $\pi'_{m,\Delta} := \widetilde{\pi}_{m,\Delta} \circ \varrho_{m,\Delta}$. Then the desired representation $\pi_{\omega,\Delta} : \mathcal{B}_\Delta \to \mathcal{B}\big(l^2(\widehat{G}_\Delta, H_{\omega,\Delta})\big)$ is given for $f \in l^2(\widehat{G}_\Delta, H_{\omega,\Delta})$ by

$$[\pi_{\omega,\Delta}(a)f](g) = \pi'_{m,\Delta}[\alpha_{g,\Delta}(a)]f(g), \quad [\pi_{\omega,\Delta}(U_{h,\Delta})f](g) = f(gh),$$

where $a \in \mathcal{A}_\Delta$ and $h, g \in \widehat{G}_\Delta$. Given $b \in \mathcal{B}$ and $\Delta \in \mathfrak{R}_G(M)$, we set $b_\Delta := P(\Delta)b$.

Theorem 5.5. *If assumptions* (A1), (A2), (A7) *are fulfilled, then an operator* $b \in \mathcal{B}$ *is invertible (left invertible, right invertible) on the space* H *if and only if for every* $\Delta \in \mathcal{R}$ *and every* $\omega \in \Omega_\Delta$, *the operator* $\pi_{\omega,\Delta}(b_\Delta)$ *is invertible (left invertible, right invertible) on the space* $l^2(\widehat{G}_\Delta, H_{\omega,\Delta})$ *and, in the case of infinite* Ω_Δ *or infinite* $\mathcal{R}$,

$$\sup \big\{\big\|(\pi_{\omega,\Delta}(b_\Delta))^{-1}\big\| : \omega \in \Omega_\Delta, \ \Delta \in \mathcal{R}\big\} < \infty$$

(respectively, there is a uniformly bounded family $\{\pi^l_{\omega,\Delta}(b_\Delta)\}_{\omega \in \Omega_\Delta,\, \Delta \in \mathcal{R}}$ *of left inverse operators with self-adjoint projections* $\pi^l_{\omega,\Delta}(b_\Delta)\pi_{\omega,\Delta}(b_\Delta)$, *or there is a uniformly bounded family* $\{\pi^r_{\omega,\Delta}(b_\Delta)\}_{\omega \in \Omega_\Delta,\, \Delta \in \mathcal{R}}$ *of right inverse operators with self-adjoint projections* $\pi^r_{\omega,\Delta}(b_\Delta)\pi_{\omega,\Delta}(b_\Delta))$.

Proof. An operator $b \in \mathcal{B}$ is invertible (left invertible, right invertible) on the space H if and only if for every $\Delta \in \mathcal{R}$ the operator $b_\Delta \in \mathcal{B}_\Delta$ is invertible (left invertible, right invertible) on the space $P(\Delta)H$ and, in the case of infinite $\mathcal{R}$, the norms $\|b_\Delta^{-1}\|$ (respectively, $\|(b_\Delta^* b_\Delta)^{-1}\|$, $\|(b_\Delta b_\Delta^*)^{-1}\|$) are uniformly bounded on $\mathcal{R}$. Since for every C^*-algebra $\mathcal{B}_\Delta$ assumptions (A1)–(A3) are satisfied, Theorem 5.5 directly follows from Theorem 4.1. $\qquad\square$

Recall that (A1) and (A7) imply that $\widehat{G}_\Delta$-orbits of points $m \in \widetilde{\Delta}$ coincide with their G-orbits $G(m)$ for every $\Delta \in \mathcal{R}$. Furthermore, if the C^*-algebra $\mathcal{Z}$ is separable, so are the C^*-algebras $Z_\Delta = P(\Delta)Z$. Therefore, by analogy with Theorem 5.5, one can prove the following criterion on the basis of Theorem 4.9.

Theorem 5.6. *Suppose* (A1), (A2), (A7) *hold, the set* $\mathcal{R}$ *is finite and, for every* $\Delta \in \mathcal{R}$ *such that* Ω_Δ *is infinite, either*

$(A5)_\Delta$ *the* C^*-*algebra* $\mathcal{Z}$ *is separable and* $\bigcap_{m\in\omega} J_{m,\Delta} = \bigcap_{m\in\overline{\omega}} J_{m,\Delta}$ *for every orbit* $\omega \in \Omega_\Delta$ *such that* $\overline{\omega} = \omega'$; *or*

$(A6)_\Delta$ *for every point* $m \in \widetilde{\Delta}$, *every net* $m_s \in \widetilde{\Delta}$ *which tends to* m *and does not contain a subnet consisting of points of one and the same* G-*orbit, and every operator* $a \in \bigcap_{\tau\in G(m)} J_{\tau,\Delta}$ *there exists a subnet* $m_v \to m$ *such that* $\lim_v \|\pi_{\omega_v,\Delta}(a)\| = 0$ *where* $\omega_v = G(m_v)$.

Then an operator $b \in \mathcal{B}$ *is invertible (left invertible, right invertible) on the space* H *if and only if for every* $\Delta \in \mathcal{R}$ *and every* $\omega \in \Omega_\Delta$ *the operator* $\pi_{\omega,\Delta}(b_\Delta)$ *is invertible (left invertible, right invertible) on the space* $l^2(\widehat{G}_\Delta, H_{\omega,\Delta})$.

Theorem 5.6 implies the following analogue of Corollary 4.10.

Corollary 5.7. *Suppose* (A1), (A2), (A7) *hold, the set* $\mathcal{R}$ *is finite, the* C^*-*algebra* $\mathcal{Z}$ *is separable and* $\omega' \neq \overline{\omega}$ *for every* $\omega \in \Omega_\Delta$ *and every* $\Delta \in \mathcal{R}$ *having infinite set* Ω_Δ. *Then an operator* $b \in \mathcal{B}$ *is invertible (left invertible, right invertible) on the space* H *if and only if for every* $\Delta \in \mathcal{R}$ *and every* $\omega \in \Omega_\Delta$ *the operator* $\pi_{\omega,\Delta}(b_\Delta)$ *is invertible (left invertible, right invertible) on the space* $l^2(\widehat{G}_\Delta, H_{\omega,\Delta})$.

Finally, let $(A3)_\Delta$ mean (A3) with $\mathcal{A}, \mathcal{Z}, G, M$ replaced by $\mathcal{A}_\Delta, \mathcal{Z}_\Delta, \widehat{G}_\Delta, \widetilde{\Delta}$, respectively, and let $M_{0,\Delta}$ be a closed subset of the compact $\widetilde{\Delta}$ such that the point m_ν in $(A3)_\Delta$ belongs to the G-orbit of the set $M_{0,\Delta}$. Set $\Omega_{0,\Delta} := \Omega(M_{0,\Delta})$.

Remark 5.8. From Theorem 4.12 it follows that Theorem 5.5 remains valid if to replace Ω_Δ by $\Omega_{0,\Delta}$.

Theorem 4.15 immediately implies the following.

Theorem 5.9. *If* (A1), (A2), (A7) *hold, the set* $\mathcal{R}$ *is finite and, for every* $\Delta \in \mathcal{R}$ *such that* Ω_Δ *is infinite, assumption* $(A6)_\Delta$ *is satisfied under the replacement of* $\widetilde{\Delta}$ *by* $M_{0,\Delta}$, *then an operator* $b \in \mathcal{B}$ *is invertible (left invertible, right invertible) on the space* H *if and only if for every* $\Delta \in \mathcal{R}$ *and every* $\omega \in \Omega_{0,\Delta}$ *the operator* $\pi_{\omega,\Delta}(b_\Delta)$ *is invertible (left invertible, right invertible) on the space* $l^2(\widehat{G}_\Delta, H_{\omega,\Delta})$.

Remark 5.10. Presented results allow us to study the Fredholmness (n-normality, d-normality) of operators $b \in \mathcal{B}(H)$ on a Hilbert space H, which is equivalent to the invertibility (left invertibility, right invertibility) of the cosets $b + \mathcal{K}(H)$ in the Calkin algebra $\mathcal{B}(H)/\mathcal{K}(H)$ where $\mathcal{K}(H)$ is the ideal of compact operators in $\mathcal{B}(H)$.

References

[1] G.R. Allan, *Ideals of vector-valued functions*. Proc. London Math. Soc., 3rd ser. **18** (1968), 193–216.

[2] A.B. Antonevich, *Linear Functional Equations. Operator Approach.* Operator Theory: Advances and Applications **83**, Birkhäuser Verlag, Basel 1996. Russian original: University Press, Minsk 1988.

[3] A. Antonevich and A. Lebedev, *Functional Differential Equations: I. C^*-Theory.* Pitman Monographs and Surveys in Pure and Applied Mathematics **70**, Longman Scientific & Technical, Harlow 1994.

[4] G.M. Adel'son-Vel'skii and Yu.A. Shreider, *The Banach mean on groups.* Uspekhi Mat. Nauk **12** (1957), no. 6, 131–136 [Russian].

[5] M.A. Bastos, C.A. Fernandez, and Yu.I. Karlovich, *C^*-algebras of integral operators with piecewise slowly oscillating coefficients and shifts acting freely.* Integral Equations and Operator Theory **55** (2006), 19–67.

[6] A. Böttcher and Yu.I. Karlovich, *Carleson Curves, Muckenhoupt Weights, and Toeplitz Operators.* Progress in Mathematics **154**, Birkhäuser Verlag, Basel 1997.

[7] A. Böttcher, Yu.I. Karlovich, and B. Silbermann, *Singular integral equations with PQC coefficients and freely transformed argument.* Math. Nachr. **166** (1994), 113–133.

[8] A. Böttcher, Yu.I. Karlovich, and I.M. Spitkovsky, *The C^*-algebra of singular integral operators with semi-almost periodic coefficients.* J. Funct. Analysis **204** (2003), 445–484.

[9] A. Böttcher and B. Silbermann, *Analysis of Toeplitz Operators.* Akademie-Verlag, Berlin 1989 and Springer-Verlag, Berlin 1990.

[10] O. Bratteli and D.W. Robinson, *Operator Algebras and Quantum Statistical Mechanics. I: C^*- and W^*-algebras, Symmetry Groups, Decomposition of States.* Springer-Verlag, New York 1979.

[11] J. Dixmier, *C^*-Algebras.* North-Holland Publishing Company, Amsterdam 1977.

[12] R.G. Douglas, *Banach Algebra Techniques in Operator Theory.* Academic Press, New York 1972.

[13] I. Gohberg and N. Krupnik, *One-Dimensional Linear Singular Integral Equations, Vols. 1 and 2.* Birkhäuser, Basel 1992; Russian original: Shtiintsa, Kishinev 1973.

[14] V.Ya. Golodets, *Crossed products of von Neumann algebras.* Russian Math. Surveys, **26** (1971), no. 5, 1–50.

[15] F.P. Greenleaf, *Invariant Means on Topological Groups and Their Representations.* Van Nostrand-Reinhold, New York 1969.

[16] Yu.I. Karlovich, *The local-trajectory method of studying invertibility in C^*-algebras of operators with discrete groups of shifts.* Soviet Math. Dokl. **37** (1988), 407–411.

[17] Yu.I. Karlovich, *C^*-algebras of operators of convolution type with discrete groups of shifts and oscillating coefficients.* Soviet Math. Dokl. **38** (1989), 301–307.

[18] Yu.I. Karlovich, *Algebras of Convolution Type Operators with Discrete Groups of Shifts and Oscillating Coefficients.* Doctoral dissertation, Math. Inst. Georgian Acad. Sci., Tbilisi, 1991 [Russian].

[19] Yu.I. Karlovich and B. Silbermann, *Local method for nonlocal operators on Banach spaces*. Toeplitz Matrices and Singular Integral Equations, The Bernd Silbermann Anniversary Volume, Operator Theory: Advances and Appl. **135** (2002), 235–247.

[20] Yu.I. Karlovich and B. Silbermann, *Fredholmness of singular integral operators with discrete subexponential groups of shifts on Lebesgue spaces*. Math. Nachr. **272** (2004), 55–94.

[21] A.A. Kirillov, *Elements of the Theory of Representations*. Springer-Verlag, Berlin 1976.

[22] V.G. Kravchenko and G.S. Litvinchuk, *Introduction to the Theory of Singular Integral Operators with Shift*. Mathematics and Its Applications **289**, Kluwer Academic Publishers, Dordrecht 1994.

[23] A.V. Lebedev, *On certain C^*-methods used for investigating algebras associated with automorphisms and endomorphisms*. Deposited in VINITI, No. 5351-B87, Minsk 1987.

[24] G.J. Murphy, *C^*-algebras and Operator Theory*. Academic Press, Boston 1990.

[25] M.A. Naimark, *Normed Algebras*. Wolters-Noordhoff Publishing, Groningen, The Netherlands 1972.

[26] G.K. Pedersen, *C^*-Algebras and Their Automorphism Groups*. Academic Press, London 1979.

[27] B.A. Plamenevsky, *Algebras of Pseudodifferential Operators*. Kluwer Academic Publishers, Dordrecht 1989.

[28] M. Reed and B. Simon, *Methods of Modern Mathematical Physics. I: Functional Analysis*. Academic Press, New York 1980.

[29] V.N. Semenyuta and A.V. Khevelev, *A local principle for special classes of Banach algebras*. Izv. Severo-Kavkazskogo Nauchn. Tsentra Vyssh. Shkoly, Ser. Estestv. Nauk **1** (1977), 15–17 [Russian].

[30] I.B. Simonenko, *A new general method of studying linear operator equations of the type of singular integral equations. Parts I, II*. Izv. Akad. Nauk SSSR, Ser. Mat. **29** (1965), 567–586; 757–782 [Russian].

[31] I.B. Simonenko and Chin Ngok Min, *Local Method in the Theory of One-Dimensional Singular Integral Equations with Piecewise Continuous Coefficients. Noetherity*. University Press, Rostov on Don 1986 [Russian].

Yu.I. Karlovich
Facultad de Ciencias
Universidad Autónoma del Estado de Morelos
Av. Universidad 1001, Col. Chamilpa,
C.P. 62209 Cuernavaca, Morelos, México
e-mail: `karlovich@buzon.uaem.mx`

Submitted: September 30, 2005

Operator Theory:
Advances and Applications, Vol. 170, 167–186
© 2006 Birkhäuser Verlag Basel/Switzerland

Boundedness in Lebesgue Spaces with Variable Exponent of the Cauchy Singular Operator on Carleson Curves

Vakhtang Kokilashvili, Vakhtang Paatashvili and Stefan Samko

Abstract. We prove the boundedness of the singular integral operator S_Γ in the spaces $L^{p(\cdot)}(\Gamma, \rho)$ with variable exponent $p(t)$ and power weight ρ on an arbitrary Carleson curve under the assumptions that $p(t)$ satisfy the log-condition on Γ. The curve Γ may be finite or infinite.

We also prove that if the singular operator is bounded in the space $L^{p(\cdot)}(\Gamma)$, then Γ is necessarily a Carleson curve. A necessary condition is also obtained for an arbitrary continuous coefficient.

Mathematics Subject Classification (2000). Primary 47B38; Secondary 42B20, 45P05.

Keywords. Weighted generalized Lebesgue spaces, variable exponent, singular operator, Carleson curves.

1. Introduction

Let $\Gamma = \{t \in \mathbb{C} : t = t(s),\ 0 \leq s \leq \ell \leq \infty\}$ be a simple rectifiable curve with arc-length measure $\nu(t) = s$. In the sequel we denote

$$\gamma(t, r) := \Gamma \cap B(t, r), \quad t \in \Gamma, \ r > 0, \tag{1.1}$$

where $B(t, r) = \{z \in \mathbb{C} : |z - t| < r\}$. We also denote for brevity

$$\nu(\gamma(t, r)) = |\gamma(t, r)|.$$

We remind that a curve is called Carleson curve (regular curve), if there exists a constant $c_0 > 0$ not depending on t and r, such that

$$|\gamma(t, r)| \leq c_0 r \tag{1.2}$$

We consider the singular integral operator

$$S_\Gamma f(t) = \frac{1}{\pi i} \int_\Gamma \frac{f(\tau)}{\tau - t} d\nu(\tau) \tag{1.3}$$

on Carleson curves Γ and prove that the operator S is bounded in weighted spaces $L^{p(\cdot)}(\Gamma, w)$, $w(t) = \prod_{k=1}^{n} |t - t_k|^{\beta_k}$, $t_k \in \Gamma$ with variable exponent $p(t)$ (see definitions in Section 2), under the assumption that $p(t)$ satisfies the standard log-condition. The curve Γ may be finite or infinite. In the latter case we assume also that $p(t)$ satisfies the log-condition at infinity.

2. Definitions

Let p be a measurable function on Γ such that $p : \Gamma \to (1, \infty)$. In what follows we assume that p satisfies the conditions

$$1 < p_- := \operatorname*{ess\,inf}_{t \in \Gamma} p(t) \leq \operatorname*{ess\,sup}_{t \in \Gamma} p(t) =: p_+ < \infty, \tag{2.1}$$

$$|p(t) - p(\tau)| \leq \frac{A}{\ln \frac{1}{|t - \tau|}}, \qquad t \in \Gamma, \tau \in \Gamma, \qquad |t - \tau| \leq \frac{1}{2}. \tag{2.2}$$

Observe that condition (2.1) may be also written in the form

$$|p(t) - p(\tau)| \leq \frac{2\ell A}{\ln \frac{2\ell}{|t - \tau|}}, \qquad t, \tau \in \Gamma, \tag{2.3}$$

where ℓ is the length of the curve.

In the case where Γ is an infinite curve, we also assume that p satisfies the following condition at infinity

$$|p(t) - p(\tau)| \leq \frac{A_\infty}{\ln \frac{1}{\left|\frac{1}{t} - \frac{1}{\tau}\right|}}, \qquad \left|\frac{1}{t} - \frac{1}{\tau}\right| \leq \frac{1}{2}, \qquad |t| \geq L, |\tau| \geq L \tag{2.4}$$

for some $L > 0$.

By $\mathcal{P} = \mathcal{P}(\Gamma)$ we denote the class of exponents p satisfying condition (2.1) and by $\mathbb{P} = \mathbb{P}(\Gamma)$ the class of those p for which the maximal operator M is bounded in the space $L^{p(\cdot)}(\Gamma)$.

The generalized Lebesgue space with variable exponent is defined via the modular

$$I_\Gamma^p (f) := \int_\Gamma |f(t)|^{p(t)} \, d\nu(\tau)$$

by the norm

$$\|f\|_{p(\cdot)} = \inf \left\{ \lambda > 0 : I_\Gamma^p \left(\frac{f}{\lambda} \right) \leq 1 \right\}.$$

Observe that

$$\|f\|_{p(\cdot)} = \|f^a\|_{\frac{p(\cdot)}{a}}^{\frac{1}{a}} \tag{2.5}$$

for any $0 < a \leq \inf p(t)$.

By $L^{p(\cdot)}(\Gamma, w)$ we denote the weighted Banach space of all measurable functions $f : \Gamma \to \mathbb{C}$ such that

$$\|f\|_{L^{p(\cdot)}(\Gamma,w)} := \|wf\|_{p(\cdot)} = \inf\left\{\lambda > 0 : \int_{\Gamma} \left|\frac{w(t)f(t)}{\lambda}\right|^{p(t)} d\nu(t) \le 1\right\} < \infty. \tag{2.6}$$

We denote $p'(t) = \frac{p(t)}{p(t)-1}$.

From the Hölder inequality for the $L^{p(\cdot)}$-spaces

$$\left|\int_{\Gamma} u(\tau)v(\tau)\, d\nu(\tau)\right| \le k\|u\|_{L^{p(\cdot)}(\Gamma)}\|v\|_{L^{p'(\cdot)}(\Gamma)}, \qquad \frac{1}{p(\tau)} + \frac{1}{p'(\tau)} \equiv 1,$$

where $k = 1 + \frac{1}{p_-} + \frac{1}{(p')_-} = 1 + \frac{1}{p_-} - \frac{1}{p_+} < 2$, it follows that

$$\left|\int_{\Gamma} u(t)v(t)\, d\nu(t)\right| \le k\|u\|_{L^{p'}(\Gamma,\frac{1}{w})}\|v\|_{L^p(\Gamma,w)}, \tag{2.7}$$

and for the conjugate space $\left[L^{p(\cdot)}(\Gamma, w)\right]^*$ we have

$$\left[L^{p(\cdot)}(\Gamma, w)\right]^* = L^{p'(\cdot)}(\Gamma, 1/w) \tag{2.8}$$

which is an immediate consequence of the fact that $\left[L^{p(\cdot)}(\Gamma)\right]^* = L^{p'(\cdot)}(\Gamma)$ under conditions (2.1), see [13], [16].

The following value

$$\frac{1}{p_\gamma} = \frac{1}{|\gamma|}\int_{\gamma} \frac{d\nu(t)}{p(t)}, \quad \gamma \subset \Gamma \tag{2.9}$$

will be used, introduced for balls in $\mathbb{R}^n$ by L. Diening [4]. Here $\gamma = \gamma(t,r)$, $t \in \Gamma$, $r > 0$, is any portion of the curve Γ.

By $\chi_\gamma(\tau) = \begin{cases} 1, & \tau \in \gamma \\ 0, & \tau \in \Gamma\backslash\gamma \end{cases}$ we denote the characteristic function of a portion γ of the curve Γ.

3. The main statements

In the sequel we consider the power weights of the form

$$w(t) = \prod_{k=1}^{n} |t - t_k|^{\beta_k}, \qquad t_k \in \Gamma \tag{3.1}$$

in the case of finite curve and the weights

$$w(t) = |t - z_0|^{\beta} \prod_{k=1}^{n} |t - t_k|^{\beta_k}, \qquad t_k \in \Gamma, \quad z_0 \notin \Gamma \tag{3.2}$$

in the case of infinite curve.

Theorem A. *Let*

  i) Γ *be a simple Carleson curve;*

  ii) *p satisfy conditions (2.1), (2.2) and also (2.4) in the case Γ is an infinite curve.*

Then the singular operator S_Γ is bounded in the space $L^{p(\cdot)}(\Gamma, w)$ with weight (3.1) or (3.2), if and only if

$$-\frac{1}{p(t_k)} < \beta_k < \frac{1}{p'(t_k)}, \qquad k = 1, \ldots, n, \tag{3.3}$$

and also

$$-\frac{1}{p_\infty} < \beta + \sum_{k=1}^{n} \beta_k < \frac{1}{p'(\infty)} \tag{3.4}$$

in the case Γ is infinite.

Remark 3.1. *From (2.4) it follows that there exists $p_\infty = \lim_{\substack{|t| \to \infty \\ t \in \Gamma}} p(t)$ and* $|p(t) - p_\infty| \leq \frac{A_\infty}{\ln |t|}$, $|t| \geq \max\{L, 2\}$.

For constant p Theorem A is due to G. David [3] in the non-weighted case, for the weighted case with constant p see [2]. For earlier results on the subject we refer to [9]. The statement of Theorem A for variable $p(\cdot)$ was proved in [11] in the case of finite Lyapunov curves or curves of bounded rotation without cusps.

Theorem B. *Let Γ be a finite rectifiable curve. Let $p : \Gamma \to [1, \infty)$ be a continuous function. If the singular operator S_Γ is bounded in the space $L^{p(\cdot)}(\Gamma)$, then the curve Γ has the property*

$$\sup_{\substack{t \in \Gamma \\ r > 0}} \frac{|\gamma(t, r)|}{r^{1-\varepsilon}} < \infty \tag{3.5}$$

for every $\varepsilon > 0$. If $p(t)$ satisfies the log-condition (2.2), then property (3.5) holds with $\varepsilon = 0$, that is, Γ is a Carleson curve.

Observe that Theorem B for the case of constant p was proved in [15].

Theorem C. *Let assumptions i)–ii) of Theorem A be satisfied, and let $a \in C(\Gamma)$. In the case where Γ is an infinite curve starting and ending at infinity, we assume that $a \in C(\dot{\Gamma})$, where $\dot{\Gamma}$ is the compactification of Γ by a single infinite point, that is, $a(t(-\infty)) = a(t(+\infty))$. Then under conditions (3.3)–(3.4), the operator*

$$(S_\Gamma a I - a S_\Gamma)f = \frac{1}{\pi i} \int_\Gamma \frac{a(\tau) - a(t)}{\tau - t} f(\tau) d\nu(\tau)$$

is compact in the space $L^{p(\cdot)}(\Gamma, w)$ with weight (3.1)–(3.2).

Theorems A, B and C are proved in Sections 6, 7 and 8, respectively.

4. Preliminaries

We base ourselves on the following result for maximal operators on Carleson curves. Let

$$Mf(t) = \sup_{r>0} \frac{1}{\nu\{\gamma(t,r)\}} \int_{\gamma(t,r)} |f(\tau)| d\nu(\tau) \qquad (4.1)$$

be the maximal operator on functions defined on a curve Γ in the complex plane. The following statements are valid.

Proposition 4.1. *Let*

   i) Γ *be a simple Carleson curve of a finite length;*
   ii) p *satisfy conditions* (2.1)–(2.2).

Then the maximal operator M is bounded in the space $L^{p(\cdot)}(\Gamma, w)$ with weight (3.1), *if and only if*

$$-\frac{1}{p(t_k)} < \beta_k < \frac{1}{p'(t_k)}, \qquad k = 1, \ldots, n. \qquad (4.2)$$

Proposition 4.2. *Let*

   i) Γ *be an infinite simple Carleson curve;*
   ii) p *satisfy conditions* (2.1)–(2.2) *and let there exist a circle $B(0, R)$ such that $p(t) \equiv p_\infty = \text{const}$ for $t \in \Gamma \backslash (\Gamma \cap B(0, R))$.*

Then the maximal operator M is bounded in the space $L^{p(\cdot)}(\Gamma, w)$, with weight (3.2), *if and only if*

$$-\frac{1}{p(t_k)} < \beta_k < \frac{1}{p'(t_k)} \qquad and \qquad -\frac{1}{p_\infty} < \beta + \sum_{k=1}^{n} \beta_k < \frac{1}{p'_\infty}. \qquad (4.3)$$

The Euclidean space versions of Propositions 4.1 and 4.2 for variable exponents were proved in [11] and [8], respectively. The proof of Propositions 4.1 and 4.2 for Carleson curves follows similar ideas, but needs some modifications. The proofs of Propositions 4.1 and 4.2 for the case of Carleson curves will be given in another publication.

We will also make use of the following Kolmogorov theorem, see [12], [3], [7].

Theorem 4.3. *Let Γ be a Carleson curve of a finite length. Then for any $s \in (0, 1)$*

$$\left(\frac{1}{|\Gamma|} \int_{\Gamma} |S_\Gamma f(t)|^s d\nu(t) \right)^{\frac{1}{s}} \leq c \frac{1}{|\Gamma|} \int_{\Gamma} |f(t)| d\nu(t). \qquad (4.4)$$

Theorem 4.3 is a consequence of the fact that the singular operator on Carleson curves has weak (1,1)-type:

$$\nu\{t \in \Gamma : |S_\Gamma f(t)| > \lambda\} \leq \frac{c}{\lambda} \int_{\Gamma} |f(t)| d\nu(t)$$

the latter being proved in [3].

Proposition 4.4. *Let $p(t)$ satisfy condition (2.1) and the maximal operator M be bounded in $L^{p(\cdot)}(\Gamma)$. Then there exists a constant $C > 0$ such that*

$$\|\chi_\gamma\|_{p(\cdot)} \leq C|\gamma|^{\frac{1}{p_\gamma}} \qquad \text{for all} \qquad \gamma = \gamma(t, r) \subset \Gamma \tag{4.5}$$

where p_γ is the mean value (2.9).

Proposition 4.4 was proved in [4], Lemma 3.4, for balls in the Euclidean space and remains the same for arcs γ on Carleson curves. For completeness of presentation we expose this proof in the appendix.

5. Auxiliary statements

Let

$$\mathcal{M}^{\#} f(t) = \sup_{r>0} \frac{1}{|\gamma(t, r)|} \int_{\gamma(t,r)} |f(\tau) - f_{\gamma(t,r)}| \, d\nu(\tau), \qquad t \in \Gamma \tag{5.1}$$

where $f_{\gamma(t,r)} = \frac{1}{|\gamma(t,r)|} \int_{\gamma(t,r)} f(\tau) \, d\nu(\tau)$, be the sharp maximal function on the curve Γ.

Theorem 5.1. *Let Γ be an infinite Carleson curve. Let $p(t)$ satisfy conditions (2.1)– (2.2) and $p(t) = p_\infty$ outside some circle $B(t_0, R)$. Let $w(t) = |t - t_0|^\beta$, $t_0 \in \mathbf{C}$, where*

$$-\frac{1}{p(t_0)} < \beta < \frac{1}{p'(t_0)} \quad \text{and} \quad -\frac{1}{p_\infty} < \beta < \frac{1}{p'_\infty} \qquad \text{if} \qquad t_0 \in \Gamma$$

and $-\frac{1}{p_\infty} < \beta < \frac{1}{p'_\infty}$ if $t_0 \notin \Gamma$. Then for $f \in L^{p(\cdot)}(\Gamma, w)$

$$\|f\|_{L^{p(\cdot)}(\Gamma,w)} \leq c \|\mathcal{M}^{\#} f\|_{L^{p(\cdot)}(\Gamma,w)}. \tag{5.2}$$

Proof. As is known, $\|f\|_{L^{p(\cdot)}} \sim \sup_{\|g\|_{L^{p'(\cdot)}} \leq 1} \left| \int_\Gamma f(t)g(t) \, d\nu(t) \right|$, see [13], Theorem 2.3 or [16], Theorem 3.5. Therefore,

$$\|fw\|_{L^{p(\cdot)}} \leq c \sup_{\|g\|_{L^{p'(\cdot)}} \leq 1} \left| \int_\Gamma f(t)g(t)w(t) \, d\nu(t) \right|.$$

We make use of the inequality

$$\int_\Gamma |f(t)g(t)| \, d\nu(t) \leq \int_\Gamma M^{\#} f(t) M g(t) \, d\nu(t) \tag{5.3}$$

where $f \in L^{p(\cdot)}(\Gamma), g \in L^{p'(\cdot)}(\Gamma)$, which is known for the Euclidean space, see [5], Lemma 3.5, and is similarly proved for infinite Carleson curves. We obtain

$$\|fw\|_{L^{p(\cdot)}} \leq c \sup_{\|g\|_{L^{p'(\cdot)}} \leq 1} \left| \int_\Gamma w(t) M^{\#} f(t) [w(t)]^{-1} M(gw) \, d\nu(t) \right|$$

and then

$$\|fw\|_{L^{p(\cdot)}} \le c \sup_{\|g\|_{L^{p'(\cdot)}} \le 1} \|wM^{\#}f\|_{L^{p(\cdot)}} \|w^{-1}M(gw)\|_{L^{p'(\cdot)}}$$

by the Hölder inequality. Since $-\frac{1}{p'(t_0)} < -\beta < \frac{1}{p(t_0)}$, we may apply Proposition 4.2 for the space $L^{p'(\cdot)}$ with β replaced by $-\beta$ and conclude that

$$\|fw\|_{L^{p(\cdot)}} \le C \sup_{\|g\|_{L^{p'(\cdot)}} \le 1} \|wM^{\#}f\|_{L^{p(\cdot)}} \|g\|_{L^{p'(\cdot)}} \le C\|wM^{\#}f\|_{L^{p(\cdot)}}$$

which proves (5.2). $\qquad\square$

6. Proof of Theorem A

6.1. General remark

Remark 6.1. It suffices to prove Theorem A for a single weight $|t - t_0|^{\beta}$ where $t_0 \in \Gamma$ in the case Γ is finite and t_0 may belong or not belong to Γ when Γ is infinite.

Indeed, in the case of a finite curve let $\Gamma = \bigcup_{k=1}^{n} \Gamma_k$ where Γ_k contains the point t_k in its interior and does not contain $t_j, j \neq k$ in its closure. Then

$$\|f\|_{L^{p(\cdot)}\left(\Gamma, \prod_{k=1}^{n} |t-t_k|^{\beta_k}\right)} \sim \sum_{k=1}^{n} \|f\|_{L^{p(\cdot)}(\Gamma_k, |t-t_k|^{\beta_k})} \tag{6.1}$$

whenever $1 \le p_- \le p_+ < \infty$. This equivalence follows from the easily checked modular equivalence

$$I_{\Gamma}^{p}\left(f(t) \prod_{k=1}^{n} |t - t_k|^{\beta_k}\right) \sim \sum_{k=1}^{n} I_{\Gamma_k}^{p}\left(f(t)|t - t_k|^{\beta_k}\right),$$

since

$$c_1 \le \|f\|_{p(\cdot)} \le c_2 \implies c_3 \le I_{\Gamma}^{p}(f) \le c_4,$$
$$C_1 \le I_{\Gamma}^{p}(f) \le C_2 \implies C_3 \le \|f\|_{p(\cdot)} \le C_4 \tag{6.2}$$

with $c_3 = \min\left(c_1^{p_-}, c_1^{p_+}\right), c_4 = \max\left(c_2^{p_-}, c_2^{p_+}\right), C_3 = \min\left(C_1^{\frac{1}{p_-}}, C_1^{\frac{1}{p_+}}\right)$ and $C_4 = \max\left(C_2^{\frac{1}{p_-}}, C_2^{\frac{1}{p_+}}\right)$.

Similarly, in the case of an infinite curve

$$\|f\|_{L^{p(\cdot)}\left(\Gamma, |t-z_0|^{\beta} \prod_{k=1}^{n} |t-t_k|^{\beta_k}\right)} \sim \|f\|_{L^{p(\cdot)}(\Gamma_{\infty}, |t-z_0|^{\beta})} + \sum_{k=1}^{n} \|f\|_{L^{p(\cdot)}(\Gamma_k, |t-t_k|^{\beta_k})} \tag{6.3}$$

where Γ_{∞} is a portion of the curve outside some large circle, so that Γ_{∞} does not contain the points $t_k, k = 1, \ldots, n$.

Then, because of (6.1) and (6.3), the statement of Remark 6.1 is obtained by introduction of the standard partition of unity $1 = \sum_{k=1}^{n} a_k(t)$, where $a_k(t)$ are smooth functions equal to 1 in a neighborhood $\gamma(t_k, \varepsilon)$ of the point t_k and equal to 0 outside its neighborhood $\gamma(t_k, 2\varepsilon)$ (and similarly in a neighborhood of infinity in the case Γ is infinite), so that $a_k(t)|t - t_j|^{\pm\beta_j} \equiv 0$ in a neighborhood of the point t_k, if $k \neq j$.

6.2. Auxiliary results

We start with proving the following statement known for singular integrals in the Euclidean space (T. Alvarez and C. Pérez, [1]).

Proposition 6.2. *Let Γ be a simple Carleson curve. Then the following pointwise estimate is valid*

$$\mathcal{M}^{\#}\left(|S_{\Gamma}f|^s\right)(t) \leq c[Mf(t)]^s, \qquad 0 < s < 1, \tag{6.4}$$

where the constant $c > 0$ may depend on Γ and s, but does not depend on $t \in \Gamma$ and f.

To prove Proposition 6.2, we need – following ideas in [1] – the following technical lemma.

Lemma 6.3. *Let Γ be a simple Carleson curve, $z_0 \in \Gamma$ and $\gamma_r = \gamma(z_0, r)$ and*

$$H_{r,z_0}(t) = \frac{1}{|\gamma_r|^2} \int_{\gamma_r} \int_{\gamma_r} \left| \frac{1}{z - t} - \frac{1}{\tau - t} \right| d\nu(z)d\nu(\tau). \tag{6.5}$$

Then for any locally integrable function f the pointwise estimate holds

$$\sup_{r>0} \int_{t\in\Gamma:|t-z_0|>2r} |f(t)| H_{r,z_0}(t) d\nu(t) \leq CMf(z_0) \tag{6.6}$$

where $C > 0$ does not depend on f and z_0.

Proof. We have

$$H_{r,z_0}(t) = \frac{1}{|\gamma_r|^2} \int_{\gamma_r} \int_{\gamma_r} \frac{|\tau - z|}{|z - t| \cdot |\tau - t|} d\nu(z)d\nu(\tau).$$

For $|t - z_0| > 2r$ we have

$$|z - t| \geq |t - z_0| - |z - z_0| \geq |t - z_0| - r \geq \frac{1}{2}|t - z_0|$$

and similarly $|\tau - t| \geq \frac{1}{2}|t - z_0|$ so that

$$H_{r,z_0}(t) \leq \frac{Cr}{|t - z_0|^2}$$

where the constant $C > 0$ depends only on the length of the curve Γ. Then

$$\sup_{r>0} \int_{t\in\Gamma:|t-z_0|>2r} |f(t)| H_{r,z_0}(t)\,d\nu(t) \le c \sup_{r>0} \sum_{k=0}^{m} \int_{2^k r < |t-z_0| < 2^{k+1}r} \frac{r|f(t)|}{|t-z_0|^2}\,d\nu(t)$$

with $m = m(r)$. Hence

$$\sup_{r>0} \int_{t\in\Gamma:|t-z_0|>2r} |f(t)| H_{r,z_0}(t)\,d\nu(t) \le 2c \sup_{r>0} \sum_{k=0}^{m} \frac{1}{2^k} \frac{1}{2^{k+1}r} \int_{|t-z_0|<2^{k+1}r} |f(t)|\,d\nu(t)$$

$$\le 2c M f(z_0) \sum_{k=0}^{m} \frac{1}{2^k} \le c_1 M f(z_0). \qquad \square$$

We will also need the following technical lemma

Lemma 6.4. *Let f be an integrable function on Γ, $f_\gamma = \frac{1}{|\gamma|} \int_\gamma f(\tau)\,d\nu(\tau)$. Then*

$$\frac{1}{|\gamma|} \int_\gamma |f(\tau) - f_\gamma|\,d\nu(\tau) \le \frac{2}{|\gamma|} \int_\gamma |f(\tau) - C|\,d\nu(\tau) \qquad (6.7)$$

for any constant C on the right-hand side.

Proof. The proof is well known:

$$\frac{1}{|\gamma|} \int_\gamma |f(\tau) - f_\gamma|\,d\nu(\tau) \le \frac{1}{|\gamma|^2} \int_\gamma \int_\gamma |f(\tau) - f(\sigma)|\,d\nu(\tau)\,d\nu(\sigma)$$

$$\le \frac{1}{|\gamma|^2} \int_\gamma \int_\gamma (|f(\tau) - C| + |C - f(\sigma)|)\,d\nu(\tau)\,d\nu(\sigma)$$

$$= \frac{2}{|\gamma|} \int_\gamma |f(\tau) - C|\,d\nu(\tau). \qquad \square$$

Proof of Proposition 6.2. To prove estimate (6.4), according to Lemma 6.4 it suffices to show that for any locally integrable function f and any $0 < s < 1$ there exists a positive constant A such that

$$\left(\frac{1}{|\gamma|} \int_\gamma \left| |S_\Gamma f(\xi)|^s - A^s \right| d\nu(\xi) \right)^{\frac{1}{s}} \le C M f(z_0), \qquad \gamma = \gamma(z_0, r) \qquad (6.8)$$

for almost all $z_0 \in \Gamma$, where $C > 0$ does not depend on f and z_0. We set $f = f_1 + f_2$, where $f_1 = f \cdot \chi_{\gamma(z_0,2r)}$ and $f_2 = f \cdot \chi_{\Gamma \setminus \gamma(z_0,2r)}$. We take

$$A = (S_\Gamma f_2)_\gamma = \frac{1}{|\gamma|} \int_\gamma |S_\Gamma f_2(\xi)|\,d\nu(\xi).$$

Then, taking into account that $||a|^s - |b^s|| \leq |a - b|^s$, for $0 < s < 1$, we have

$$\left(\frac{1}{|\gamma|} \int_\gamma \left| |S_\Gamma f(\xi)|^s - A^s \right| d\nu(\xi) \right)^{\frac{1}{s}} \leq c \left(\frac{1}{|\gamma|} \int_\gamma \left| S_\Gamma f_1(\xi) \right|^s d\nu(\xi) \right)^{\frac{1}{s}} +$$

$$+ c \left(\frac{1}{|\gamma|} \int_\gamma \left| |S_\Gamma f_2(\xi)| - A \right|^s d\nu(\xi) \right)^{\frac{1}{s}} =: c(I_1 + I_2).$$

For I_1 by (4.4) we obtain

$$I_1 \leq \frac{1}{|\gamma|} \int_\gamma |f_1(\xi)| d\nu(t) \leq \frac{1}{|\gamma|} \int_\gamma |f(t)| d\nu(\xi) \leq Mf(z_0). \tag{6.9}$$

For I_2, by Jensen inequality and Fubini theorem after easy estimations we get

$$I_2 \leq \frac{1}{|\gamma|} \int_\gamma \left| (S_\Gamma f_2)(\xi) - \frac{1}{|\gamma|} \int_\gamma (S_\Gamma f_2)(\tau) d\nu(\tau) \right| d\nu(\xi) \leq \int_{\Gamma \setminus \gamma(z_0, 2r)} |f(t)| H_{r, z_0}(t) d\nu(t),$$

where $H_{r, z_0}(t)$ is the function defined in (6.5). Therefore, by Lemma 6.3, $I_2 \leq CMf(z_0)$ which completes the proof. $\qquad\square$

6.3. Proof of Theorem A itself. Sufficiency part

According to Remark 6.1, we consider the case of a single weight $|t - t_0|^\beta$ where t_0 may be not belonging to Γ in case Γ is infinite.

I). The case of infinite curve and p constant at infinity. First we consider the case where Γ is an infinite curve and we additionally suppose at this step that $p(t) \equiv \text{const} = p_\infty$ outside some large ball $B(0, R)$.

Let $0 < s < 1$. Observe that

$$\|S_\Gamma f\|_{L^{p(\cdot)}(\Gamma, w)} = \||S_\Gamma f|^s\|^{\frac{1}{s}}_{L^{\frac{p(\cdot)}{s}}(\Gamma, w^s)}.$$

Then by Theorem 5.1 we have

$$\|S_\Gamma f\|_{L^{p(\cdot)}(\Gamma, w)} \leq C \left\| \mathcal{M}^{\#}(|S_\Gamma f|^s) \right\|^{\frac{1}{s}}_{L^{\frac{p(\cdot)}{s}}(\Gamma, w)}$$

for s sufficiently close to 1. Indeed, Theorem, 5.1 is applicable in this case, because $\frac{p(t)}{s}$ satisfies conditions (2.1)–(2.2) and when s is sufficiently close to 1, then the exponent $s\beta(t_0)$ of the weight w^s satisfies the conditions $-\frac{1}{\frac{p(t_0)}{s}} < s\beta(t_0) < \frac{1}{\frac{p'(t_0)}{s}}$, required by Theorem 5.1. Therefore, by Proposition 6.2 we get

$$\|S_\Gamma f\|_{L^{p(\cdot)}(\Gamma, w)} \leq c \|(Mf)^s\|^{\frac{1}{s}}_{L^{\frac{p(\cdot)}{s}}(\Gamma, w)} = c \|Mf\|_{L^{p(\cdot)}(\Gamma, w)}.$$

It remains to apply Proposition 4.2 to obtain $\|S_\Gamma f\|_{L^{p(\cdot)}(\Gamma, w)} \leq c \|f\|_{L^{p(\cdot)}(\Gamma, w)}$.

II). The case of finite curve and p constant on some arc. At the next step we consider the case of finite curve under the additional assumption that there exists an arc $\gamma \subset \Gamma$ with $|\gamma| > 0$ on which $p(t) \equiv \mathrm{const}$.

First we observe that the singular integral may be considered in the form

$$S_\Gamma f(t) = \frac{1}{\pi i} \int_\Gamma \frac{f(\tau)}{\tau - t}\, d\tau \tag{6.10}$$

instead of (1.3), since $d\tau = \tau'(s)d\nu(\tau)$ and $|\tau'(s)| = 1$ on Carleson curves so that $\|f\tau'\|_{L^{p(\cdot)}(\Gamma, w)} = \|f\|_{L^{p(\cdot)}(\Gamma, w)}$.

The case considered now is reduced to the previous case **I)** by the change of variables. Let $z_0 \in \gamma$ be any point of γ (different from t_0 if $t_0 \in \gamma$). Without loss of generality we may assume that $z_0 = 0$. Let

$$\Gamma_* = \{t \in \mathbf{C} : t = \frac{1}{\tau},\ \tau \in \Gamma\} \qquad \text{and} \qquad \widetilde{p}(t) = p\left(\frac{1}{t}\right), \quad t \in \Gamma_*,$$

so that Γ_* is an infinite curve and $\widetilde{p}(z)$ is constant on Γ_* outside some large circle. By the change of variables $\frac{1}{\tau} = w$ and $\frac{1}{t} = z$ we get

$$(S_\Gamma f)(t) = -z(S_{\Gamma_*}\psi)(z), \qquad z \in \Gamma_* \tag{6.11}$$

where $\psi(w) = \frac{1}{w}f\left(\frac{1}{w}\right)$. The following lemma is valid where the equivalence $A \sim B$ means that $c_1 A \leq B \leq c_2 A$ with c_1 and c_2 not depending on A and B.

Lemma 6.5. *The following modular equivalence holds*

$$I_\Gamma^p\left(|t - t_0|^\beta f(t)\right) \sim I_{\Gamma_*}^{\widetilde{p}}\left(\rho(t)\psi(t)\right) \tag{6.12}$$

where $\rho(t) = |t|^\nu\, |t - t_0^*|^\beta$ *with* $t_0^* = \frac{1}{t_0} \in \Gamma_*$ *and*

$$\nu = 1 - \beta - \frac{2}{\widetilde{p}(\infty)} = 1 - \beta - \frac{2}{p(0)}. \tag{6.13}$$

Proof. Indeed,

$$I_\Gamma^p\left(|t - t_0|^\beta f(t)\right) = \int_\Gamma |t - t_0|^{\beta p(t)}|f(t)|^{p(t)}|dt|.$$

After the change of variables $t \to \frac{1}{t}$ we get

$$I_\Gamma^p\left(|t - t_0|^\beta f(t)\right) = \int_{\Gamma_*} \left|\frac{1}{t} - t_0\right|^{\beta \widetilde{p}(t)} \left|f\left(\frac{1}{t}\right)\right|^{\widetilde{p}(t)} \frac{|dt|}{|t^2|}$$

$$= \int_{\Gamma_*} |t_0|^{\beta \widetilde{p}(t)} \frac{|t_0^* - t|^{\beta \widetilde{p}(t)}}{|t|^{\beta \widetilde{p}(t)}} \left|f\left(\frac{1}{t}\right)\right|^{\widetilde{p}(t)} \frac{|dt|}{|t^2|}.$$

Since $|t_0|^{\beta \widetilde{p}(t)} \sim \mathrm{const}$, we obtain

$$I_\Gamma^p\left(|t - t_0|^\beta f(t)\right) \sim \int_{\Gamma_*} \frac{|t_0^* - t|^{\beta \widetilde{p}(t)}}{|t|^{\beta \widetilde{p}(t)+2}} \left|f\left(\frac{1}{t}\right)\right|^{\widetilde{p}(t)} |dt|. \tag{6.14}$$

Now we have

$$f\left(\frac{1}{t}\right) = t\psi(t).$$

Therefore, from (6.14) we get

$$I_\Gamma^p\left(|t-t_0|^\beta f(t)\right) \sim \int_{\Gamma_*} \frac{|t_0^* - t|^{\beta\tilde{p}(t)}}{|t|^{\beta\tilde{p}(t)+2}}\, |t\psi(t)|^{\tilde{p}(t)}\, |dt|.$$

Hence

$$I_\Gamma^p\left(|t-t_0|^\beta f(t)\right) \sim \int_{\Gamma_*} \left(\frac{|t_0^* - t|^\beta}{|t|^{(\beta-1)+\frac{2}{\tilde{p}(t)}}}\, |t\psi(t)|\right)^{\tilde{p}(t)} |dt|.$$

Observe that the point $z = 0$ does not pass through the origin and therefore $|t|^{(\beta-1)+\frac{2}{\tilde{p}(t)}} \sim |t|^{(\beta-1)+\frac{2}{\tilde{p}(\infty)}}$. As a result we arrive at (6.12)–(6.13). $\quad\square$

According to (6.2), from (6.12) we also have

$$\|f\|_{L^{p(\cdot)}(\Gamma,|t-t_0|^\beta)} \sim \|\psi\|_{L^{\tilde{p}(\cdot)}(\Gamma_*,\rho(t))}$$

and

$$\|S_\Gamma f\|_{L^{p(\cdot)}(\Gamma,|t-t_0|^\beta)} \sim \|S_{\Gamma_*}\psi\|_{L^{\tilde{p}(\cdot)}(\Gamma_*,\rho(t))} \tag{6.15}$$

where (6.11) was taken into account. Observe also that

$$-\frac{1}{p(t_0)} < \beta < \frac{1}{p'(t_0)} \qquad \Longleftrightarrow \qquad -\frac{1}{\tilde{p}(t_0^*)} < \nu < \frac{1}{\tilde{p}'(t_0^*)}. \tag{6.16}$$

Obviously, $\tilde{p}(t)$ satisfies conditions (2.1)–(2.2). Since $\tilde{p}(t)$ is constant at infinity, according to part **I)** and Remark 6.1, the operator S_{Γ_*} is bounded in the space $L^{\tilde{p}(\cdot)}(\Gamma_*,\rho(t))$, the required conditions on the weight $\rho(t)$ being satisfied by (6.16) and by the fact that $\beta + \nu = 1 - \frac{2}{\tilde{p}(\infty)}$ is automatically in the interval $\left(-\frac{1}{\tilde{p}(\infty)}, \frac{1}{\tilde{p}'(\infty)}\right)$. Then the operator S_Γ is bounded in the space $L^{p(\cdot)}(\Gamma,|t-t_0|^\beta)$ by (6.15).

III). The general case of finite curve. Let $\gamma_1 \subset \Gamma$ and $\gamma_2 \subset \Gamma$ be two disjoint non-empty arcs of Γ, $\overline{\gamma_1} \cap \overline{\gamma_2} = \emptyset$. According to the part **II)** the operator $wS_\Gamma\frac{1}{w}$ with $w(t) = |t-t_0|^\beta, t_0 \in \Gamma$, is bounded in the space $L^{p_1(\cdot)}(\Gamma)$, if

1) $p_1(t)$ satisfies conditions (2.1)–(2.2),
2) $p_1(t)$ is constant at γ_1,
3) $-\frac{1}{p_1(t_0)} < \beta < \frac{1}{p_1'(t_0)}$

and similarly in the space $L^{p_2(\cdot)}(\Gamma)$, if

1') $p_2(t)$ satisfies conditions (2.1)–(2.2),
2') $p_2(t)$ is constant at γ_2,
3') $-\frac{1}{p_2(t_0)} < \beta < \frac{1}{p_2'(t_0)}$

Aiming to make use of the Riesz interpolation theorem, we observe that the following statement is valid (see its proof in Appendix 2).

Lemma 6.6. *Given a function $p(t)$ satisfying conditions (2.1)–(2.2), there exist arcs $\gamma_2 \subset \Gamma$ and $\gamma_1 \subset \Gamma$ such that $p(t)$ may be represented in the form*

$$\frac{1}{p(t)} = \frac{\theta}{p_1(t)} + \frac{1 - \theta}{p_2(t)}, \qquad \theta = \frac{1}{2}, \tag{6.17}$$

where $p_j(t)$, $j = 1, 2$, satisfy the above conditions 1), 2) and 1'), 2'), respectively and

$$\frac{1}{p_j(t_0)} = \frac{1}{p(t_0)}, \qquad j = 1, 2,$$

so that conditions 3) and 3'(are also satisfied whenever they are satisfied for $p(t)$.

In view of Lemma 6.6, the boundedness of the singular operator in $L^{p(\cdot)}(\Gamma)$ with given p follows from the Riesz-Thorin interpolation theorem for the spaces $L^{p(\cdot)}(\Gamma)$ (proved in [14]).

IV). The general case of infinite curve. Obviously, the after the step **III)** the general case of infinite curve, that is, the case where Γ is infinite and p is not necessarily constant outside some circle, is reduced to the case of finite curve by mapping the infinite curve Γ onto a finite curve Γ_* in the same way as it was done in the step **II)**. What is important to note is that thanks to conditions (2.2) and (2.4), the new exponent $\tilde{p}(t), t \in \Gamma_*$ is log-continuous on the curve Γ_*.

Remark 6.7. We emphasize the following. We had to prove the boundedness of the singular operator on an infinite curve under additional assumption that $p(t)$ is constant at infinity, then by the change of variables we could cover the case of a finite curve with $p(t)$ constant at any arc. After that we could use the interpolation theorem to get a result on boundedness on a finite curve without the assumption on $p(t)$ to be constant an an arc. After that it remained to use change of variables to get the general result for an infinite curve. This order is essential. Indeed, a seeming possibility to treat first the general case of a finite curve and then cover the case of infinite curve by the change of variables, is not applicable, because the initial step in the whole proof was based on Theorem 5.1, which was proved for infinite curves.

6.4. Proof of the necessity part of Theorem A

The proof of the necessity is in fact the same as in the case of smooth curves, see [10], p. 153. We dwell on the main points. Let Γ be a finite curve. From the boundedness of S_Γ in $L^{p(\cdot)}(\Gamma, \rho)$ it follows that $S_\Gamma f(t)$ exists almost everywhere for an arbitrary $f \in L^{p(\cdot)}(\Gamma, \rho)$. Thus ρ should be such that $f \in L^1(\Gamma)$ for arbitrary $f \in L^{p(\cdot)}(\Gamma, \rho)$. The function $f = f\rho\rho^{-1}$ belongs to $L^1(\Gamma)$ for arbitrary $f \in L^{p(\cdot)}(\Gamma, \rho)$ if and only if $\rho^{-1} \in L^{q(\cdot)}$. Then the function $\rho^{-1}(t) = |t - t_0|^{-\beta}$, $t_0 \in \Gamma$, belongs to $L^{q(\cdot)}(\Gamma)$ if and only if $\beta < \frac{1}{q(t_0)}$. Indeed, by the log-condition we have

$$|t - t_0|^{-\beta q(t)} \sim |t - t_0|^{-\beta q(t_0)}.$$

On the other hand, since Γ is a Carleson curve, from $|t - t_0|^{-\beta q(t_0)} \in L^1$ we have $\beta < \frac{1}{q(t_0)}$.

The necessity of the condition $-\frac{1}{p(s_0)} < \beta$ follows from the duality argument.

In a similar way, with slight modifications the case of infinite curve and weight fixed to infinity, is treated.

7. Proof of Theorem B

We start with the following remark.

Remark 7.1. If the operator S_Γ is bounded in $L^{p(\cdot)}(\Gamma)$ and γ is a measurable subset of Γ, then the operator $S_\gamma = \chi_\gamma S_\Gamma \chi_\gamma$ is bounded in $L^{p(\cdot)}(\gamma)$ and $\|S_\gamma\|_{L^{p(\cdot)}(\gamma)} \leq \|S_\Gamma\|_{L^{p(\cdot)}(\Gamma)}$ (we denote the restriction of $p(\cdot)$ onto γ by the same symbol $p(\cdot)$).

7.1. Auxiliary lemmas

Lemma 7.2. *For every point* $t \in \Gamma$ *and every* $\rho \in \left(0, \frac{1}{6}\operatorname{diam}\Gamma\right)$ *there exists a function* $\varphi_t := \varphi_{t,\rho}(\tau)$ *such that*

$$I_p\left(S_\Gamma\varphi_t\right) \geq m\left(\frac{|\gamma(t,\rho)|}{\rho}\right)^{p_- - 1} I_p\left(\varphi_t\right), \tag{7.1}$$

where $m > 0$ *is a constant not depending on* t *and* ρ.

Proof. Let us fix the point $t = t_0$ and consider circles centered at t_0 of the radii $\rho, 2\rho$ and 3ρ and 8 rays with the angle $\frac{\pi}{4}$, one of them being parallel to the axis of abscissas. These rays split the circle $|z - t_0| < \rho$ and the annulus $2\rho < |z - t_0| < 3\rho$ into 16 parts. It suffices to treat only those partes which lie in a semiplane, for example, in the upper semiplane. We denote these parts of the circle $|z - t_0| < \rho$ by $\Gamma_k := \Gamma_{k,t_0,\rho}$ and the parts of the annulus $2\rho < |z - t_0| < 3\rho$ by $\gamma_k := \gamma_{k,t_0,\rho}$, respectively, $k = 1, 2, 3, 4$, counting them, e.g., counter clockwise. These rays may be chosen so that there exists a pair k_0, j_0 such that

$$|\Gamma_{k_0}| \geq \frac{1}{8}|\gamma(t_0,\rho)| \quad \text{and} \quad |\gamma_{j_0}| \geq \frac{1}{8}\rho. \tag{7.2}$$

Without loss of generality we may take $k_0 = 1$.

Let

$$\varphi_{t_0} = \varphi_{t_0,\rho}(t) = \begin{cases} 1, & t \in \Gamma_1 \\ 0, & t \in \Gamma\backslash\Gamma_1 \end{cases} \tag{7.3}$$

We have to estimate the integral

$$I_p(S_\Gamma\varphi_{t_0,\rho}) = \int\limits_\Gamma \left|\int\limits_\Gamma \frac{\varphi_{t_0}(\tau)}{\tau - t}\, d\nu(\tau)\right|^{p(t)} d\nu(t). \tag{7.4}$$

Let $\tau - t = |\tau - t|e^{i\alpha(\tau,t)}$. We have

$$I_p(S_\Gamma\varphi_{t_0}) \geq \int\limits_{\gamma_{j_0}} \left|\int\limits_{\Gamma_1} \frac{\cos\alpha(\tau,t) - i\sin\alpha(\tau,t)}{|\tau - t|}\, d\nu(\tau)\right|^{p(t)} d\nu(t). \tag{7.5}$$

Let first $j_0 = 1$. We put $M_1 = (\rho, 0)$ and $M_2 = \left(2\rho\cos\frac{\pi}{4}, 2\rho\sin\frac{\pi}{4}\right)$. It is easily seen that $\max|\alpha(\tau, t)| \leq \beta_1 < \frac{\pi}{2}$, where β_1 is the angle between the vector $\overline{M_2M_2}$ and the axis of abscissas. Similarly it can be seen that

$$
\begin{array}{llll}
\text{if} & \tau \in \Gamma_1, t \in \gamma_2, & \text{then} & \frac{\pi}{4} \leq \alpha(\tau, t) \leq \pi - \beta_2, \\
\text{if} & \tau \in \Gamma_1, t \in \gamma_3, & \text{then} & \frac{\pi}{2} \leq \alpha(\tau, t) \leq \pi - \beta_3, \\
\text{if} & \tau \in \Gamma_1, t \in \gamma_4, & \text{then} & \frac{3\pi}{4} \leq \alpha(\tau, t) \leq \pi + \beta_4
\end{array}
$$

where $\beta_2 = \operatorname{arctg} 2$, $\beta_3 = \operatorname{arctg}\frac{1}{3}$ and $\beta_4 = \operatorname{arctg}\frac{2\sqrt{2}-1}{7}$. Therefore, when $\tau \in \Gamma_1$ and $t \in \gamma_{j_0}, j_0 = 1, 2, 3, 4$, then

$$
\text{either} \quad |\cos\alpha(\tau, t)| \geq m_0 > 0, \quad \text{or} \quad |\sin\alpha(\tau, t)| \geq m_0 > 0.
$$

Moreover, when $\tau \in \Gamma_1$ and $t \in \gamma_2$ or $t \in \gamma_4$, then $\cos\alpha(\tau, t)$ preserves the sign and when $\tau \in \Gamma_1$ and $t \in \gamma_2$ or $t \in \gamma_3$, then $\sin\alpha(\tau, t)$ preserves the sign. Consequently, from (7.5) we get

$$
I_p(S_\Gamma\varphi_{t_0}) \geq \int\limits_{\gamma_{j_0}} \max\left(\left|\operatorname{Re}\int\limits_{\Gamma_1}\frac{\varphi_{t_0}(\tau)d\nu(\tau)}{\tau - t}\right|^{p(t)}, \left|\operatorname{Im}\int\limits_{\Gamma_1}\frac{\varphi_{t_0}(\tau)d\nu(\tau)}{\tau - t}\right|^{p(t)}\right).
$$

Hence

$$
I_p(S_\Gamma\varphi_{t_0}) \geq \int\limits_{\gamma_{j_0}}\left|\int\limits_{\Gamma_1}\frac{m_0}{3\rho}d\nu(\tau)\right|^{p(t)} d\nu(t) \geq \left(\frac{m_0}{3}\right)^{p+}\int\limits_{\gamma_{j_0}}\left(\frac{|\Gamma_1|}{\rho}\right)^{p(t)} d\nu(t).
$$

Then by (7.2)

$$
I_p(S_\Gamma\varphi_{t_0}) \geq \left(\frac{m_0}{3\cdot 8}\right)^{p+}\int\limits_{\gamma_{j_0}}\left(\frac{|\gamma(t, \rho)|}{\rho}\right)^{p(t)} d\nu(t) \geq m_1\left(\frac{|\gamma(t, \rho)|}{\rho}\right)^{p-}|\gamma_{j_0}|.
$$

Since $|\gamma(t, \rho)| \geq I_p(\varphi_t) = |\Gamma_1|$ and $|\gamma_{j_0}| \geq \frac{\rho}{8}$, we obtain

$$
I_p(S_\Gamma\varphi_{t_0}) \geq \frac{m_1}{8}\left(\frac{|\gamma(t, \rho)|}{\rho}\right)^{p-}\frac{\rho}{|\gamma(t, \rho)|}|\Gamma_1| = m\left(\frac{\nu(\gamma(t, \rho))}{\rho}\right)^{p--1} I_p(\varphi_{t_0})
$$

which proves (7.1) with (7.3). $\qquad\square$

We denote for brevity

$$
\alpha(f) = \alpha_\gamma(f) = \begin{cases} p_+, & \text{if } \|f\|_{p(\cdot)} \geq 1, \\ p_-, & \text{if } \|f\|_{p(\cdot)} < 1, \end{cases}
$$

and

$$
\beta(f) = \beta_\gamma(f) = \begin{cases} p_-, & \text{if } \|f\|_{p(\cdot)} \geq 1, \\ p_+, & \text{if } \|f\|_{p(\cdot)} < 1, \end{cases}
$$

so that $\alpha(f) + \beta(f) \equiv p_+ + p_-$ and

$$
\|f\|_{p(\cdot)}^{\beta(f)} \leq I_p(f) \leq \|f\|_{p(\cdot)}^{\alpha(f)}. \tag{7.6}
$$

Lemma 7.3. *If the operator S_Γ is bounded in the space $L^{p(\cdot)}(\Gamma)$, then for every $t \in \Gamma$ the estimate holds*

$$\frac{|\gamma(t,\rho)|}{\rho} \le c_\Gamma |\gamma(t,\rho)|^{\delta_\Gamma(t)} \tag{7.7}$$

where $\delta_\Gamma(t) = \frac{1}{p_- - 1}\left(\frac{\alpha(S_\Gamma \varphi_t)}{\beta(\varphi_t)} - 1\right)$, $C_\Gamma = \left(\frac{8\|S_\Gamma\|_{L^{p(\cdot)}(\Gamma)}^{p_+}}{m}\right)^{\frac{1}{p_- - 1}}$ and the function φ_t and the constant were defined in (7.1).

Proof. Let $K = \|S_\Gamma\|_{L^{p(\cdot)}}$ for brevity. By the boundedness $\|S_\Gamma f\|_{L^{p(\cdot)}} \le K\|f\|_{L^{p(\cdot)}}$ and property (7.6) we have

$$I_p(S_\Gamma f) \le K^{\alpha(S_\Gamma f)}\|f\|_{L^{p(\cdot)}}^{\alpha(S_\Gamma f)} \le K^{\alpha(S_\Gamma f)}[I_p(f)]^{\frac{\alpha(S_\Gamma f)}{\beta(f)}}.$$

We choose $f = \varphi_t$ with φ_t from Lemma 7.2 and take (7.1) and (7.3) into account, which yields

$$K^{\alpha(S_\Gamma f)}[I_p(\varphi_t)]^{\frac{\alpha(S_\Gamma f)}{\beta(S_\Gamma \varphi_t)}} \ge I_p(S_\Gamma \varphi_t) \ge m\left(\frac{|\gamma(t,\rho)|}{\rho}\right)^{p_- - 1} I_p(\varphi_t)$$

$$\ge \frac{m}{8}\left(\frac{|\gamma(t,\rho)|}{\rho}\right)^{p_- - 1} |\gamma(t,\rho)|. \tag{7.8}$$

We observe that in the first term in this chain of inequalities we have $I_p(\varphi_t) \le |\gamma(t,\rho)|$ and then (7.8) yields (7.7). $\qquad\square$

7.2. Proof of Theorem B itself

Let $\gamma = \gamma(t,3\rho) = \Gamma \cap \{z : |z - t| < 3\rho\}$. According to Remark 7.1, the operator S_γ is boundeed in $L^{p(\cdot)}(\gamma)$. Then by Lemma 7.3 we obtain

$$\frac{|\gamma(\xi,\rho)|}{\rho} \le c_\gamma |\gamma(\xi,\rho)|^{\delta_\gamma(\xi)} \le c_\Gamma |\gamma(\xi,\rho)|^{\delta_\gamma(\xi)}, \qquad \xi \in \gamma, \tag{7.9}$$

where C_Γ is the same as in Lemma 7.3 and

$$\delta_\gamma(\xi) = \frac{1}{p_-(\gamma) - 1}\left(\frac{\alpha(S_\gamma \varphi_\xi)}{\beta(\varphi_\xi)} - 1\right)$$

with $p_-(\gamma) = \min_{\tau \in \gamma} p(\tau)$. Depending on the values $\|S_\gamma \varphi_\xi\|_{L^{p(\cdot)}(\gamma)}$ and $\|\varphi_\xi\|_{L^{p(\cdot)}(\gamma)}$, the exponent $\delta_\gamma(\xi)$ may take only three values $0, -\delta_1$ and δ_2, where

$$\delta_1 = \frac{p_+(\gamma) - p_-(\gamma)}{p_+(\gamma)}\frac{1}{p_-(\gamma) - 1}, \qquad \delta_2 = \frac{p_+(\gamma) - p_-(\gamma)}{p_-(\gamma)}\frac{1}{p_-(\gamma) - 1}$$

(in fact, according to (7.9) only two values 0 and $-\delta_2$ are possible, since $\frac{\gamma(\xi,\rho)}{\rho} \ge 1$). Therefore, when ρ is small, $|\delta_\gamma(\xi)|$ also has small values:

$$|\delta_\gamma(\xi)| \le \lambda \omega(p, 6\rho), \qquad \lambda = \frac{1}{(p_-(\Gamma) - 1)p_-(\Gamma)}, \tag{7.10}$$

where $\omega(p, h)$ is the continuity modulus of the function p, since $p(t)$ is continuous on the compact set Γ and consequently is uniformly continuous.

Let $\rho_1 < 1$ be sufficiently small such that $\lambda\omega(p, 6\rho_1) < \varepsilon$.

From (7.9) we have $|\gamma(\xi, \rho)|^{1-\delta_\gamma(\xi)} \leq C_\Gamma \rho$ and then

$$|\gamma(\xi, \rho)| < C_\Gamma^{\frac{1}{1-\delta(\xi)}} \rho^{\frac{1}{1-\delta_\gamma(\xi)}} < C_\Gamma^{\frac{1}{1-\varepsilon}} \rho^{\frac{1}{1+\varepsilon}} \qquad \text{for} \qquad \rho < \rho_1 \qquad (7.11)$$

(where we took into account that $C_\Gamma > 1$ and $\rho \leq \rho_1 < 1$). Thus, (3.5) has been proved.

Let now $p(t)$ satisfy the log-condition (2.2). For the function

$$\psi_\xi(\rho) = |\gamma(\xi, \rho)|^{\delta_\gamma(\xi)}$$

by (7.11) we have

$$|\ln \psi_\xi(\rho)| = |\delta_\gamma(\xi) \ln |\gamma(\xi, \rho)|| \leq \lambda\omega(p, 6\rho) \left(\frac{\ln C_\Gamma}{1 - \varepsilon} + \frac{|\ln \rho|}{1 + \varepsilon} \right).$$

In view of (7.10) and (2.3) we then obtain

$$|\ln \psi_\xi(\rho)| \leq \frac{\lambda A}{1 - \varepsilon} \frac{\ln \frac{C_\Gamma}{\rho}}{\ln \frac{\ell}{3\rho}}, \qquad \rho < \min\left\{ \rho_1, \frac{\ell}{6} \right\}. \qquad (7.12)$$

It is easy to see that $\dfrac{\ln \frac{C_\Gamma}{\rho}}{\ln \frac{\ell}{3\rho}}$ is bounded for small ρ, so that $|\ln \psi_\xi(\rho)| \leq C < \infty$. Since $\frac{|\gamma(\xi,\rho)|}{\rho} \leq C_\Gamma \psi_\xi(\rho)$ by (7.9), we get $\frac{|\gamma(\xi,\rho)|}{\rho} \leq C_\Gamma e^C$, which means that Γ is a Carleson curve.

8. Proof of Theorem C

Theorem C is derived from Theorem A, which is standard. Indeed, it is known that any function $a(t)$ continuous on Γ may be approximated in $C(\Gamma)$ by a rational function $r(t)$, whatsoever Jordan curve Γ we have, as is known from the Mergelyan's result, see for instance, [6], p. 169. Therefore, since the singular operator S is bounded in $L_w^{p(\cdot)}(\Gamma)$ by Theorem A, we obtain that the commutator $aS - SaI$ is approximated in the operator norm in $L_w^{p(\cdot)}(\Gamma)$ by the commutator $rS - SrI$ which is finite-dimensional operator, and consequently compact in $L_w^{p(\cdot)}(\Gamma)$. Therefore, $aS - SaI$ is compact.

9. Appendices

9.1. Appendix 1: Proof of Proposition 4.4

Let $f(\tau) = \chi_\gamma(\tau)|\gamma|^{-\frac{1}{p(\tau)}}$, $\gamma = \gamma(t,r)$, so that $\|f\|_{p(\cdot)} = 1$. For all $z \in \gamma$ we have

$$CMf(z) \geq \frac{1}{|\gamma|} \int_\gamma f(\tau)\, d\nu(\tau) = \frac{1}{|\gamma|} \int_\gamma |\gamma|^{-\frac{1}{p(\tau)}}\, d\nu(\tau) \tag{9.1}$$

$$\text{for any } \gamma = \gamma(t,r).$$

Since the function $\Phi(x) = a^{-x}, x \in \mathbb{R}^1_+$, is convex for any $a > 0$, by Jensen's inequality

$$\Phi\left(\frac{1}{|\gamma|} \int_\gamma |f(\tau)|d\nu(\tau)\right) \leq \frac{1}{|\gamma|} \int_\gamma \Phi(|f(\tau)|)\, d\nu(\tau) \tag{9.2}$$

we obtain

$$CMf(z) \geq |\gamma|^{-\frac{1}{|\gamma|} \int_\gamma \frac{d\nu(\tau)}{p(\tau)}} = |\gamma|^{-\frac{1}{p_\gamma}}, \qquad z \in \gamma.$$

Hence $\|\chi_\gamma(z)|\gamma|^{-\frac{1}{p_\gamma}}\|_{p(\cdot)} \leq C\|Mf\|_{p(\cdot)}$ and by the boundedness of the maximal operator we obtain that $\|\chi_\gamma(z)|\gamma|^{-\frac{1}{p_\gamma}}\|_{p(\cdot)} \leq C$, which yields (4.5).

9.2. Appendix 2: Proof of Lemma 6.6

We have to prove the following. *Let Γ be a Carleson curve and $a(t)$ any function on Γ, satisfying the* log*-condition and such that*

$$0 < d \leq a(t) \leq D < 1 \qquad on \qquad \Gamma. \tag{9.3}$$

Then there exist non-intersecting non-empty arcs γ_1 and γ_2 on Γ such that

$$a(t) = \frac{b(t) + c(t)}{2} \qquad with \qquad b(t) \equiv 0 \ \ on \ \ \gamma_1 \qquad and \qquad c(t) \equiv 0 \ \ on \ \ \gamma_2 \tag{9.4}$$

and $b(t)$ and $c(t)$ are log*-continuous on Γ, satisfy the same condition (9.3) and $b(t_0) = c(t_0) = a(t_0)$.*

We will take γ_1 and γ_2 so that $t_0 \notin \overline{\gamma_1 \cup \gamma_2}$ and construct the functions $b(t)$ and $c(t)$ as follows

$$b(t) = \begin{cases} A & , \ t \in \gamma_1 \\ \ell(t) & , \ t \in \Gamma\backslash(\gamma_1 \cup \gamma_2) \\ 2a(t) - B & , \ t \in \gamma_2 \end{cases} \tag{9.5}$$

where $A, B \in (0,1)$ are some constants. The link $\ell(t)$ between the values of $b(t)$ on γ_1 and on γ_2 may be constructed for instance in the following way: at each of the components of the set $\Gamma\backslash(\gamma_1 \cup \gamma_2)$ it is introduced as the linear interpolation between the number A and the values of $2a(t) - B$ at the endpoints of this component, if it does not belong to it, and as the piece-wise linear interpolation between

A the value $a(t_0)$ and the values of $2a(t) - B$ at the endpoints of this component, if it contains t_0. Then

$$c(t) = 2a(t) - b(t) = \begin{cases} 2a(t) - A & , \ t \in \gamma_1 \\ 2a(t) - \ell(t) & , \ t \in \Gamma \backslash (\gamma_1 \cup \gamma_2) \\ B & , \ t \in \gamma_2 \end{cases} \tag{9.6}$$

Obviously, $b(t)$ and $c(t)$ are log-continuous on Γ. Checking condition (9.3) for $b(t), c(t)$, we only have to verify this condition for $2a(t) - A$ on γ_1 and for $2a(t) - B$ on γ_2. To this end, we have to choose A and B so that

$$2a(t) - 1 < A < 2a(t) \quad \text{on} \quad \overline{\gamma_1}, \quad 2a(t) - 1 < B < 2a(t) \quad \text{on} \quad \overline{\gamma_2}$$

Let $a_-(\gamma_i) = \inf_{t \in \gamma_i} a(t)$ and $a_+(\gamma_i) = \sup_{t \in \gamma_i} a(t)$, $i = 1, 2$. It suffices to choose A and B in the intervals

$$A \in \left(\max\{0, 2a_+(\gamma_1) - 1\}, \min\{2a_-(\gamma_1), 1\} \right),$$

$$B \in \left(\max\{0, 2a_+(\gamma_2) - 1\}, \min\{2a_-(\gamma_2), 1\} \right)$$

These intervals are non-empty, if $a_+(\gamma_i) - a_-(\gamma_i) > \frac{1}{2}, i = 1, 2$. Obviously, γ_i may be chosen sufficiently small so that the last condition is satisfied.

References

[1] T. Alvarez and C. Pérez. Estimates with A_∞ weights for various singular integral operators. *Boll. Un. Mat. Ital*, A (7) 8(1):123–133, 1994.

[2] A. Böttcher and Yu. Karlovich. *Carleson Curves, Muckenhoupt Weights, and Toeplitz Operators*. Basel, Boston, Berlin: Birkhäuser Verlag, 1997, 397 pages.

[3] G. David. Opérateurs intégraux singuliers sur certaines courbes du plan complexe. *Ann. Sci. École Norm. Sup. (4)*, 17(1):157–189, 1984.

[4] L. Diening. Riesz potential and Sobolev embeddings on generalized Lebesgue and Sobolev spaces $L^{p(\cdot)}$ and $W^{k,p(\cdot)}$. *Mathem. Nachrichten*, 268:31–43, 2004.

[5] L. Diening and M. Ružička. Calderon-Zygmund operators on generalized Lebesgue spaces $L^{p(x)}$ and problems related to fluid dynamics. *J. Reine Angew. Math*, 563:197–220, 2003.

[6] D. Gaier. *Vorlesungen über Approximation im Komplexen*. Birkhäuser, Basel, Boston, Stuttgart, 1980, 174 pages.

[7] S. Hofmann. Weighted norm inequalities and vector valued inequalities for certain rough operators. *Indiana Univ. Math. J.*, 42(1):1–14, 1993.

[8] M. Khabazi. Maximal operators in weighted $L^{p(x)}$ spaces. *Proc. A. Razmadze Math. Inst.*, 135:143–144, 2004.

[9] B.V. Khvedelidze. The method of Cauchy type integrals in discontinuous boundary problems of the theory of holomorphic functions of one variable (Russian). In *Collection of papers "Contemporary problems of mathematics" (Itogi Nauki i Tekhniki)*, t. 7, pages 5–162. Moscow: Nauka, 1975.

[10] V. Kokilashvili and S. Samko. Singular Integrals in Weighted Lebesgue Spaces with Variable Exponent. *Georgian Math. J.*, 10(1):145–156, 2003.

[11] V. Kokilashvili and S. Samko. Maximal and fractional operators in weighted $L^{p(x)}$ spaces. *Revista Matematica Iberoamericana*, 20(2):495–517, 2004.

[12] A.N. Kolmogorov. Sur les fonctions harmoniques conjugées et les séries de Fourier. *Fund. Math.*, 7:24–29, 1925.

[13] O. Kováčik and J. Rákosník. On spaces $L^{p(x)}$ and $W^{k,p(x)}$. *Czechoslovak Math. J.*, 41(116):592–618, 1991.

[14] J. Musielak. *Orlicz spaces and modular spaces*, volume 1034 of *Lecture Notes in Mathematics*. Springer-Verlag, Berlin, 1983.

[15] V.A. Paatashvili and G.A. Khuskivadze. Boundedness of a singular Cauchy operator in L ebesgue spaces in the case of nonsmooth contours. *Trudy Tbiliss. Mat. Inst. Razmadze Akad. Nauk Gruzin. SSR*, 69:93–107, 1982.

[16] S.G. Samko. Differentiation and integration of variable order and the spaces $L^{p(x)}$. Proceed. of Intern. Conference "Operator Theory and Complex and Hypercomplex Analysis", 12–17 December 1994, Mexico City, Mexico, Contemp. Math., Vol. 212, 203–219, 1998.

Vakhtang Kokilashvili
A.Razmadze Mathematical Institute Tbilisi
Georgia

and

International Black Sea University
e-mail: kokil@rmi.acnet.ge

Vakhtang Paatashvili
A. Razmadze Mathematical Institute Tbilisi
Georgia
e-mail: paata@rmi.acnet.ge

Stefan Samko
University of Algarve
Portugal
e-mail: ssamko@ualg.pt

Operator Theory:
Advances and Applications, Vol. 170, 187–203

On the Averaging Method for the Problem of Heat Convection in the Field of Highly-Oscillating Forces

V.B. Levenshtam

Dedicated to the 70th birthday of Professor I.B. Simonenko

Abstract. In the paper have been proved two theorems an averaging of convection problem and on stability or instability its periodic solutions.

Mathematics Subject Classification (2000). Primary 99Z99; Secondary 00A00.

Keywords. Averaging method, convection, stability.

1. Introduction

The foundations of the averaging method [1] were established in the works of Lagrange, Laplace and Gauss devoted to space mechanics. The classical theory of the method for the mechanical vibrational systems with a finite number of degrees of freedom was created by Van der Pol, Krylov and Bogoljubov [2]–[4]. The mathematical basis of the main algorithms of the theory was developed by Bogoljubov [4]. Since the 50s the averaging method is used (as a rule, without any mathematical justification) for continuum mechanics too.

The problems of thermal convective fluid flows arising under high-frequency vertical [5] and oblique [6] vibrations or vibrations in a weightlessness [7, 8] were studied with the help of the averaging method. The works [9]–[12] are devoted to justify mathematically the averaging method applied to such problems. The present paper links on the latest two works, where existence and local uniqueness of a certain periodic (with respect to time) convective flow is proved, under some conditions, for the frequency equal to a high vibration frequency. In [11, 12] the stability conditions have been formulated for this periodic flaw. Here, for completeness, these results are cited with a new proof. In [11, 12] instability of the periodic regime is studied, however the proof of the result is performed under an essential additional restriction (the spectrum of the respective linearized stationary averaging problem does not contain imaginary values). Here we give a complete proof of the result about instability. Besides, a detailed investigation of a conditional stability of the periodic flow is carried out at the present work. Some methods of Simonenko [9] and Yudovich [13] are essentially used here.

2. Statement of the results

Let Ω denote a bounded domain of the Euclidean space E^3 with $C^{2+\alpha}$-smooth boundary $\dot\Omega$, $\alpha \in (0,1)$; $\bar\Omega = \Omega \cup \dot\Omega$. Let us consider the problem about $2\pi\omega^{-1}$-periodic (with respect to time) solution of the following system of differential equations

$$\frac{\partial \bar u'}{\partial t} - \nu\Delta\bar u' + \nabla p' = -(\bar u', \nabla)\bar u' + \left[\bar a_0(x) + \omega \sum_{0<|k|\leq m} \bar a_k(x)e^{ik\omega t}\right] T'$$

$$+ \sum_{j=1}^{3} \frac{\partial}{\partial x_j}\bar b_j(x,\omega t),$$

$$\frac{\partial T'}{\partial t} - \chi\Delta T' = -(\bar u', \nabla T') + b_0(x,\omega t),$$

$$\operatorname{div}\bar u' = 0, \quad \bar u'|_{\dot\Omega} = 0, \quad T'|_{\dot\Omega} = h(x).$$

(2.1)

Here m – a positive integer, ν, $\chi > 0$; $\bar a_k$ – vector-functions with values from the three-dimensional complex space C^3, $\bar a_k \in C^1(\bar\Omega)$, $k \neq 0$, $\bar a_0 \in C(\Omega)$, $\bar a_k$ and $\bar a_{-k}$ are complex conjugate; b_j – vector-functions with values from R^3 and b_0, b_1 – the functions with values from R^1, b_j and b_0 are defined on the set $\bar\Omega \times R^1$, being continuous and 2π-periodic (with respect to τ) mappings to vector and scalar spaces $L_q(\Omega)$, $q > 3$ respectively; $h \in W_q^2(\Omega)$; ω – a large asymptotic parameter.

As noted in [11, 12] the problems of vibrational fluid convection, described in the introduction, is a particular case of the problem (2.1).

Following to Krylov–Bogoljubov idea (see Eq. (21.31) in [4]) and to Ref. [10], let us represent the function $\bar u'$ from (2.1) as

$$\bar u'(x,t) = \bar u(x,t) + \bar v(x,t),$$

where $\bar v(x,t) = \bar w(x,t,t)$, and the vector-function $\bar w(x,t,\tau)$ is a $2\pi\omega^{-1}$-periodic (with respect to t) solution of the problem

$$\frac{\partial \bar w(x,t,\tau)}{\partial t} - \nu\Pi\Delta\bar w = \omega\Pi \sum_{0<|k|\leq m} \bar a_k(x)e^{ik\omega t}T'(x,\tau), \quad \operatorname{div}\bar w = 0, \quad \bar w|_{\dot\Omega} = 0.$$

Here Π denotes the well-known Weyl projector (see for instance [13]) acting from the space L_q to the subspace S_q of solinoidal ($\operatorname{div}\bar a = 0$) vectors, which is an orthoprojector if $q = 2$.

For convenience, let us introduce the following notations

$$A_3(\bar{u}, T') = \chi \Delta T' - (\bar{u}, \nabla T'), \quad N(t)r = \sum_{0<|k|\leq m} (ik)^{-1} e^{ikt} \Pi \bar{a}_k r,$$

$$M_\omega(t)r = \bar{v}(x,t) - N(\omega t)r,$$

where $\bar{v}(x,t)$ – a $2\pi\omega^{-1}$-periodic (with respect to t) solution of the system

$$\frac{\partial \bar{v}}{\partial t} = \nu \Pi \Delta \bar{v} + \omega \Pi \sum_{0<|k|\leq m} \bar{a}_k e^{ik\omega t} r, \quad \operatorname{div} \bar{v} = 0, \quad \bar{v}|_{\dot{\Omega}} = 0.$$

Here $r \in L_q(\Omega)$. Further on the role of $r(x)$ is often played by the elements of the family of functions $T'(x,t)$ with t as parameter. Then the system (2.1) is reduced to the following form

$$\frac{\partial \bar{u}}{\partial t} - \nu \Delta \bar{u} + \nabla p = \bar{f}_1'(\bar{u}, T', \omega t) + \bar{\varphi}_{\omega,1}'(\bar{u}, T', t) + \sum_{j=1}^{3} \frac{\partial}{\partial x_j} \bar{b}_j(\cdot, \omega t)$$

$$\frac{\partial T'}{\partial t} - \chi \Delta T' = f_2'(\bar{u}, T', \omega t) + \varphi_{\omega,2}'(\bar{u}, T', t) + b_0(\cdot, \omega t), \quad \operatorname{div} \bar{u} = 0 \qquad (2.2)$$

$$\bar{u}|_{\dot{\Omega}} = 0, \qquad T'|_{\dot{\Omega}} = h.$$

Here

$$\bar{f}_1'(\bar{u}, T', t) = \bar{F}_1'(\bar{u}, \bar{T}') + \tilde{\bar{f}}_1'(\bar{u}, T', t), \quad \bar{F}_1'(\bar{u}, T') = -(\bar{u}, \nabla)\bar{u}$$

$$- \sum_{0<|k|\leq m} k^{-2} (\Pi \bar{a}_k T', \nabla) \Pi \bar{a}_{-k} T' + \bar{a}_0 T' + \sum_{0<|k|\leq m} k^{-2} \Pi \bar{a}_k (\Pi \bar{a}_{-k} T', \nabla T'),$$

$$\tilde{\bar{f}}_1'(\bar{u}, T', t) = -(N(t)T', \nabla)\bar{u} - (\bar{u}, \nabla)N(t)T'$$

$$- \sum_{\binom{0<|k|,|s|\leq m}{k+s\neq 0}} (ks)^{-1} e^{i(k+s)t} (\Pi \bar{a}_k T', \nabla) \Pi \bar{a}_s T' - N(t)A_3(\bar{u}, T')$$

$$+ \sum_{\binom{0<|k|,|s|\leq m}{k+s\neq 0}} (ks)^{-1} e^{i(k+s)t} \Pi \bar{a}_k (\Pi \bar{a}_s T', \nabla T'),$$

$$\bar{\varphi}_{\omega,1}'(\bar{u}, T', t) = -(M_\omega(t)T', \nabla)\bar{u} - (\bar{u}, \nabla)M_\omega(t)T' - (M_\omega(t)T', \nabla)M_\omega(t)T'$$

$$- (N(\omega t)T', \nabla)M_\omega(t)T' - (M_\omega(t)T', \nabla)N(\omega(t)T' - M_\omega(t)A_3(\bar{u}, T')$$

$$+ M_\omega(t)(M_\omega(t)T', \nabla T') + M_\omega(t)(N(\omega t)T', \nabla T') + N(\omega t)(M_\omega(t)T', \nabla T'),$$

$$f_2'(\bar{u}, T', t) = F_2'(\bar{u}, T') + \tilde{f}_2(\bar{u}, T', t), \quad F_2'(\bar{u}, T') = -(\bar{u}, \nabla T'),$$

$$\tilde{f}_2'(\bar{u}, T', t) = -(N(t)T', \nabla T'), \quad \varphi_{\omega,2}'(\bar{u}, T', t) = -(M_\omega(t)T', \nabla T').$$

Let $\bar{B}_j(x)$ and $B_0(x)$ denote averaged, over τ, values of $\bar{b}_j(x,\tau)$ and $b_0(x,\tau)$:

$$\bar{B}_j(x) = (2\pi)^{-1} \int_0^{2\pi} \bar{b}_j(x,\tau)d\tau, \quad B_0(x) = (2\pi)^{-1} \int_0^{2\pi} b_0(x,\tau)d\tau.$$

We denote as $\overset{\circ}{S}{}_q^2$ (as $\overset{\circ}{W}{}_q^2$) a closure, in the norm of $W_q^2(\Omega)$, of the set of smooth vector-functions from S_q (from L_q) vanishing on $\dot{\Omega}$. It should be noted that W_q^k (and L_q) denote not only usual Sobolev spaces of scalar functions but also spaces of three- and four-dimensional vector-functions, with their components being from those spaces.

Let the averaged system (2.2)

$$\frac{\partial \bar{v}}{\partial t} - \nu \Delta \bar{v} + \nabla q' = \bar{F}_1(\bar{v}, \tau') + \sum_{j=1}^{3} \frac{\partial}{\partial x_j} \bar{B}_j(x), \quad \operatorname{div} \bar{v} = 0$$

$$\frac{\partial \tau'}{\partial t} - \chi \Delta \tau' = \bar{F}_2'(\bar{v}, \tau') + B_0, \quad \bar{v}|_{\dot{\Omega}} = 0, \quad \tau'|_{\dot{\Omega}} = h$$

(2.3)

have a stationary solution $(\overset{\circ}{v}, \overset{\circ}{q}', \overset{\circ}{\tau}') \in \overset{\circ}{S}{}_q^2 \times \overset{\circ}{W}{}_q^1 \times \overset{\circ}{W}{}_q^2$.

Let us introduce the three linear operators A_1, A_2, A. The first two of them act from S_q and L_q, respectively, as

$$A_1 \bar{u} = -\nu \Pi \Delta \bar{u}, \qquad u \in D(A_1) = \overset{\circ}{S}{}_q^2$$
$$A_2 T = -\chi \Delta T, \qquad T \in D(A_2) = \overset{\circ}{W}{}_q^2$$

where $D(B)$ denotes a domain of definition of the operator B. Operator A acts in the space $S_q \times L_q$, $q > 3$ as

$$A\begin{pmatrix} \bar{v} \\ \tau \end{pmatrix} = \begin{pmatrix} A_1 \bar{v} + B_1 \bar{v} + C_1 \tau \\ A_2 \tau + B_2 \tau + C_2 \bar{v} \end{pmatrix}, \quad \begin{pmatrix} \bar{v} \\ \tau \end{pmatrix} \in D(A) = \overset{\circ}{S}{}_q^2 \times \overset{\circ}{W}{}_q^2.$$

Here

$$B_1 = \Pi B_1', \quad B_1' \bar{v} = (\overset{\circ}{v}, \nabla)\bar{v} + (\bar{v}, \nabla)\overset{\circ}{v}, \quad C_1 = \Pi C_1',$$

$$C_1' \tau = \sum_{0<|k|\leq m} k^{-2}(\Pi \bar{a}_k \overset{\circ}{\tau}', \nabla)\Pi \bar{a}_{-k}\tau + \sum_{0<|k|\leq m} k^{-2}(\Pi \bar{a}_k \tau', \nabla)\Pi \bar{a}_{-k} \overset{\circ}{\tau}'$$

$$- \sum_{0<|k|\leq m} k^{-2}\bar{a}_k(\Pi \bar{a}_{-k}\tau, \nabla \overset{\circ}{\tau}') \sum_{0<|k|\leq m} k^{-2}\bar{a}_k(\Pi \bar{a}_{-k} \overset{\circ}{\tau}', \nabla \tau) - \bar{a}_0 \tau,$$

$$B_2 \tau = (\overset{\circ}{v}, \nabla \tau), \quad C_2 \bar{v} = (\bar{v}, \nabla \overset{\circ}{\tau}').$$

$C_{(-\infty,\infty)}(E)$ $(E -$ Banach space$)$ denotes the space of the continuous and uniformly bounded E-valued vector-functions, defining on the axis $(-\infty, \infty)$, with usual *sup*-norm.

Theorem 2.1. *Let the spectrum Λ of the operator A do not contain zero. Then for some value $\omega > 0$ the following statements are valid:*

1. *A positive number r_0 exists such that at $\omega > \omega_0$ the system (2.1) has a $2\pi\omega^{-1}$-periodic (with respect to t) solution $(\bar{u}_\omega', p_\omega', T_\omega')$, with the component $(\bar{u}_\omega', T_\omega')$ of the solution being unique in the sphere*

$$\|T' - \overset{\circ}{\tau}'\|_{C_{(-\infty,\infty)}(W_q^1)} + \left\| \bar{u}_\omega' - \left(\overset{\circ}{v} + \sum_{0<|k|\leq m} (ik)^{-1} e^{ik\omega t} \Pi \bar{a}_k \overset{\circ}{\tau}' \right) \right\|_{C_{(-\infty,\infty)}(S_q)} \leq r_0$$

and the following relation holding for this solution

$$\lim_{\omega \to \infty} \left(\|T'_\omega - \overset{\circ}{\tau}{}'\|_{C_{(-\infty,\infty)}(W_q^1)} \right.$$
$$\left. + \left\| \bar{u}'_\omega - \left(\overset{\circ}{v} + \sum_{0<|k|\leq m} (ik)^{-1} e^{ik\omega t} \Pi \bar{a}_k \overset{\circ}{\tau}{}' \right) \right\|_{C_{(-\infty,\infty)}(S_q)} \right) = 0.$$

2. *If the spectrum* Λ *is contained in the right complex half-plane, then the solution* $(\bar{u}'_\omega, T'_\omega)$ *is exponentially stable in the norm of* $S_q \times W_q^1$, *uniformly with respect to* $\omega > \omega_0$ *and to the initial data.*

3. *If* Λ *contains at list one point at the left half-plane, then the solution* $(\bar{u}'_\omega, T'_\omega)$ *is unstable in the norm of* $S_q \times W_q^1$.

Prior to formulate Theorem 2.2, devoted to study conditional stability of the solution $(\bar{u}'_\omega, T'_\omega)$, we first introduce some preliminary concepts and new notations.

Let $(\bar{v}_\omega, q_\omega, T'_\omega)$ denote a $2\pi\omega^{-1}$ – periodic (with respect to t) solution of the system (2.2) connected with the solution $(\bar{u}'_\omega, p'_\omega, T'_\omega)$, cited in Theorem 2.1, by the change of variables described above. Namely, $\bar{v}_\omega = \bar{u}'_\omega - \sum\limits_{0<|k|\leq m} (ikI - \omega^{-1}A_1)^{-1} \Pi \bar{a}_k(x) e^{ik\omega t} T'(x,t)$, where I is a unit operator in S_q.

The system (2.2) may be linearized near the solution $(\bar{v}_\omega, q_\omega, T'_\omega)$. Let us apply a projector Π to the first equation (i.e., the Navier–Stokes equations) of the obtained system. Then we come to the system

$$\begin{aligned} \frac{\partial \bar{u}}{\partial t} - \nu \Delta \bar{u} &= \Pi[D_1(\omega,t)](\bar{u},\tau), \qquad \operatorname{div} \bar{u} = 0 \\ \frac{\partial \tau}{\partial t} - \chi \Delta \tau &= [D_2(\omega,t)](\bar{u},\tau), \qquad \bar{u}|_{\dot{\Omega}} = 0, \quad \tau_{\dot{\Omega}} = 0. \end{aligned} \tag{2.4}$$

Here $[D_1(\omega,t)](\bar{u},\tau)$, $[D_2(\omega,t)](\bar{u},\tau)$ – linear differential expressions with $2\pi\omega^{-1}$-periodic (with respect to t) coefficients

$$[D_k(\omega,t)](\bar{u},\tau) = f'_k(\bar{u}+\bar{v}_\omega, \tau+T'_\omega) + \varphi'_{\omega,k}(\bar{u}+\bar{v}_\omega, \tau+T'_\omega, t) - f'_k(\bar{v}_\omega, T'_\omega)$$
$$- \varphi'_{\omega,k}(\bar{v}_\omega, T'_\omega, t) - f'_k(\bar{u},\tau) - \varphi'_{\omega,k}(\bar{u},\tau) - \delta_k^1(N(t)\chi\Delta\tau + M_\omega(t)\chi\Delta\tau),$$

$k = 1, 2$. Here δ_k^i is the Kronecker delta, and in notations of the vector quantities $(k = 1)$ the sign of vector is omitted.

Let $t_\infty > 0$ denote some quantity so that $\exp(\lambda t_\infty) \neq 1$ when $\lambda \in \Lambda$, and let $t_\omega = [t_\infty \omega(2\pi)^{-1}]2\pi\omega^{-1}$. For arbitrary real numbers s_1, s_2 U_{s_1,s_2} denotes the translation operator over trajectories of the system (2.4), which maps the value of the solution at the time s_1 to the value of the solution at the time s_2. In particular, u_{s_1,s_1+t_ω} denotes the monodromy operator.

Theorem 2.2. *Let the spectrum* $\Lambda = \Lambda_- \cup \Lambda_+$, *where the set* Λ_- *is located in the left, and* Λ_+ *in the right complex half-plane. Then the positive numbers* r_0, σ_0, C_0 *exist such that for any* $s \in R$ *some finite-dimensional manifold* Y_ω^+ *and a manifold* *of finite codimension Y_ω^- are defined in the sphere $\|(\bar{u},\tau)\|_{S_q\times W_q^1} \le r_0$, with the following properties:*

1. *If $(\bar{u}_0,\tau_0) \in Y_\omega^-$ then for the vector $(\bar{u}_s',T_s')$, where $\bar{u}_s' = \bar{u}_\omega'(\cdot,s) + \bar{u}_0 + \sum_{0<|k|\le m} (ikI - \omega^{-1}A_1)^{-1}\Pi\bar{a}_k e^{ik\omega s}\tau_0,\ T_s' = T_\omega'(\cdot,s) + \tau_0$, a unique solution $(\bar{u}_\omega^s, T_\omega^s)$ of the initial-boundary problem exists for the equation (2.1) with the initial condition $\bar{u}_\omega(\cdot,s) = \bar{u}_s',\ T_\omega^s(\cdot,s) = T_s'$, defined on the ray $t \ge s$, with the following estimate*

$$\|\bar{u}_\omega'(\cdot,t) - \bar{u}_\omega^s(\cdot,t)\|_{S_q} + \|T_\omega'(\cdot,t) - T_\omega^s(\cdot,t)\|_{W_q^1}$$
$$\le c_0 e^{-\sigma_0(t-s)}\left(\|\bar{u}_\omega'(\cdot,s) - \bar{u}_s'\|_{S_q} + \|T_\omega'(\cdot,s) - T_s'\|_{W_q^1}\right).$$

2. *If the initial value (u_s',T_s') is outside (in the norm of $S_q \times W_q^1$) a vicinity of the vector $(u_\omega'(\cdot,s),T_\omega'(\cdot,s))$ with the radius r_0 , and it cannot be represented in the form stated in 1), then the following estimate is valid at some time $t_1 > s$*

$$\|\bar{u}_\omega'(\cdot,t_1) - \bar{u}_\omega^s(\cdot,t_1)\|_{S_q} + \|T_\omega'(\cdot,t_1) - T_\omega^s(\cdot,t_1)\|_{W_q^1} > r_0. \qquad (2.5)$$

3. *If $(\bar{u}_0,\tau_0) \in Y_\omega^-$ and the initial value $(\bar{u}_s',T_s')$ has the form cited in 1), then the respective solution $(\bar{u}_\omega^s,T_\omega^s)$ of the equation (2.1) is defined for all $t \le s$, with*

$$\|\bar{u}_\omega'(\cdot,t) - \bar{u}_\omega^s(\cdot,t)\|_{S_q} + \|T_\omega'(\cdot,t) - T_\omega^s(\cdot,t)\|_{W_q^1}$$
$$\le c_0 e^{\sigma_0(t-s)}\left(\|\bar{u}_\omega'(\cdot,s) - \bar{u}_s'\|_{S_q} + \|T_\omega'(\cdot,s) - T_s'\|_{W_q^1}\right), \quad t \le s.$$

4. *If the initial value $(\bar{u}_s',T_s')$ is outside (in the norm of $S_q \times W_q^1$) a vicinity of the vector $(u_\omega'(\cdot,s),T_\omega'(\cdot,s))$ with the radius r_0 , and it cannot be represented in the form stated in 3, then the respective solution is either undefined for all $t \le s$ or the inequality (2.5) holds at some time $t_1 < s$.*

5. *The manifolds Y_ω^- and Y_ω^+ are tangent (at the origin) to subspaces, invariant respectively the monodromy operator $U_{s,s+t_\omega}$, and corresponding to its spectral subdomains in and out of the unite circle, respectively.*

3. Proof of the theorems

$1°$. Acting to the first (vector) equation of the system (2.2) by the projector Π and performing then the change

$$\bar{u} \to \bar{u} + \overset{\circ}{v}, \qquad T' \to T + \overset{\circ}{\tau}',$$

one comes to the system

$$\frac{\partial z}{\partial t} + Az = f(z,t,\omega), \qquad z|_{\dot{\Omega}} = 0. \qquad (3.1)$$

Here

$$z = (\bar{u},T) \in S_q \times L_q, \quad q > 3, \quad f = (\Pi\bar{f}_1, f_2),$$

$$\bar{f}_1(\bar{u}, T, \omega) = \bar{f}_1'(\bar{u} + \overset{\circ}{v}, T + \overset{\circ'}{\tau}, \omega t) - \bar{F}_1'(\overset{\circ}{v}, \overset{\circ'}{\tau}) + B_1\bar{u} + C_1 T$$

$$+ \bar{\varphi}_{\omega,1}'(\bar{u} + \overset{\circ}{v}, T + \overset{\circ'}{\tau}, \omega t) + \sum_{j=1}^{3} \frac{\partial}{\partial x_j}\bar{b}_j(\cdot, \omega t),$$

$$\bar{f}_2(\bar{u}, T, \omega) = \bar{f}_2'(\bar{u} + \overset{\circ}{v}, T + \overset{\circ'}{\tau}, \omega t) - \bar{F}_2'(\overset{\circ}{v}, \overset{\circ'}{\tau}) + B_2 T + C_2\bar{u}$$

$$+ \bar{\varphi}_{\omega,2}'(\bar{u} + \overset{\circ}{v}, T + \overset{\circ'}{\tau}, \omega t) + b_0(\cdot, \omega t).$$

Let us introduce some notations. Let $W_{q,0}^1$ denote a subspace of the space $W_q^1(\Omega)$ which consists of the functions $h(x)$ vanishing on $\dot{\Omega}(h|_{\dot{\Omega}} = 0)$. Let the brief B_q denote the Cartesian product $S_q \times W_{q,0}^1$.

Let E be a Banach space, and $C(E) \equiv C_{[0,t_\infty]}(E)$ denote the space of the continuous vector-functions $u\colon [0, t_\infty] \to E$ with a max-norm. Let $C_\gamma(E), \gamma \in (0,1)$ denote a usual Hölder space of the vector-functions $u \in C(E)$ satisfying the Hölder condition with the index γ. At last, we denote as $\hat{C}_\gamma(E)$ the weighted Hölder space containing the vector-functions $u \in C(E)$ which satisfy the condition

$$\|u\|_{\hat{C}_\gamma(E)} = \|u\|_{C(E)} + \sup_{0<t_1<t_2\leq\infty}\left(t_1^\gamma \frac{\|u(t_2) - u(t_1)\|_E}{(t_2 - t_1)^\gamma}\right) < \infty.$$

The following multiplicative inequality is trivially proved

$$\|u\|_{\hat{C}_{\gamma_1}(E)} \leq \|u\|_{\hat{C}_{\gamma_2}(E)}^{\gamma_1/\gamma_2}\left(2\|u\|_{C(E)}\right)^{1-\gamma_1/\gamma_2}$$

where $0 < \gamma_1 < \gamma_2$, $u \in \hat{C}_{\gamma_2}(E)$.

According to [14]–[16] the operator-A generates an analytical semigroup e^{-tA} both in the space $S_q \times L_q$ and in the space B_q, $q > 3$.

Statement 3.1. *For $q \geq p > 3$ a certain constant c exists such that at arbitrary $z = (\bar{u}, T) \in B_p$ for all $t \in (0, t_\infty)$, $0 < t_1 < t_2 \leq t_\infty$ and $\gamma \in (0,1)$ the following estimates are valid*

$$\left\|\frac{\partial^k}{\partial x_i^k}e^{-tA}z\right\|_{B_q} \leq ct^{-\beta-k/2}\|z\|_{B_p} \tag{3.2}$$

$$\frac{\|\frac{\partial^k}{\partial x_i^k}(e^{-t_2 A} - e^{-t_1 A})z\|_{B_q}}{(t_2 - t_1)^\gamma} \leq ct_1^{-\beta-\gamma-k/2}\|z\|_{B_p} \tag{3.3}$$

$$\left\|e^{-tA}\tilde{\Pi}\frac{\partial}{\partial x_i}z\right\|_{B_q} \leq ct^{-1/2-\beta}\|z\|_{B_p} \tag{3.4}$$

$$\frac{\|(e^{-t_2 A} - e^{-t_1 A})\tilde{\Pi}\frac{\partial}{\partial x_i}z\|_{B_q}}{(t_2 - t_1)^\gamma} \leq ct_1^{-\beta-\gamma-1/2}\|z\|_{B_p} \tag{3.5}$$

where $\tilde{\Pi}z = (\Pi\bar{u}, T)$, $\beta = (3/2)(1/p - 1/q)$, $i = 1, 2, 3$, $k = 0, 1$.

The proof of Statement 3.1 is based on the following estimates for resolvent of the operator A.

Lemma 3.2. *Real numbers λ_0, c and $\varphi \in (0, \pi/2)$ exist such that the spectrum Λ of the operator A is in the sector $\Sigma_{\lambda_0,\varphi} \equiv \{\lambda \in C : |\arg(\lambda - \lambda_0)| \le \varphi\}$ with the boundary $\Gamma_{\lambda_0,\varphi} \equiv \Gamma$, and for all $\lambda \notin \Sigma_{\lambda_0,\varphi}$ and all vector-functions z from the sets cited below the following estimates hold*

$$\left\|\frac{\partial^k}{\partial x_i^k}(\lambda I - A)^{-1} z\right\|_{L_q} \le C(1 + |\lambda|)^{\beta+k/2-1}\|z\|_{L_p}, \qquad z \in S_p \times L_p \qquad (3.6)$$

$$\left\|\frac{\partial^k}{\partial x_i^k}(\lambda I - A)^{-1} z\right\|_{B_q} \le C(1 + |\lambda|)^{\beta+k/2-1}\|z\|_{B_p}, \qquad z \in B_p \qquad (3.7)$$

$$\left\|(\lambda I - A)^{-1}\tilde{\Pi}\frac{\partial}{\partial x_i}z\right\|_{L_q} \le C(1 + |\lambda|)^{\beta-1/2}\|z\|_{L_p}, \qquad z \in S_p \times L_p \qquad (3.8)$$

$$\left\|(\lambda I - A)^{-1}\tilde{\Pi}\frac{\partial}{\partial x_i}z\right\|_{B_q} \le C(1 + |\lambda|)^{\beta-1/2}\|z\|_{B_p}, \qquad z \in B_p \qquad (3.9)$$

where $i = 1, 2, 3$, $k = 0, 1$.

Proof of Lemma 3.2. The estimates (3.6) and (3.8) are obtained in [10]. The estimates (3.7) and (3.9) can be proved by analogy, with the use of relations (3.7)–(3.13) of [10]. With this, additionally to inequalities (3.11)–(3.13) of [10] the estimate

$$\|(\lambda I_2 - A_2)^{-1}T\|_{W_{q,0}^1} < \frac{c}{1 + |\lambda|}\|T\|_{W_{q,0}^1}$$

should be taken into account, where $\varphi_0 \le |\arg \lambda| \le \pi$ for arbitrary fixed $\varphi_0 \in (0, \pi/2)$, I_2 – a unit operator in $W_{q,0}^1$, $T \in W_{q,0}^1$ and $C \equiv C(\varphi_0)$ is a constant independent on the quantities λ and T. This estimate is derived with the help of some methodology described in [13, p. 51–52] for the estimates of resolvent of certain differentiable operators in the space W_p^k. $\square$

Proof of Statement 3.1. We will derive here the estimate (3.5) only, since the estimates (3.2)–(3.4) are proved by analogy. For the semigroup e^{-tA}, according to (3.6) the following representation [17] is valid

$$e^{-tA} = \frac{1}{2\pi i}\int_\Gamma e^{-\lambda t}(\lambda I - A)^{-1}d\lambda. \qquad (3.10)$$

Obviously, the estimate (3.5) may be proved only for smooth finitary vector-functions $z \in B_{p,0}$. For such z, according to (3.10), we have

$$\left\|(e^{-t_2 A} - e^{-t_1 A})\tilde{\Pi}\frac{\partial}{\partial x_i}z\right\|_{B_q} = \left\|\int_{t_1}^{t_2}\frac{d}{d\tau}e^{-\tau A}\tilde{\Pi}\frac{\partial}{\partial x_i}z\,d\tau\right\|_{B_q}$$

$$= \left\|\int_{t_1}^{t_2}d\tau\int_\Gamma \lambda e^{-\lambda \tau}(\lambda I - A)^{-1}\tilde{\Pi}\partial_{x_i}z\,d\lambda\right\|_{B_q}$$

$$\leq c_1 \int\limits_{t_1}^{t_2} \exp -\lambda_0 \tau d\tau \int\limits_{0}^{\infty} e^{-r\tau \cos\varphi} \frac{rdr}{(1+r)^{1/2-\beta}} \|z\|_{B_p} \leq c_2 \int\limits_{t_1}^{t_2} \tau^{-3/2-\beta} d\tau \|z\|_{B_p}$$

where the constants c_1, c_2 do not depend on t_1, t_2, z. Now, applying the Hölder inequality, we obtain

$$c_2 \int\limits_{t_1}^{t_2} \frac{d\tau}{\tau^{3/2+\beta}} \|z\|_{B_p} \leq c_2 (t_2-t_1)^\gamma \left(\int\limits_{t_1}^{t_2} \frac{d\tau}{\tau^{(3/2+\beta)/(1-\gamma)}} \right)^{1-\gamma} \|z\|_{B_p}$$

$$\leq c_3 (t_2-t_1)^\gamma t_1^{-\beta-\gamma-1/2} \|z\|_{B_p}. \qquad \square$$

$2°$. Let $q > 3$, $\gamma \in (0, (1-3/q)/2)$. Let us introduce the space $H_q^\gamma \equiv \hat{C}_\gamma(S_q) \times \hat{C}_\gamma(W_{q,0}^1)$ and the operator $N \colon H_q^\gamma \times B_q \times (0,\infty] \to H_q^\gamma$ acting as follows

$$[N(z,z_0,\omega)](t) = \begin{cases} \displaystyle\int\limits_0^t e^{-(t-s)A} f[z(s),s,\omega]ds + e-tAz_0, & \omega \neq \infty \\[6pt] \displaystyle\int\limits_0^t e^{-(t-s)A} F[z(s)]ds + e-tAz_0, & \omega = \infty \end{cases}$$

Here $F(z) \equiv (\Pi\bar{F}_1(\bar{u},T), \bar{F}_2(\bar{u},T))$, $\bar{F}_1(\bar{u},T) = \bar{F}_1'(\bar{u}+\overset{\circ}{v}, T+\overset{\circ'}{\tau}) - \bar{F}_1'(\overset{\circ}{v},\overset{\circ'}{\tau}) + B_1\bar{u} + C_1 T + \sum\limits_{j=1}^{3} (\partial)(\partial x_j)\bar{B}_j$, $F_2(\bar{u},T) = F_2'(\bar{u}+\overset{\circ}{v}, T+\overset{\circ'}{\tau}) - F_2'(\overset{\circ}{v},\overset{\circ'}{\tau}) + B_2 T + C_2 + B_0$.

Lemma 3.3. *Operator N is continuous and continuously differentiable with respect to variables z, z_0.*

Proof. Let $(D_z N)(z,z_0,\omega)$ and $(D_{z_0} N)(z,z_0,\omega)$ denote the Fréchet differentials of the operator N with respect to z and z_0. It follows from the estimates (3.2), (3.3) at $q = p$, $k = 0$ that

$$[(D_{z_0} N)(z,z_0,\omega)]\xi_0 = e^{-tA}\xi_0. \tag{3.11}$$

Prior to define the differential $D_z N$ let us denote as $[(D_z f_1)(z,t,\omega)]\xi$ and $[(D_z f_2)(z,t,\omega)]\xi$ ($z = (\bar{u},T)$, $\xi = (\bar{y},S)$) the linear (with respect to ξ) parts of the differential expressions $\bar{f}_1(z+\xi,t,\omega) - \bar{f}_1(z,t,\omega)$ and $f_2(z+\xi,t,\omega) - f_2(z,t,\omega)$. Let $[(D_z f)(z,t,\omega)]\xi$ denote the vector-function $\Pi[(D_z\bar{f}_1)(z,t,\omega)]\xi$, $[(D_z f_2)(z,t,\omega)]\xi$. By analogy, the vector-function $[(D_z F)(z)]\xi$ is introduced.

Let us prove that

$$[(D_z N)(z,z_0,\omega)](\xi) = \begin{cases} \displaystyle\int\limits_0^t e^{-(t-s)A} f[(D_z f)(z,s,\omega)]\xi ds, & \omega \neq \infty \\[6pt] \displaystyle\int\limits_0^t e^{-(t-s)A} [(D_z F)(z)]\xi ds, & \omega = \infty. \end{cases} \tag{3.12}$$

First of all, we state the two simple relations:

$$(\bar{a}, \nabla)\bar{b} = \sum_{i=1}^{3} \frac{\partial}{\partial x_i}(a_i \bar{b}), \quad \bar{a} = (a_1, a_2, a_3) \in S_q \cap W_q^1, \quad \bar{b} \in W_q^1 \qquad (3.13)$$

$$M_\omega(t)r = -\sum_{0<|k|\leq m} (ik)^{-1} e^{ik\omega t} A_1 (ik\omega I_1 - A_1)^{-1} \times \Pi a_k r, \quad r \in L_q. \qquad (3.14)$$

It follows from Eq. (3.14), with the fact that the operator A_1 generates an analytical semigroup [13, 15, 16], the following estimate

$$\|M_\omega(t)r\|_{C(S_q)} \leq C\|r\|_{L_q}, \qquad (3.15)$$

where the constant C does not depend on $r \in L_q$ and $\omega \gg 1$.

With the help of relations (3.13), (3.15), the Hölder inequality

$$\|ab\|_{L_{p/2}} \leq \|a\|_{L_p}\|b\|_{L_p}, \quad a, b \in L_p, \quad p > 2 \qquad (3.16)$$

and Statement 3.1, the following estimate is proved

$$\left\| \int_0^t e^{-(t-s)A}[(D_z f)(z, s, \omega)]\xi ds \right\|_{C_\gamma(B_q)} \leq C\|\xi\|_{H_q^\gamma}, \qquad (3.17)$$

where $z, \xi \in H_q^\gamma$, and the constant C does not depend on ξ and ω. Analogous estimate can be stated by the same technique for $D_z F$.

As an example, we derive here the inequality (3.17). Let us operate with the typical term $\varphi \equiv (M_\omega(t)(M_\omega(t)T', \nabla S) - (\bar{y}, \nabla T))$ instead of $[(D_z f)(z, t, \omega)]\xi$. Let us restrict the consideration by the estimate of the quantity

$$\chi = (t_2 - t_1)^{-\gamma} \left\| \int_0^{t_1} \left[e^{-(t_2-s)A} - e^{-(t_1-s)A} \right] \varphi ds \right\|_{B_p}$$

$$+ (t_2 - t_1)^{-\gamma} \left\| \int_{t_1}^{t_2} e^{-(t_2-s)A} \varphi ds \right\|_{B_p} \equiv \chi_1 + \chi_2,$$

where $0 < t_1 < t_2 \leq \infty$. We have, due to (3.15), (3.16), (3.2) and (3.3)

$$\chi_1 \leq c_1 \int_0^t (t_1 - s)^{-3/2q-\gamma} ds (\|T'\|_{L_q}\|\nabla S\|_{L_q} + \|\nabla T\|_{L_q}\|\bar{y}\|_{S_q}) \leq c_1\|\xi\|_{B_q},$$

$$\chi_2 \leq (t_2 - t_1)^{-\gamma} c_1 \int_{t_1}^{t_2} (t_2 - s)^{-3/2q} ds \left(\|T\|_{L_q} + \|\nabla T\|_{L_q}\right)\left(\|\nabla S\|_{L_q} + \|\bar{y}\|_{S_q}\right)$$

$$\leq c_1\|\xi\|_{B_q}, \quad c_1 = \text{const.}$$

$3°$. It is known that solution of the Eq. (3.1) satisfying the initial condition $z(0) = z_0$, is a fixed point of the operator $M: H_q^\gamma \times B_q \times (0, +\infty] \to H_q^\gamma$ acting by the law

$$M(z, z_0, \omega) \equiv z - N(z, z_0, \omega). \qquad (3.18)$$

It follows from Lemma 3.3, together with the evident equalities $M(0,0,+\infty) = 0$, $(D_z M)(0,0,+\infty) = I$ ($D_z M$ – the Fréchet differential of the operator M with respect to z, I – the unit operator in H_q^γ), and from the theory of implicit functions, that a number $\omega_1 > 0$ and a vicinity O_1 of the origin in B_q exist, such that at $\omega > \omega_1$ and $z_0 \in O_1$ a solution $Z(z_0,\omega)$ of the problem (3.18) exists, with the mapping $Z: O_1 \times (\omega_1, +\infty] \to H_q^\gamma$ being continuous and continuously-differentiable with respect to z_0.

Let us denote the value of the solution $Z(z_0,\omega)$ at the point t as $[Z(z_0,\omega)](t)$. The operator $U(z_0,\omega,t) = [Z(z_0,\omega)](t)$ provides a continuous and continuously-differentiable (with respect to z_0) mapping of the space $O_1 \times (\omega_1,+\infty] \times (0, t_\infty]$ to the space B_q. Note that the spectrum of the Fréchet differential $D_{z_0} U(0,+\infty,t) = e^{-tA}$ consists of the numbers $e^{-\lambda t}$, $\lambda \in \Lambda$ and the number 0.

According to the theorem of implicit function, for arbitrary sphere $O_2(O_1)$ of a sufficiently small radius $r > 0$ with the center in the origin, such numbers $\omega_2 > \omega_1$ and $\delta > 0$ exist that at $\omega > \omega_2$ and $t \in (t_\infty - \delta, t_\infty)$ the equation

$$z_0 = U(z_0,\omega,t) \tag{3.19}$$

has a unique solution z_0^ω in the sphere O_2. The equation (3.19) at $t = t_\omega \equiv [t_\infty \omega (2\pi)^{-1}] 2\pi \omega^{-1}$ defines initial data of the t_ω-periodic (with respect to t) solutions. These solutions, due to their local uniqueness in the space $C_{(-\infty,\infty)}(B_q)$ and to a $2\pi\omega^{-1}$-periodicity (with respect to t) of the right-hand side of the system (3.1), are also $2\pi\omega^{-1}$-periodic. Statement 1 of Theorem 2.2 follows from this fact.

4°. In this and the next sections we prove Statements 2 and 3 of Theorem 2.1 and Theorem 2.2. We set here $s = 0$, to reduce the volume of the writing.

Let $z_\omega(t)$ is a $2\pi\omega^{-1}$-periodic (with respect to t) solution of the system (3.1) corresponding to the initial condition z_0^ω, which exists as stated in the previous section. In accordance with the above notations, $z_\omega = (\bar{v}_\omega - \overset{\circ}{v}, T'_\omega - \overset{\circ'}{\tau})$. Let us apply in (3.1) the change $z = y + z_\omega$. One then obtains

$$\frac{\partial y}{\partial t} + Ay = \widehat{f}\,(y,t,\omega), \tag{3.20}$$

where $\widehat{f}\,(y,t,\omega) = f(y + z_\omega, t, \omega) - f(z_\omega, t, \omega)$. Let us set $\widehat{F}\,(y,t) = F(y) - F(0)$, and introduce the operator $\widehat{N}: H_q^\gamma \times B_q \times (0, \infty] \to H_q^\gamma$, $\gamma \in (0,1)$, $q > 3$ by the formula

$$\widehat{N}\,(y, y_0, \omega) = \begin{cases} \displaystyle\int_0^t e^{-(t-s)A}\, \widehat{f}\,[y(s),s,\omega)]ds + e^{-tA}y_0, & \omega \neq \infty \\[2em] \displaystyle\int_0^t e^{-(t-s)A}\, \widehat{F}\,[y(s)]ds + e^{-tA}y_0, & \omega = \infty. \end{cases}$$

Lemma 3.4. *Operator* $\widehat{N}$ *is continuous and continuously-differentiable with respect to the variables* y, y_0.

By analogy to Section 3° we can state existence of the number $\omega > 0$ and a vicinity O_1 of the origin in B_q such that at $\omega > \omega_1$ and $y_0 \in O_1$ some solution $y = \hat{U}(y_0, \omega)$ of the equation $y = \widehat{N}(y, y_0, \omega)$ exists, with the map $\hat{U} : O_1 \times (\omega_1, +\infty) \to H_q^\gamma$ being continuous and continuously-differentiable with respect to y_0. We denote the value of this solution at the point $t \in (0, t_\infty)$ as $\widehat{U}(y_0, \omega, t)$, i.e., $\widehat{U}(y_0, \omega, t) = [\hat{U}(y_0, \omega)](t)$.

Obviously, $U(\cdot, \omega, t)$ at $\omega \neq \infty$ is a translation operator along trajectories of the system (3.20). It can be stated, in the same way like in 3°, that the operator $\widehat{U}$ represents a continuous and continuously-differentiable (with respect to y_0) mapping of the space $O_1 \times (\omega_1, +\infty] \times (0, t_\infty]$ to B_q. Besides, the following lemma is valid.

Lemma 3.5. *The following relations take place*

$$(D_{y_0} \widehat{U})(0, \omega, t) = U_{0,t}, \quad \omega \neq \infty; \qquad (D_{y_0} \widehat{U})(0, \infty, t) = e^{-tA}.$$

Proof. First of all, let $\widehat{U}(y, y_0, \omega, t)$ denote the value of the function $\widehat{U}(y, y_0, \omega)$ at the time t. The theorem about implicit functions involves

$$[(D_{y_0} \widehat{U})(0, \omega, t)]u_0 = -[(D_y \widehat{N})(0, 0, \omega, t)]^{-1}[(D_{y_0} \widehat{N})(0, 0, \omega, t)]u_0. \qquad (3.21)$$

According to equations (3.11), (3.12)

$$[(D_{y_0} \widehat{N})(0, 0, \omega, t)]u_0 = e^{-tA}u_0 \qquad (3.22)$$

$$(D_y \widehat{N})(0, 0, \omega, t)]\xi = \begin{cases} \int_0^t e^{-(t-s)A}[(D_y f)(z_\omega, s, \omega)]\xi(s)ds, & \omega \neq \infty \\ \\ \int_0^t e^{-(t-s)A}[(D_y F)(0, s, \omega)]\xi(s)ds, & \omega = \infty. \end{cases} \qquad (3.23)$$

If one rewrites the system (2.4) so that the left-hand side consists of the expression $d\hat{z}/dt - A\hat{z}$, $\hat{z} = (\bar{u}, \tau)$, then passes from the so obtained system to the integral equation (with the kernel $e^{-(t-s)A}$), and then applies to the later a theorem of implicit function, thus as a result the translation operator $U_{0,t}$ will be represented at the same form as $(D_{y_0} \widehat{U})(0, \omega, t)]\xi$ (see Eqs. (3.21)–(3.23)). The lemma is proved.

Lemma 3.6. *For all* $t > 0$ *the operators* $(D_{y_0} \widehat{U})(0, \omega, t)]\xi$ *at sufficiently large* ω *and the operator* e^{-tA}, *all acting in the space* B_q, $q > 3$, *are completely continuous.*

Proof. Obviously, one-to-one operator A acts continuously from the space $D(A) = \overset{\circ}{S}{}^2_q$ to $S_q \times L_q$. It follows from this fact, in accordance with the Banach theorem, the coercive inequality

$$\|z\|_{\overset{\circ}{S}{}^2_q \times \overset{\circ}{W}{}^2_q} \le c\|Az\|_{S_q \times L_q}, \quad z \in D(A), \quad c = \mathrm{const.}$$

From this inequality and the estimate

$$\|Ae^{-tA}z\|_{S_q \times L_q} \le ct^{-1}\|z\|_{S_q \times L_q}$$

well known in group theory, one can state that the operator e^{-tA} transfers arbitrary bounded set in B_q to a bounded set in $\overset{\circ}{S}{}^2_q \times \overset{\circ}{W}{}^2_q$, which is compact in B_q, due to an enclosure theorem. Thus complete continuity of the operator e^{-tA} is proved. Then the complete continuity of the operators $(D_{y_0} \widehat{U})(0, \omega, t)$, $\omega \gg 1$ is evident, due to their closeness to the operator e^{-tA}. Lemma 3.6 is proved.

It should be noted that the proof of the stated fact could be omitted. Indeed, the operator A^{-1} is completely continuous in B_q. So the semigroup of operators, whose resolvent is completely continuous, consists itself of completely continuous operators.

5°. The spectrum of the operator $e^{-t_\omega A}$, $\omega \gg 1$ acting in B_q is given as $e^{-\lambda t_\omega}$, $\lambda \in \Lambda$ and the value zero. Hence, under conditions 2 of Theorem 2.1 the spectrum of the operator $e^{-t_\omega A}$, $\omega \gg 1$ is inside a unit circle; under conditions 3 of Theorem 2.1 it, additionally to some points of the closed unit circle, contains at least one point out of the circle; under conditions of Theorem 2.2 the spectrum of the operator $e^{-t_\omega A}$ is a union of spectral subdomains σ_ω^+ and σ_ω^- such that σ_ω^+ is inside the unit circle of the complex plane, and σ_ω^- outside of it. The stated closeness between operators $e^{-t_\omega A}$ and $(D_{y_0} \widehat{U})(0, \omega, t_\omega)$, $\omega \gg 1$ involves closeness of their respective spectral projectors, so the cited properties of the spectrum of the operator $e^{-t_\omega A}$ are valid for the spectrum of the operator $(D_{y_0} \widehat{U})(0, \omega, t_\omega)$ too. Now the statements of the points 2, 3 in Theorem 2.1 and Theorem 2.2 follow, respectively, from Lemmas 3.1, 4.1 and 5.1, Chapter 3 of the Ref. [13].

Let us accent briefly on some details in the proof of Theorem 2.2 only. Let γ_1 and γ_2 be smooth positively oriented contours containing the sets σ_ω^+ and σ_ω^-, with γ_1 being entirely inside and γ_2 outside of the unit circle. Obviously, the contours γ_1 and γ_2 belong to the resolvent set of the operator $(D_{y_0} \widehat{U})(0, \omega, t_\omega)$ for sufficiently large ω. Let us denote the spectral sets of this operator, located inside the contours γ_1 and γ_2 as s_ω^+ and s_ω^-, respectively. Due to a closeness (in a usual spectral topology $\mathrm{Hom}(B_q, B_q)$) of the spectral projectors

$$P_\omega^+ = -\frac{1}{2\pi i} \int_{\gamma_1} (\lambda I - e^{-t_\omega A})^{-1} d\lambda$$

and

$$Q_\omega^+ = -\frac{1}{2\pi i} \int_{\gamma_1} [\lambda I - (D_{y_0} \widehat{U})(0, \omega, t_\omega)]^{-1} d\lambda, \quad \omega \gg 1,$$

the spectral subspaces P_ω^+ and Q_ω^+ are isomorphic. Therefore, the total multiplicity of eigenvalues (which are in fact non-trivial points of the spectrum, due to Lemma 3.6) of the operator $e^{-t_\omega A}$, located in σ_ω^+, coincides with the total multiplicity of the eigenvalues of the operator $(D_{y_0} \widehat{U})(0, \omega, t_\omega)$, and so

$$s_\omega^+ \neq \varnothing. \tag{3.24}$$

When proving Theorem 2.2, we need to attract the following simple

Lemma 3.7. *There are some vicinity of the origin v_0 in B_q, $q > 3$ and the positive values ω_0, c_0 such that for $\omega > \omega_0$, arbitrary $y_0 \in v_0$ and arbitrary positive integer n a solution of equation (3.20) with the initial condition $y(nt_\omega) = y_0$ exists, with the following estimate holding*

$$\|y(t)\|_{B_q} \leq c_0, \qquad t \in [nt_\omega, (n+1)t_\omega]. \tag{3.25}$$

Proof. Let v_1 and ω_1 be the same as in definition of the operator $\widehat{U}$. Then there are a vicinity of the origin $v_0 \subset v_1$ and the numbers $\omega_0 \geq \omega_1$, c_0 such that for all $y_0 \in v_0$ and $\omega > \omega_0$ a solution $\widehat{U}(y_0, \omega, t)$ of equation (3.20) exists on the segment $t \in [0, t_\infty]$, under the initial condition $u(0) = y_0$, and the following estimate is valid

$$\| \widehat{U}(y_0, \omega, t) \|_{B_q} \leq c_0, \quad t \in [0, t_\infty]. \tag{3.26}$$

The results of Lemma 3.6 follow from the last estimate, due to a t_ω-periodicity (with respect to t) of the coefficients of equation (3.20). In particular, the inequality (3.26) involves the estimate (3.25).

6°. The following statement proposed by Yudovich (see [13], Chapter 3, Lemma 5.2) is of a great importance, when proving Theorem 2.2. We cite it here, to complete the present work.

Let us consider at the arbitrary Banach space B the equation

$$x = Nx,$$

where N – operator defined at some vicinity D_{r_0} of the origin in the space B, continuous and continuously-differentiable in this vicinity. Let U denote the Fréchet differential $(DN)(0)$ of the operator N at the origin. Suppose $N0 = 0$. Then Lemma 7 of Ref. [13] states the following result. Let the spectrum of the operator U can be represented as a union of nonintersecting closed sets σ_1 and σ_2, with $|\sigma_1| > 1$, $|\sigma_2| < 1$. Then in a certain neighborhood $D_r \subset D_{r_0}$ of the origin in the space B there are defined invariant (with respect to operator N) manifolds Y_1 and Y_2, which are tangent to invariant subspaces of the operator U, corresponding to the spectral sub-domains σ_1 and σ_2 respectively. With all this: 1) if $x_0 \in Y_2$, then the successive approximations $N_{x_0}^n \notin D_r$ converge to zero at $n \to \infty$, and for a certain $\rho_0 \in (0, 1)$ the estimates $\|N^n x_0\| \leq C\rho_0^n \|x_0\|$ ($C = $ const, $n = 0, 1, 2, \ldots$)

are valid; 2) if $x_0 \notin Y_2$, then $N^n x_0 \notin D_r$ for some n; 3) for arbitrary $x_0 \in Y_2$ the inverse map $N^{-1} x_0 \in Y_1$ is defined and $N^{-n} x_0 \to 0$ at $n \to \infty$, with the following estimate holding $\|N^{-n} x_0\| \leq C \rho_0^n \|x_0\|$, where $\rho_0 \in (0, 1)$, $C = \text{const}$, $n = 0, 1, 2, \ldots$; 4) if $x_0 \notin Y_1$ then for some n either the elements N_{-n} are not defined or $N^{-n} x_0 \notin D_r$.

Proof of Theorem 2.2. Let the neighborhood v_1 of the origin in B_q and $\omega_1 > 0$ be the same as in $4°$. Then at $\omega > \omega_1$ a continuous and continuously-differentiable operator $V_\omega \colon v_1 \to v_1$ is defined, such that $V_\omega(y_0) = \widehat{U}\ (y_0, \omega, t_\omega)$. Let us consider the equation $y_0 = V_\omega(y_0)$. Obviously, $V_\omega(0) = 0$. In accordance with $5°$ the spectrum of the Fréchet differential $W_\omega \equiv (D_{y_0} V_\omega)(0)$ is a union of the spectral subdomains s_ω^+ and s_ω^- such that $|s_\omega^+| > 1$ and $|s_\omega^-| < 1$, with (see Eq. (3.24)) $s_\omega^+ \neq \oslash$ and consists of a finite number of eigenfrequencies of finite multiplicities. Let X_ω^+ and X_ω^- – invariant (with respect to operator W_ω) subspaces, corresponding to the spectral subdomains s_ω^+ and s_ω^-, respectively. It is clear that X_ω^+ – a finite-dimensional space. According to Lemma 3.7, there is a positive $r > 0$ such that in the sphere $S_r \equiv \{y \in B_q \colon \|y\|_{B_q} \leq r\}$ some invariant (with respect to operator V_ω) manifolds Y_ω^+ and Y_ω^- are defined, with the first of them being of a finite dimension, and the second one having a finite co-dimension, which are tangent at the origin to the subspaces X_ω^+ and X_ω^-, respectively. In addition, there are such numbers $\rho_0 \in (0, 1)$ and $C > 0$ that a) if $y_0 \in Y_\omega^-$ then the successive approximations $V_\omega^n(y_0)$ satisfy the estimates $\|V_\omega^n(y_0)\|_{B_q} \leq C \rho_0^n \|y_0\|_{B_q}$; b) if $y_0 \notin Y_\omega^-$ then $V_\omega^n(y_0) \notin S_r$ for some n; c) for arbitrary $y_0 \notin Y_\omega^+$ the inverse map $V_\omega^{-1}(y_0)$ is defined with $\|V_\omega^{-n}(y_0)\|_{B_q} \leq C \rho_0^n \|y_0\|_{B_q}$; d) if $y_0 \notin Y_\omega^+$ then for a certain n either the elements V_ω^{-n} are not defined or $V^{-n}(y_0) \notin S_r$.

Let us finish to prove Theorem 2.2. We start from Statement 1). Let $y_0 = (\bar{u}_0, \tau_0) \in Y_\omega^-$. Then, due to Statement a) and to Lemma 3.7, a solution $\hat{y}_0(x, t) = \hat{\bar{u}}_0(x, t)$, $\hat{\tau}_0(x, t)$, $t \geq 0$ of the equation (3.20) exists, which satisfies the initial condition $\hat{y}_0(x, 0) = y_0$, with the following estimate

$$\|\hat{y}_0(\cdot, t)\|_{B_q} \leq c_1 e^{-\sigma_0 t} \|y_0\|_{B_q} \tag{3.27}$$

where $c_1 = \text{const}$, $\sigma_0 = -\ln \rho_0$. Recalling the change of variables when passing from the system (2.1) to the system (3.20), one comes to the conclusion that the vector-function $(\bar{u}_\omega^0, T_\omega^0)$, where

$$\bar{u}_\omega^0(x, t) = \bar{u}_\omega'(x, t) + \hat{\bar{u}}_0(x, t) + \sum_{0 < k \leq m} (ikI - \omega^{-1} A_1)^{-1} \Pi \bar{a}_k(x) e^{ik\omega t} \hat{\tau}_0(x, t)$$

$$T_\omega^0(x, t) = T_\omega'(x, t) + \hat{\tau}_0(x, t), \tag{3.28}$$

is a solution of the system (2.1) with the initial condition

$$\bar{u}_0'(x) = \bar{u}_\omega'(x, 0) + u_0(x) + \sum_{0 < k \leq m} (ikI - \omega^{-1} A_1)^{-1} \Pi \bar{a}_k(x) \tau_0(x)$$

$$T_0'(x) = T_\omega'(x, 0) + \tau_0(x). \tag{3.29}$$

The following estimate follows from equations (3.27)–(3.29), (3.14)–(3.15)

$$\|\bar{u}'_\omega(\cdot,t) - \bar{u}^0_\omega(\cdot,t)\|_{S_q} + \|T'_\omega(\cdot,t) - T^0_\omega(\cdot,t)\|_{W^1_q} = c_1 e^{-\sigma_0 t}(\|\bar{u}'_\omega(\cdot,0) - \bar{u}'_0(\cdot)\|_{S_q}$$
$$+ \|T'_\omega(\cdot,0) - T'_0(\cdot)\|_{W^1_q}), \quad c_1 = \text{const.}$$

Statement 1) of Theorem 2.2 is proved.

Let us prove Statement 2). Let $\bar{u}, \tau_0) \notin Y^-_\omega$. Then, due to b), for the solution $\bar{y}_0$ of equation (3.20) with such an initial condition, at a certain time $t_1 \equiv t(\bar{u}_0, \tau_0)$ the relation

$$\bar{y}_0(\cdot, t_1) \notin S_r \tag{3.30}$$

is valid. Statement 2) of Theorem 2.2 follows from (3.30). Indeed, if inequality (2.5) for sufficiently small $r_0 > 0$ and arbitrary $t_1 > 0$ is not valid then, due to relations (3.28)–(3.29) and (3.14)–(3.15), the vectors $\bar{y}_0(\cdot, t_1)$ belong to the sphere S_r, that contradicts equation (3.30).

Statements 3), 4) of Theorem 2.2 are proved by analogy. Statement 5) was proved above. So, Theorem 2.2 is proved.

Acknowledgements

The work was performed under financial support of the Russian Foundation for the Fundamental Researches (Grants 96–01–01417 and 98–01–00136).

References

[1] N.N. Bogoljubov, Yu.A. Mitropolsky, *Asymptotic Methods in the Theory of Nonlinear Oscillations*. Nauka, Moscow, 1974 (in Russian).

[2] B. Van der Pol, *A theory of the amplitude of free and forced triode vibrations*. The Radio Review. London **1** (1920), 701–710.

[3] N.M. Krylov, N.N. Bogoljubov, *Introduction to Nonlinear Mechanics*. Izdatelstvo AN Ukr. SSR, Kiev (1937) (in Russian).

[4] N.N. Bogoljubov, *On Some Statistical Methods in Mathematical Physics*. Izdatelstvo AN Ukr. SSR, Kiev (1945) (in Russian).

[5] S.M. Zen'kovskaja, I.B. Simonenko, *On the influence of a high-frequency vibration on appearance of convection*. Izvestiya AN SSSR. Mechanics of Fluids and Gas, No. 5 (1966) (in Russian).

[6] S.M. Zen'kovskaja, *On convection in a fluid layer under oblique vibrations*, VINITI, No. 2437–78 (1978) (in Russian).

[7] G.Z. Gershuni, E.M. Zhuhovitsky, *On free heat convection in oscillating field under the conditions of weightlessness*, Soviet Doklady **249** (1979), No. 3.

[8] G.Z. Gershuni, E.M. Zhuhovitsky, *On convective fluid instability in oscillating field*, Izvestiya AN SSSR. Mechanics of Fluids and Gas, No. 4 (1981) (in Russian).

[9] I.B. Simonenko, *On the averaging method for the problem of convection in the high-frequency field of oscillating force, and some other parabolic equations*, Mat. Sbornik **87** (1972), No. 2.

[10] V.B. Levenshtam, *On the averaging methods for the convection problem under high-frequency vibrations*, Siberian Mat. Journ. **34** (1993), No. 2.

[11] V.B. Levenshtam, *On the averaging method for the problem of heat vibrational convection*, Russian Doklady **349** (1996), No. 5.

[12] V.B. Levenshtam, *On the averaging method in the convection problem under high-frequency oblique vibrations*, Siberian Mat. Journ. **37** (1996), No. 5.

[13] V. Yudovich, *The Linearization Method in Hydrodynamical Stability Theory*, Transl. Math. Monographs. No. 74, Amer. Math. Soc., Providence (1989).

[14] M.Z. Solomyak, *Application of the semigroup theory to a study of differential equations in the Banach spaces*, Soviet Doklady **125** (1958), No. 5.

[15] P.E. Sobolevsky, *Study of the Navier-Stokes equations by methods of the theory of parabolic equations in the Banach spaces*, Soviet Doklady **156** (1964), No. 4.

[16] P.E. Sobolevsky, *On a coercive inequality for abstract parabolic equations*, Soviet Doklady **157** (1964), No. 1.

[17] M.A. Krasnoselsky, P.P. Zabreiko, E.I. Pustil'nik, P.E. Sobolevsky, *Integral Operators in the Spaces of Integrable Functions*. Nauka, Moscow, 1966 (in Russian).

V.B. Levenshtam
South Scientific Centre of
Russian Academy of Science
Zorge Str., 5
344090 Rostov-on-Don, Russia
e-mail: `vleven@math.rsu.ru`

Operator Theory:
Advances and Applications, Vol. 170, 205–228
© 2006 Birkhäuser Verlag Basel/Switzerland

Finite Sections of Band-dominated Operators with Almost Periodic Coefficients

Vladimir S. Rabinovich, Steffen Roch and Bernd Silbermann

Dedicated to I.B. Simonenko on occasion of his seventieth birthday.

Abstract. We consider the sequence of the finite sections $R_n A R_n$ of a band-dominated operator A on $l^2(\mathbb{Z})$ with almost periodic coefficients. Our main result says that if the compressions of A onto $\mathbb{Z}^+$ and $\mathbb{Z}^-$ are invertible, then there is a distinguished subsequence of $(R_n A R_n)$ which is stable. Moreover, this subsequence proves to be fractal, which allows us to establish the convergence in the Hausdorff metric of the singular values and pseudoeigenvalues of the finite section matrices.

Mathematics Subject Classification (2000). Primary 65J10; Secondary 47B36.

Keywords. Band operator, Almost Mathieu operator, finite sections, spectral behavior.

1. Introduction

Given a non-empty subset $\mathbb{I}$ of the set $\mathbb{Z}$ of the integers, let $l^2(\mathbb{I})$ stand for the Hilbert space of all sequences $(x_n)_{n \in \mathbb{I}}$ of complex numbers with $\sum_{n \in \mathbb{I}} |x_n|^2 < \infty$. We identify $l^2(\mathbb{I})$ with a closed subspace of $l^2(\mathbb{Z})$ in the natural way, and we write $P_\mathbb{I}$ for the orthogonal projection from $l^2(\mathbb{Z})$ onto $l^2(\mathbb{I})$.

The set of the non-negative integers will be denoted by $\mathbb{Z}^+$, and we write P in place of $P_{\mathbb{Z}^+}$ and Q in place of the complementary projection $I - P$. Thus, $Q = P_{\mathbb{Z}^-}$ where $\mathbb{Z}^-$ refers to the set of all negative integers. For $k \in \mathbb{Z}$, define the *shift operator*

$$U_k : l^2(\mathbb{Z}) \to l^2(\mathbb{Z}), \quad (x_n) \mapsto (y_n) \text{ with } y_n = x_{n-k}.$$

Further, each function $a \in l^\infty(\mathbb{Z})$ induces a *multiplication operator*

$$a : l^2(\mathbb{Z}) \to l^2(\mathbb{Z}), \quad (x_n) \mapsto (a_n x_n).$$

The first two authors are supported by CONACYT project 43432.

Notice that the shifted multiplication operator $U_{-k}aU_k$ is a multiplication operator again:

$$(U_{-k}aU_k x)_n = (aU_k x)_{n+k} = a_{n+k}x_n.$$

Definition 1.1. *A function $a \in l^\infty(\mathbb{Z})$ is called* almost periodic *if the set of all multiplication operators $U_{-k}aU_k$ with $k \in \mathbb{Z}$ is relatively compact in the norm topology of $L(l^2(\mathbb{Z}))$ or, equivalently, in the norm topology of $l^\infty(\mathbb{Z})$. We denote the set of all almost periodic functions on $\mathbb{Z}$ by $AP(\mathbb{Z})$, and we write $\mathcal{A}_{AP}(\mathbb{Z})$ for the norm closure in $L(l^2(\mathbb{Z}))$ of the set of all operators*

$$A = \sum_{k=-K}^{K} a_k U_k \quad \text{with } a_k \in AP(\mathbb{Z}).$$

The operators in $\mathcal{A}_{AP}(\mathbb{Z})$ are called band-dominated operators with almost periodic coefficients.

Is is easy to see that $AP(\mathbb{Z})$ and $\mathcal{A}_{AP}(\mathbb{Z})$ are C^*-subalgebras of $l^\infty(\mathbb{Z})$ and $\mathcal{A}_{AP}(\mathbb{Z})$, respectively.

For each positive integer n, set

$$P_n := P_{\{0, 1, \dots, n-1\}} \quad \text{and} \quad R_n := P_{\{-n, -n+1, \dots, n-1\}}.$$

The projections R_n converge *-strongly to the identity operator on $l^2(\mathbb{Z})$, and the projections P_n converge *-strongly to the identity operator on $l^2(\mathbb{Z}^+)$ when considered as acting on $l^2(\mathbb{Z}^+)$ and to the projection P when considered as acting on $l^2(\mathbb{Z})$. For each operator $A \in \mathcal{A}_{AP}(\mathbb{Z})$, we consider the sequences $(R_n A R_n)$ and $(P_n PAP P_n)$ of its finite sections. These sequences converge *-strongly to A and PAP, respectively. Hence, they can be viewed as approximation methods for these operators. The finite sections sequences $(R_n A R_n)$ resp. $(P_n PAP P_n)$ are said to be *stable* if the operators $R_n A R_n : \operatorname{im} R_n \to \operatorname{im} R_n$ resp. $P_n PAP P_n : \operatorname{im} P_n \to \operatorname{im} P_n$ are invertible for sufficiently large n and if the norms of their inverses are uniformly bounded.

The stability of the finite section method for band-dominated operators with *arbitrary l^∞-coefficients* has been studied in [15, 16]. The crucial observation employed there is that the stability of the sequence $(R_n A R_n)$ is equivalent to the Fredholmness of an associated band-dominated operator which can be treated by means of the limit operators method. The resulting criterion says that the sequence $(R_n A R_n)$ is stable if and only if the operator A is invertible and if a whole family of so-called limit operators associated with that sequence is uniformly invertible. Similarly, the stability of $(P_n PAP P_n)$ is equivalent to the invertibility of PAP plus the uniform invertibility of an associated limit operator family. The precise statements can be found in [15, 16, 18].

In the present paper we will show if $A \in \mathcal{A}_{AP}(\mathbb{Z})$ and if the operators PAP and QAQ are invertible then one can always find a *subsequence* of $(R_n A R_n)$ resp. of $(P_n PAP P_n)$ which is stable. Moreover, this subsequence can be effectively determined in many situations. Thus, the uniform invertibility of the (in general, infinite) family of limit operators is replaced by the invertibility of the single operator QAQ.

The motivation to consider suitable subsequences of $(R_n A R_n)$ comes from a special class of band-dominated operators with almost periodic coefficients: the *block Laurent operators with continuous generating function*. These are the operators on $l^2(\mathbb{Z})$ with matrix representation $(a_{i-j})_{i,j\in\mathbb{Z}}$ with respect to the standard basis of $l^2(\mathbb{Z})$ where a_j is the jth Fourier coefficient of a continuous function $a : \mathbb{T} \to \mathbb{C}^{l\times l}$,

$$a_j := \frac{1}{2\pi} \int_0^{2\pi} a(e^{it}) e^{-ijt}\, dt.$$

The block Laurent operator with generating function a will be denoted by $L(a)$. Since every continuous function on $\mathbb{T}$ can be uniformly approximated by a polynomial, block Laurent operators with continuous generating function are band-dominated operators with l-periodic (hence, almost periodic) coefficients. If $L(a)$ is a block Laurent operator, then the operator

$$T(a) := PL(a)P : l^2(\mathbb{Z}^+) \to l^2(\mathbb{Z}^+)$$

is called the associated *block Toeplitz operator with generating function a*.

Let, for simplicity, $l = 2$ and write the jth Fourier coefficient a_j of the continuous function $a : \mathbb{T} \to \mathbb{C}^{2\times 2}$ as

$$a_j = \begin{pmatrix} a_{00}^j & a_{01}^j \\ a_{10}^j & a_{11}^j \end{pmatrix}.$$

Then the standard finite sections sequence $(P_n PAP P_n)$ for the block Toeplitz operator $A = T(a)$ starts with

$$\left(a_{00}^0\right),\quad \begin{pmatrix} a_{00}^0 & a_{01}^0 \\ a_{10}^0 & a_{11}^0 \end{pmatrix},\quad \begin{pmatrix} a_{00}^0 & a_{01}^0 & a_{00}^{-1} \\ a_{10}^0 & a_{11}^0 & a_{10}^{-1} \\ a_{00}^1 & a_{01}^1 & a_{00}^0 \end{pmatrix},\quad \begin{pmatrix} a_{00}^0 & a_{01}^0 & a_{00}^{-1} & a_{01}^{-1} \\ a_{10}^0 & a_{11}^0 & a_{10}^{-1} & a_{11}^{-1} \\ a_{00}^1 & a_{01}^1 & a_{00}^0 & a_{01}^0 \\ a_{10}^1 & a_{11}^1 & a_{10}^0 & a_{11}^0 \end{pmatrix},\ \dots$$

These finite sections do not completely reflect the 2×2-block structure of the operator $T(a)$. It is thus much more natural to consider the *subsequence* $(P_{2n} PAP P_{2n})$ of $(P_n PAP P_n)$ which starts with

$$\begin{pmatrix} a_{00}^0 & a_{01}^0 \\ a_{10}^0 & a_{11}^0 \end{pmatrix},\quad \begin{pmatrix} a_{00}^0 & a_{01}^0 & a_{00}^{-1} & a_{01}^{-1} \\ a_{10}^0 & a_{11}^0 & a_{10}^{-1} & a_{11}^{-1} \\ a_{00}^1 & a_{01}^1 & a_{00}^0 & a_{01}^0 \\ a_{10}^1 & a_{11}^1 & a_{10}^0 & a_{11}^0 \end{pmatrix},\ \dots$$

where each finite section is a 2×2-block Toeplitz matrix, too. In fact, it is the sequence $(P_{ln} PT(a)P P_{ln})$ which is usually referred to as *the* finite sections sequence for the $l \times l$-block Toeplitz operator $T(a)$ rather than the sequence $(P_n PT(a)P P_n)$ itself. The stability of the sequence $(P_{ln} PT(a)P P_{ln})$ for a block Toeplitz operator $T(a)$ with continuous generating function is well understood (see [10, 7, 8], for instance). It is stable if and only if the operators $PL(a)P$ and $QL(a)Q$ are invertible. The same results holds for the stability of the finite sections sequence

$(R_{ln}L(a)R_{ln})$, simply because the operators $R_{ln}L(a)R_{ln}$ and $P_{2ln}PT(a)PP_{2ln}$ possess the same matrix representation with respect to the standard basis of $l^2(\mathbb{Z})$.

The paper is organized as follows. We start with some simple observations concerning band-dominated operators with almost periodic coefficients and their limit operators. Thereby we will learn how to choose a distinguished subsequence of the sequences $(R_n AR_n)$ and $(P_n PAPP_n)$ such that the above-mentioned results hold. Then we will prove the stability results. We will *not* derive them from the stability theorem for the finite sections method for general band-dominated operators from [15, 16, 18]. Rather we prefer to show that these results follow in a completely elementary way from basic properties of band-dominated operators with almost periodic coefficients, in the very same manner as the stability of the finite sections method for Toeplitz operators with (scalar-valued) continuous generating functions has been proved in [13], Theorem 4.45 (see also [5] and Section 1.3.3 in [11]).

We will have occasion to observe that many properties of band-dominated operators with almost periodic coefficients are unexpected close to those of block Laurent operators with continuous generating function (= band-dominated operators with periodic coefficients). Thus, for readers which are familiar with Toeplitz and Hankel operators, it might be helpful to introduce the following notations for every band-dominated operator A:

$$T(A) := PAP, \quad \widetilde{A} := JAJ, \quad \text{and} \quad H(A) := PAQJ$$

where J stands for the flip operator

$$J : l^2(\mathbb{Z}) \to l^2(\mathbb{Z}), \quad (x_n) \mapsto (y_n) \quad \text{with} \quad y_n := x_{-n-1}.$$

Then one has

$$T(\widetilde{A}) = PJAJP = JQAQJ \quad \text{and} \quad H(\widetilde{A}) = PJAJQJ = JQAP,$$

and equalities like $PABP = PAPBP + PAQBP$ can be written as

$$T(AB) = T(A)T(B) + H(A)H(\widetilde{B})$$

which reminds of a basic identity relating Toeplitz and Hankel operators.

Finally we would like to mention that the results of this paper can be transferred to l^p-spaces over $\mathbb{Z}$ and $\mathbb{Z}^+$ with $1 < p < \infty$ without great effort. For spectral and pseudospectral approximation on such spaces see [6] and [17], whereas the splitting property of the singular values is treated in [19]. Also the multidimensional case (l^p-spaces over subcones of $\mathbb{Z}^N$) seems to be practicable (and will be the subject of a forthcoming paper). Notice that in case of band-dominated operators with constant coefficients (discrete convolutions), the multidimensional case has been studied by Kozak and Simonenko [12].

2. Limit operators of band-dominated operators with almost periodic coefficients

We start with recalling the definition of a limit operator of a given operator. Let $\mathcal{H}$ refer to the set of all sequences $h : \mathbb{Z}^+ \to \mathbb{Z}$ which tend to infinity.

Definition 2.1. *An operator $A_h \in L(l^2(\mathbb{Z}))$ is called a* strong limit operator *of the operator $A \in L(l^2(\mathbb{Z}))$ with respect to the sequence $h \in \mathcal{H}$ if*

$$U_{-h(k)} A U_{h(k)} \to A_h \qquad as \ k \to \infty \tag{1}$$

*-strongly. The sets of all strong limit operators of a given operator A will be denoted by $\sigma_{\mathrm{op},\,s}(A)$, and we refer to this set as the strong operator spectrum *of A. Further, let $\mathcal{H}_{A,\,s}$ stand for the set of all sequences $h \in \mathcal{H}$ such that (1) holds with respect to the *-strong topology. Analogously, we call A_h a* norm limit operator *of A if (1) holds with respect to norm convergence, and we introduce the related* norm operator spectrum *$\sigma_{\mathrm{op},\,n}(A)$ of A and the corresponding class $\mathcal{H}_{A,\,n}$.*

In [14, 15, 16, 18] we have exclusively worked with limit operators in the *-strong sense (simply because the norm operator spectrum proved to be to small to be of any use in general). But for band-dominated operators with almost periodic coefficients, one can work in the norm topology as well.

Lemma 2.2. *For $A \in \mathcal{A}_{AP}(\mathbb{Z})$, one has $\sigma_{\mathrm{op},\,s}(A) = \sigma_{\mathrm{op},\,n}(A)$.*

Proof. The inclusion $\supseteq$ is obvious. The reverse inclusion holds for operators of multiplication by an almost periodic function due to the definition of the class $AP(\mathbb{Z})$. Then it holds also for band operators with almost periodic coefficients. For the proof in the general case, approximate the operator A in the norm topology by a sequence (A_n) of band operators with almost periodic coefficients. Let $g_0 :=$ $h \in \mathcal{H}_{A,\,s}$. Then there is a subsequence g_1 of g_0 which belongs to $\mathcal{H}_{A_1,\,n}$. Further, there is a subsequence g_2 of g_1 with $g_2 \in \mathcal{H}_{A_2,\,n}$. We proceed in this way and find, for every positive integer k, a subsequence g_k of g_{k-1} with $g_k \in \mathcal{H}_{A_k,\,n}$. The sequence g defined by $g(k) := g_k(k)$ is a subsequence of each sequence g_k. Thus, all limit operators $(A_k)_g$ exist with respect to norm convergence. Then also the limit operator A_g exists with respect to norm convergence, whence $A_h \in \sigma_{\mathrm{op},\,n}(A)$. $\square$

It follows in particular that $\mathcal{H}_{A,\,n}$ is not empty if $A \in \mathcal{A}_{AP}(\mathbb{Z})$.

Lemma 2.3. *Let $A \in \mathcal{A}_{AP}(\mathbb{Z})$ and $h \in \mathcal{H}_{A,\,n}$. Then $(A_h)_{-h} = A$.*

This follows immediately from

$$\|U_{h(n)} A_h U_{-h(n)} - A\| = \|A_h - U_{-h(n)} A U_{h(n)}\| \to 0.$$

Lemma 2.4. *If $A \in \mathcal{A}_{AP}(\mathbb{Z})$, then $A \in \sigma_{\mathrm{op},\,n}(A)$.*

Proof. Let h be any sequence in $\mathcal{H}_{A,\,n}$. We define a sequence $(n_k)_{k \geq 1}$ as follows. Let $n_1 = 0$. If n_k is already defined for some $k \geq 1$, then we choose $n_{k+1} > n_k$ such that

$$|h(n_{k+1}) - h(n_k)| \geq k + 1 \tag{2}$$

which is possible since $h \in \mathcal{H}$. Set $g(k) := h(n_k) - h(n_{k+1})$. Then

$$\|U_{-g(k)} A U_{g(k)} - A\|$$
$$= \|U_{h(n_{k+1})} U_{-h(n_k)} A U_{h(n_k)} U_{-h(n_{k+1})} - A\|$$
$$\leq \|U_{h(n_{k+1})} (U_{-h(n_k)} A U_{h(n_k)} - A_h) U_{-h(n_{k+1})}\| + \|U_{h(n_{k+1})} A_h U_{-h(n_{k+1})} - A\|$$
$$\leq \|U_{-h(n_k)} A U_{h(n_k)} - A_h\| + \|U_{h(n_{k+1})} A_h U_{-h(n_{k+1})} - A\| \to 0$$

as $k \to \infty$. Thus, $\lim U_{-g(k)} A U_{g(k)} = A$ in the norm. Since condition (2) ensures that $g \in \mathcal{H}$, we have $g \in \mathcal{H}_{A,n}$ and $A_g = A$. $\qquad\square$

In case of $l \times l$-block Laurent operators ($=$ band-dominated operators with l-periodic coefficients) this result is obvious: the sequence $g(k) := lk$ belongs to $\mathcal{H}_{L(a),n}$ and $L(a)_g = L(a)$.

3. Band-dominated operators with almost periodic coefficients on $l^2(\mathbb{Z}^+)$

Here we consider compressions of band-dominated operators with almost periodic coefficients onto $l^2(\mathbb{Z}^+)$. Notice that the compression of an operator of multiplication by an almost periodic function a to $l^2(\mathbb{Z}^+)$ (considered as a subspace of $l^2(\mathbb{Z})$) is no longer almost periodic unless the trivial case $a = 0$.

Definition 3.1. *Let $A \in \mathcal{A}_{AP}(\mathbb{Z})$. Then we call PAP a band-dominated operator with AP coefficients on $l^2(\mathbb{Z}^+)$. The smallest closed subalgebra of $L(l^2(\mathbb{Z}^+))$ which contains all band-dominated operators with AP coefficients on $l^2(\mathbb{Z}^+)$ will be denoted by $\mathcal{A}_{AP}(\mathbb{Z}^+)$.*

Evidently, $\mathcal{A}_{AP}(\mathbb{Z}^+)$ is a C^*-subalgebra of $L(l^2(\mathbb{Z}^+))$.

Lemma 3.2. *For $A \in \mathcal{A}_{AP}(\mathbb{Z})$, one has $\|A\| = \|PAP\|$.*

In case of periodic coefficients, this simply says that $\|L(a)\| = \|T(a)\|$.

Proof. Choose a sequence $h \in \mathcal{H}_{A,n}$ which converges to $+\infty$ and for which $A_h = A$. (Starting with a suitable sequence h in the proof of Lemma 2.4 one easily gets a sequence with these properties.) Then $h \in \mathcal{H}_{P,s}$ and $P_h = I$. Hence, $h \in \mathcal{H}_{PAP,s}$ and $(PAP)_h = A_h = A$. This implies the assertion since

$$\|A\| = \|A_h\| = \|(PAP)_h\| \leq \|PAP\| \leq \|A\|$$

where we have used the elementary estimate $\|B_h\| \leq \|B\|$ for limit operators (Proposition 1.2.2 in [16]). $\qquad\square$

Corollary 3.3. *Let $B, C \in \mathcal{A}_{AP}(\mathbb{Z})$. If $PBP = PCP$, then $B = C$.*

This follows from Lemma 3.2 with $A := B - C$. One can consider the statement of the preceding corollary as a rigidity property of band-dominated operators with AP coefficients: The restriction of an operator $A \in \mathcal{A}_{AP}(\mathbb{Z})$ onto $l^2(\mathbb{Z}^+)$ can be extended to an operator in $\mathcal{A}_{AP}(\mathbb{Z})$ in exactly one manner. The extension of a Toeplitz operator $T(a)$ is just the Laurent operator $L(a)$.

Lemma 3.4. *Let $A \in \mathcal{A}_{AP}(\mathbb{Z})$. Then*

(a) $\|A\| \leq \|A + K\|$ *for each compact operator* $K \in L(l^2(\mathbb{Z}))$;
(b) $\|PAP\| \leq \|PAP + K\|$ *for each compact operator* $K \in L(l^2(\mathbb{Z}^+))$.

Proof. Let h be as in the proof of Lemma 3.2, and let K be compact. Then, in both cases, $h \in \mathcal{H}_{K,n}$ and $K_h = 0$. Thus,

$$\|A\| = \|A_h\| = \|(A + K)_h\| \leq \|A + K\|$$

and, by Lemma 3.2,

$$\|PAP\| = \|A\| = \|A_h\| = \|(PAP + K)_h\| \leq \|PAP + K\|$$

which implies assertions (a) and (b), respectively. $\qquad\square$

Lemma 3.5. *One has*

$$\mathcal{A}_{AP}(\mathbb{Z}^+) = \{PAP + K : A \in \mathcal{A}_{AP}(\mathbb{Z}), \ K \in L(l^2(\mathbb{Z}^+)) \ compact\}, \qquad (3)$$

and each operator $B \in \mathcal{A}_{AP}(\mathbb{Z}^+)$ *can be written as* $PAP + K$ *with* $A \in \mathcal{A}_{AP}(\mathbb{Z})$ *and* K *compact in a unique way.*

The well-known analogue of (3) for Toeplitz operators ([11], Theorem 1.51) is

$$\mathcal{A}_{\mathbb{C}}(\mathbb{Z}^+) = \{T(a) + K : a \in C(\mathbb{T}), \ K \ compact\}$$

where $\mathcal{A}_{\mathbb{C}}(\mathbb{Z}^+)$ stands for the smallest closed subalgebra of $L(l^2(\mathbb{Z}^+))$ which contains all Toeplitz operators with continuous generating function ($=$ all restrictions of band-dominated operators with constant coefficients to $l^2(\mathbb{Z}^+)$).

Proof. Denote the right-hand side of (3) by $\mathcal{A}'$ for a moment. The inclusion $\mathcal{A}' \subseteq \mathcal{A}_{AP}(\mathbb{Z}^+)$ holds since $PAP \in \mathcal{A}_{AP}(\mathbb{Z}^+)$ by definition and since $K \in \mathcal{A}_{\mathbb{C}}(\mathbb{Z}^+)$ as mentioned above. For the reverse inclusion notice that the operator

$$PAPBP - PABP = -PAQBP$$

is compact for each pair of band-dominated operators A, B (since PAQ is of finite rank if A is a band operator). Hence, all finite sums of products $\sum_i \prod_j PA_{ij}P$ with band-dominated operators A_{ij} belong to $\mathcal{A}'$, and the implication $\mathcal{A}_{AP}(\mathbb{Z}^+) \subseteq \mathcal{A}'$ will follow once we have shown that $\mathcal{A}'$ is closed.

Let $(PA_nP + K_n)$ be a Cauchy sequence in $\mathcal{A}'$. By Lemma 3.2 and Lemma 3.4 (b),

$$\|A_n - A_m\| = \|P(A_n - A_m)P\| \leq \|(PA_nP + K_n) - (PA_mP + K_m)\|.$$

Thus, (A_n) is a Cauchy sequence in $\mathcal{A}_{AP}(\mathbb{Z})$. Let $A \in \mathcal{A}_{AP}(\mathbb{Z})$ denote its limit. Then PA_nP converges to PAP in the norm, which implies that (K_n) is a Cauchy sequence, too. Its limit K is compact. So we finally get that $PA_nP + K_n$ converges in the norm to $PAP + K$ which obviously is in $\mathcal{A}'$. $\qquad\square$

Lemma 3.6. *Let $A \in \mathcal{A}_{AP}(\mathbb{Z})$. Then A is invertible if and only if PAP is a Fredholm operator on $l^2(\mathbb{Z}^+)$.*

In particular, the block Laurent operator $L(a)$ with continuous generating function a is invertible if and only if the Toeplitz operator $T(a)$ is Fredholm.

Proof. If PAP is a Fredholm operator, then every strong limit operator $(PAP)_h$ of PAP is invertible (Proposition 1.2.9 in [16]). Choosing a sequence h such that $(PAP)_h = A$ gives the invertibility of A. The reverse implication holds for arbitrary band-dominated operators A since PAQ and QAP are compact. $\square$

4. Distinguished finite sections methods

Definition 4.1. *Let $A \in \mathcal{A}_{AP}(\mathbb{Z})$. By a* distinguished sequence *for A we mean a monotonically increasing sequence $h : \mathbb{Z}^+ \to \mathbb{Z}^+$ which belongs to $\mathcal{H}_{A,n}$ and for which $A_h = A$. If h is a distinguished sequence for A, then the sequences $(P_{h(n)}PAPP_{h(n)})$ and $(R_{h(n)}AR_{h(n)})$ are called the associated* distinguished finite sections methods *for PAP and A, respectively.*

Theorem 4.2. *Let $A \in \mathcal{A}_{AP}(\mathbb{Z})$ and let h be a distinguished sequence for A. Let further L be a compact operator on $l^2(\mathbb{Z}^+)$. Then the sequence $(P_{h(n)}(PAP + L)P_{h(n)})$ is stable if and only if the operators $PAP + L$ and QAQ are invertible.*

Of course, this result implies the well-known criterion for the stability of the finite sections method $(P_{ln}T(a)P_{ln})$ for the block Toeplitz operator $T(a)$ with continuous function $a : \mathbb{T} \to \mathbb{C}^{l \times l}$: This method is stable if and only if the Toeplitz operator $T(a) = PL(a)P$ itself and the associated Toeplitz operator $T(\tilde{a}) = JQL(a)QJ$ with $\tilde{a}(t) := a(1/t)$ is invertible.

In what follows we will several times make use of the following elementary lemma.

Lemma 4.3 (Kozak). *Let X be a linear space, P a projection, $Q := I - P$ and A an invertible linear operator on X. Then the operator $PAP|_{\operatorname{im} P}$ is invertible if and only if the operator $QA^{-1}Q|_{\operatorname{im} Q}$ is invertible, and*

$$(PAP)^{-1}P = PA^{-1}P - PA^{-1}Q(QA^{-1}Q)^{-1}QA^{-1}P. \tag{4}$$

Proof of Theorem 4.2. First we show that if $PAP + L$ and QAQ are invertible, then the distinguished finite sections sequence $(P_{h(n)}(PAP + L)P_{h(n)})$ is stable.

The invertibility of $PAP + L$ implies those of A by Lemma 3.6, and the invertibility of QAQ implies those of $PA^{-1}P$ by Kozak's lemma. Thus one has

$$P = PAA^{-1}P = PAPA^{-1}P + PAQA^{-1}P$$

and

$$PAP + L = (PA^{-1}P)^{-1} - PAQA^{-1}P(PA^{-1}P)^{-1} =: (PA^{-1}P)^{-1} + L - K \tag{5}$$

where $K := PAQA^{-1}P(PA^{-1}P)^{-1}$ is compact due to the compactness of PAQ.

We claim that the finite sections method $(P_{h(n)}(PA^{-1}P)^{-1}P_{h(n)})$ for the operator $(PA^{-1}P)^{-1}$ is stable if the operator QAQ is invertible. By Kozak's lemma again, the sequence $(P_{h(n)}(PA^{-1}P)^{-1}P_{h(n)})$ is stable if and only if the sequence

$(Q_{h(n)}PA^{-1}PQ_{h(n)})$ with $Q_n := I - P_n : l^2(\mathbb{Z}^+) \to l^2(\mathbb{Z}^+)$ is stable, i.e., if the operators

$$Q_{h(n)}PA^{-1}PQ_{h(n)}|_{\mathrm{im}\,Q_{h(n)}}$$

are invertible for sufficiently large n and if the norms of their inverses are uniformly bounded. This happens if and only if the operators

$$U_{-h(n)}Q_{h(n)}PA^{-1}PQ_{h(n)}U_{h(n)}|_{\mathrm{im}\,(U_{-h(n)}Q_{h(n)}U_{h(n)})}$$
$$= U_{-h(n)}Q_{h(n)}U_{h(n)}\; U_{-h(n)}A^{-1}U_{h(n)}\; U_{-h(n)}Q_{h(n)}U_{h(n)}|_{\mathrm{im}\,P}$$
$$= PU_{-h(n)}A^{-1}U_{h(n)}P|_{\mathrm{im}\,P} \tag{6}$$

are invertible for sufficiently large n and if the norms of their inverses are uniformly bounded. Since h is a distinguished sequence for A, one has

$$\|U_{-h(n)}AU_{h(n)} - A\| \to 0$$

which implies

$$\|U_{-h(n)}A^{-1}U_{h(n)} - A^{-1}\| \to 0.$$

Hence, (6) converges in the norm to $PA^{-1}P$. Since this operator is invertible as mentioned above, the operators in (6) are invertible for sufficiently large n, and their inverses are uniformly bounded. This proves the claim.

Now (5) gives

$$P_{h(n)}(PAP + L)P_{h(n)} = P_{h(n)}(PA^{-1}P)^{-1}P_{h(n)} + P_{h(n)}(L - K)P_{h(n)},$$

i.e., the sequence $(P_{h(n)}(PAP+L)P_{h(n)})$ we are interested in is a compact perturbation of the stable sequence $(P_{h(n)}(PA^{-1}P)^{-1}P_{h(n)})$. Since $(PA^{-1}P)^{-1}+L-K = PAP + L$ is an invertible operator by hypothesis, the perturbation theorem for approximation methods (Corollary 1.22 in [11]) implies the stability of the finite sections method $(P_{h(n)}(PAP + L)P_{h(n)})$.

Conversely, we have to show that the stability of that sequence implies the invertibility of the operators $PAP + L$ and QAQ. This follows in a standard way from

$$P_{h(n)}(PAP + L)P_{h(n)} \to PAP + L \qquad \text{*-strongly}$$

and

$$U_{-h(n)}P_{h(n)}(PAP + L)P_{h(n)}U_{h(n)} \to QAQ \qquad \text{*-strongly}$$

which holds for every distinguished sequence h. $\qquad\square$

Next we consider the finite section method for operators in $\mathcal{A}_{AP}(\mathbb{Z})$. We will need one more simple lemma.

Lemma 4.4. *Let $A \in \mathcal{A}_{AP}(\mathbb{Z})$, and let h be a sequence in $\mathcal{H}_{A,\,n}$ with $A_h = A$. Then $2h$ and $-h$ are sequences in $\mathcal{H}_{A,\,n}$ with $A_{2h} = A$ and $A_{-h} = A$.*

This follows easily from

$$\|U_{-2h(n)}AU_{2h(n)} - A\|$$
$$\leq \|U_{-2h(n)}AU_{2h(n)} - U_{-h(n)}AU_{h(n)}\| + \|U_{-h(n)}AU_{h(n)} - A\|$$
$$\leq 2\|U_{-h(n)}AU_{h(n)} - A\| \to 0$$

and

$$\|U_{h(n)}AU_{-h(n)} - A\| \;=\; \|U_{h(n)}(A - U_{-h(n)}AU_{h(n)})U_{-h(n)}\|$$
$$\leq \; \|A - U_{-h(n)}AU_{h(n)}\| \to 0.$$

Theorem 4.5. *Let $A \in \mathcal{A}_{AP}(\mathbb{Z})$, and let h be a distinguished sequence for A. Furthermore, let L be a compact operator on $l^2(\mathbb{Z})$. Then the sequence $(R_{h(n)}(A+L)R_{h(n)})$ is stable if and only if the operators $A+L$, PAP and QAQ are invertible.*

In case $L = 0$, the invertibility of $A + L = A$ follows from the invertibility of PAP due to Lemma 3.6. Hence, in this case, the stability of the finite section method is equivalent to the invertibility of PAP and QAQ.

Proof. The crucial observation is that

$$\|U_{h(n)}R_{h(n)}AR_{h(n)}U_{-h(n)} - P_{2h(n)}PAPP_{2h(n)}\|$$
$$= \; \|P_{2h(n)}U_{h(n)}AU_{-h(n)}P_{2h(n)} - P_{2h(n)}PAPP_{2h(n)}\|$$
$$\leq \; \|U_{h(n)}AU_{-h(n)} - A\| \to 0$$

by the preceding lemma. The same lemma states furthermore that $2h$ is a distinguished sequence for A. Thus, if PAP and QAQ are invertible, then the sequence $(P_{2h(n)}PAPP_{2h(n)})$ is stable by Theorem 4.2. Since

$$(P_{2h(n)}PAPP_{2h(n)}) \qquad \text{and} \qquad (U_{h(n)}R_{h(n)}AR_{h(n)}U_{-h(n)})$$

differ by a sequence which tends to zero in the norm, the latter sequence is stable, too. But then, clearly, the sequence $(R_{h(n)}AR_{h(n)})$ is stable. Since $A + L$ is invertible by hypothesis, the stability of the compactly perturbed sequence $(R_{h(n)}(A + L)R_{h(n)})$ follows via the perturbation theorem (Corollary 1.22 in [11]) again. The reverse implication in Theorem 4.5 follows as in the proof of Theorem 4.2. $\qquad\qquad\square$

In the following examples we are going to make the previous constructions more explicit.

Example A: Multiplication operators. For each real number $\alpha \in [0, 1)$, the function

$$a : \mathbb{Z} \to \mathbb{C}, \quad n \mapsto e^{2\pi i \alpha n} \tag{7}$$

is almost periodic. Indeed, for every integer k, $U_{-k}aU_k$ is the operator of multiplication by the function a_k with $a_k(n) = a(n + k) = e^{2\pi i \alpha k}a(n)$, i.e.,

$$U_{-k}aU_k = e^{2\pi i \alpha k}a. \tag{8}$$

Let $(U_{-k(n)}aU_{k(n)})$ by any sequence in $\{U_{-k}aU_k : k \in \mathbb{Z}\}$. Due to the compactness of $\mathbb{T}$, there are a subsequence $(e^{2\pi i \alpha k(n(r))})_{r \geq 1}$ of $(e^{2\pi i \alpha k(n)})_{n \geq 1}$ and a real number β such that

$$e^{2\pi i \alpha k(n(r))} \to e^{2\pi i \beta} \qquad \text{as } r \to \infty.$$

Thus, the functions $a_{k(n(r))} = e^{2\pi i \alpha k(n(r))} a$ converge uniformly to $e^{2\pi i \beta} a$, whence the almost periodicity of a. For the operator spectrum of the operator aI one finds

$$\sigma_{\mathrm{op},\,s}(aI) = \sigma_{\mathrm{op},\,n}(aI) = \begin{cases} \{e^{2\pi i l/q}\, a : l = 1,\, 2,\, \ldots,\, q\} & \text{if} \quad \alpha = p/q \in \mathbb{Q}, \\ \{e^{it}\, a : t \in \mathbb{R}\} & \text{if} \quad \alpha \notin \mathbb{Q}. \end{cases}$$

Here, p and q are relatively prime integers with $q > 0$. Indeed, the inclusion $\subseteq$ follows immediately from (8). The reverse inclusion is evident in case $\alpha \in \mathbb{Q}$. If $\alpha \notin \mathbb{Q}$, then it follows from a theorem by Kronecker which states that the set of all numbers $e^{2\pi i \alpha k}$ with integer k lies dense in the unit circle $\mathbb{T}$.

In case $\alpha = p/q \in \mathbb{Q}$, the sequence a is q-periodic, and $h(n) = qn$ is a distinguished sequence for the multiplication operator aI. To get a distinguished sequence h for aI in case $\alpha \notin \mathbb{Q}$, too, one has to ensure that

$$\lim_{n \to \infty} e^{2\pi i \alpha h(n)} = 1$$

(cp. (8)). For develop $\alpha \in (0,\, 1)$ into a continued fraction

$$\alpha = \lim_{n \to \infty} \cfrac{1}{b_1 + \cfrac{1}{b_2 + \cfrac{1}{\ddots \atop \displaystyle b_{n-1} + \cfrac{1}{b_n}}}}$$

with uniquely determined positive integers b_i. Write this continued fraction as p_n/q_n with positive and relatively prime integers $p_n,\, q_n$. These integers satisfy the recursions

$$p_n = b_n p_{n-1} + p_{n-2}, \qquad q_n = b_n q_{n-1} + q_{n-2} \tag{9}$$

with $p_0 = 0$, $p_1 = 1$, $q_0 = 1$ and $q_1 = b_1$, and one has for all $n \geq 1$

$$\left| \alpha - \frac{p_n}{q_n} \right| < \frac{1}{q_n q_{n+1}} < \frac{1}{q_n^2}. \tag{10}$$

These facts can be found in any book on continued fractions. From (10) we conclude that

$$|\alpha q_n - p_n| \leq q_n \left| \alpha - \frac{p_n}{q_n} \right| \leq \frac{1}{q_n} \to 0,$$

whence

$$e^{2\pi i \alpha q_n} = e^{2\pi i (\alpha q_n - p_n)} \to 1.$$

Since moreover $q_1 < q_2 < \cdots$ due to the recursion (9), this shows that the sequence $h(n) := q_n$ belongs to $\mathcal{H}_{A,\,n}$ and that $A_h = A$, i.e., h is a distinguished sequence for the operator aI with a as in (7). $\qquad\square$

Example B: Almost Mathieu operators. These are the operators $H_{\alpha,\,\lambda,\,\theta} : l^2(\mathbb{Z}) \to l^2(\mathbb{Z})$ given by

$$(H_{\alpha,\,\lambda,\,\theta} x)_n := x_{n+1} + x_{n-1} + \lambda x_n \cos 2\pi(n\alpha + \theta)$$

with real parameters α, λ and θ. Thus, $H_{\alpha,\lambda,\theta}$ is a band operator with almost periodic coefficients, and

$$H_{\alpha,\lambda,\theta} = U_{-1} + U_1 + aI \qquad \text{with} \qquad a(n) = \lambda \cos 2\pi(n\alpha + \theta).$$

For a treatment of the spectral theory of Almost Mathieu operators see [4]. As in Example A one gets

$$U_{-k} H_{\alpha,\lambda,\theta} U_k = U_{-1} + U_1 + a_k I$$

with

$$\begin{aligned}
a_k(n) &= a(n+k) = \lambda \cos 2\pi((n+k)\alpha + \theta) \\
&= \lambda(\cos 2\pi(n\alpha + \theta) \cos 2\pi k\alpha - \sin 2\pi(n\alpha + \theta) \sin 2\pi k\alpha). \qquad (11)
\end{aligned}$$

We will only consider the non-periodic case, i.e., we let $\alpha \in (0,1)$ be irrational. As in the previous example, we write α as a continued fraction with nth approximant p_n/q_n such that (10) holds. Then

$$\cos 2\pi\alpha q_n = \cos 2\pi(\alpha q_n - p_n) = \cos 2\pi q_n(\alpha - p_n/q_n) \to \cos 0 = 1$$

and, similarly, $\sin 2\pi\alpha q_n \to 0$. Further we infer from (11) that

$$|(a_{q_n}) - a)(n)| \le |\lambda| \, |1 - \cos 2\pi\alpha q_n| + |\lambda| \, |\sin \pi\alpha q_n|.$$

Hence, $a_{q_n} \to a$ uniformly. Thus, $h(n) := q_n$ defines a distinguished sequence for the Almost Mathieu operator $H_{\alpha,\lambda,\theta}$. Notice that this sequence depends on the parameter α only. Theorems 4.2 and 4.5 imply the following.

Corollary 4.6. *Let* $A := H_{\alpha,\lambda,\theta}$ *be an Almost Mathieu operator and* h *a distinguished sequence for* A. *Then the following conditions are equivalent:*

(a) *the distinguished finite sections method* $(P_{h(n)} PAP P_{h(n)})$ *for* PAP *is stable;*
(b) *the distinguished finite sections method* $(R_{h(n)} AR_{h(n)})$ *for* A *is stable;*
(c) *the operators* PAP *and* QAQ *are invertible.*

If $\theta = 0$, then the Almost Mathieu operator $A = H_{\alpha,\lambda,0}$ is flip invariant, i.e., $JAJ = A$. So we observe in this case that the third condition in Corollary 4.6 is equivalent to the invertibility of PAP alone.

For a different approach to the numerical treatment of Almost Mathieu and other operators in irrational rotation algebras see [9].

5. The algebra of the finite sections method

In what follows we fix a strongly monotonically increasing sequence $h : \mathbb{Z}^+ \to \mathbb{Z}^+$. Define

$$\mathcal{A}_{AP,h}(\mathbb{Z}) := \{A \in \mathcal{A}_{AP}(\mathbb{Z}) : h \in \mathcal{H}_{A,n} \text{ and } A_h = A\}.$$

Thus, an operator $A \in \mathcal{A}_{AP}(\mathbb{Z})$ belongs to $\mathcal{A}_{AP,h}(\mathbb{Z})$ if and only if h is a distinguished sequence for PAP. By (a slightly improved version of) Lemma 2.4, every operator $A \in \mathcal{A}_{AP}(\mathbb{Z})$ belongs to one of the sets $\mathcal{A}_{AP,h}(\mathbb{Z})$ with a suitably chosen sequence h.

It is easy to check that $\mathcal{A}_{AP,h}(\mathbb{Z})$ is a C^*-subalgebra of $L(l^2(\mathbb{Z}))$ which is moreover shift invariant, i.e., $U_{-k}AU_k$ belongs to this algebra for each $k \in \mathbb{Z}$ whenever A does. It is also clear that all Laurent operators with continuous and complex-valued generating function belong to each of the algebras $\mathcal{A}_{AP,h}(\mathbb{Z})$.

Let $\mathcal{A}_{AP,h}(\mathbb{Z}^+)$ refer to the smallest closed subalgebra of $L(l^2(\mathbb{Z}^+))$ which contains all operators PAP with $A \in \mathcal{A}_{AP,h}(\mathbb{Z})$. For instance, all Toeplitz operators with continuous and complex-valued generating function lie in this algebra. Hence, $\mathcal{A}_{AP,h}(\mathbb{Z}^+)$ also contains all compact operators, and one can show as in Lemma 3.5 that

$$\mathcal{A}_{AP,h}(\mathbb{Z}^+) = \{PAP + K : A \in \mathcal{A}_{AP,h}(\mathbb{Z}), \ K \in L(l^2(\mathbb{Z}^+)) \text{ compact}\}. \qquad (12)$$

Let $\mathcal{F}_h$ stand for the set of all bounded sequences (A_n) of matrices $A_n \in \mathbb{C}^{h(n) \times h(n)}$. Provided with pointwise defined operations and the supremum norm, $\mathcal{F}_h$ becomes a C^*-algebra. As earlier, we will identify the matrices A_n with operators acting on $\operatorname{im} P_{h(n)}$. Finally, we let $\mathcal{S}_{AP,h}(\mathbb{Z}^+)$ denote the smallest closed subalgebra of $\mathcal{F}_h$ which contains all sequences $(P_{h(n)}PAPP_{h(n)})$ with operators $A \in \mathcal{A}_{AP,h}(\mathbb{Z})$. The following result describes this algebra completely. For, introduce

$$W_n : l^2(\mathbb{Z}^+) \to l^2(\mathbb{Z}^+), \quad (x_n)_{n \geq 0} \mapsto (x_{n-1}, x_{n-2}, \ldots, x_0, 0, 0, \ldots).$$

Theorem 5.1. *The algebra $\mathcal{S}_{AP,h}(\mathbb{Z}^+)$ consists exactly of all sequences of the form*

$$(P_{h(n)}PAPP_{h(n)} + P_{h(n)}KP_{h(n)} + W_{h(n)}LW_{h(n)} + C_{h(n)}) \qquad (13)$$

with $A \in \mathcal{A}_{AP,h}(\mathbb{Z})$, $K, L \in L(l^2(\mathbb{Z}^+))$ compact and $\|C_{h(n)}\| \to 0$ as $n \to \infty$, and each sequence in $\mathcal{S}_{AP,h}(\mathbb{Z}^+)$ can be written in the form (13) in a unique way.

The Toeplitz analogue of Theorem 5.1 is well known (Theorem 1.53 in [11], for instance): the smallest closed subalgebra $\mathcal{S}_{\mathbb{C}}(\mathbb{Z}^+)$ of $\mathcal{F}_{\mathrm{id}}$ which contains all sequences $(P_n T(a) P_n)$ with a continuous function $a : \mathbb{T} \to \mathbb{C}$ consists exactly of all sequences of the form

$$(P_n T(a) P_n + P_n K P_n + W_n L W_n + C_n)$$

where a is continuous, K and L are compact, and (C_n) is a sequence tending to zero in the norm.

Proof of Theorem 5.1. First let A and B be arbitrary band-dominated operators and n a positive integer. Then

$$\begin{aligned}
P_n &PAPP_n \ P_n PBPP_n \\
&= P_n PAPBPP_n - P_n PAPQ_n PBPP_n \\
&= P_n PABPP_n - P_n PAQBPP_n - P_n PAPQ_n PBPP_n. \qquad (14)
\end{aligned}$$

Since

$$PQ_n P = U_n P U_{-n}, \qquad PW_n PJ = PU_n Q, \qquad JPW_n P = QU_{-n}P \qquad (15)$$

we obtain

$$\begin{aligned}
P_n PAPQ_n PBPP_n &= W_n J\, JPW_n P\, AU_n PU_{-n}B\, PW_n PJ\, JW_n \\
&= W_n JQU_{-n}P\, AU_n PU_{-n}B\, PU_n QJW_n.
\end{aligned} \qquad (16)$$

Further we conclude from

$$W_n JQQ_{-n}Q = 0 \qquad \text{and} \qquad QU_n QJW_n = 0$$

and from (16) that

$$P_n PAPQ_n PBPP_n = W_n JQU_{-n}\, AU_n PU_{-n}B\, U_n QJW_n.$$

Together with (14) this gives

$$\begin{aligned}
&P_n PAPP_n\, P_n PBPP_n \\
&\quad = P_n PABPP_n - P_n PAQBPP_n - W_n JQU_{-n}\, AU_n PU_{-n}B\, U_n QJW_n \\
&\quad = P_n PABPP_n + P_n KP_n - W_n JQ\, U_{-n}AU_n PU_{-n}BU_n\, QJW_n \qquad (17)
\end{aligned}$$

with a compact operator $K = -PAQBP$. Now let especially $A, B \in \mathcal{A}_{AP,h}(\mathbb{Z})$ and replace n in (17) by $h(n)$. Since

$$\|U_{-h(n)}AU_{h(n)}PU_{-h(n)}BU_{h(n)} - APB\| \to 0,$$

we obtain from (17) the identity

$$\begin{aligned}
&P_{h(n)}PAPP_{h(n)}\, P_{h(n)}PBPP_{h(n)} \\
&\quad = P_{h(n)}PABPP_{h(n)} + P_{h(n)}KP_{h(n)} + W_{h(n)}LW_{h(n)} + C_{h(n)}
\end{aligned}$$

with compact operators K and $L := -JQAPBQJ$ and with

$$\|C_{h(n)}\| = \|W_{h(n)}JQ(U_{-h(n)}AU_{h(n)}PU_{-h(n)}BU_{h(n)} - APB)QJW_{h(n)}\| \to 0.$$

Thus, the (non-closed) dense subalgebra of $\mathcal{S}_{AP,h}(\mathbb{Z}^+)$ which is generated by all sequences of the form $(P_{h(n)}PAPP_{h(n)})$ with $A \in \mathcal{A}_{AP,h}(\mathbb{Z})$ is contained in the set $\mathcal{S}'$ of all sequences of the form (13). The inclusion $\mathcal{S}_{AP,h}(\mathbb{Z}^+) \subseteq \mathcal{S}'$ will follow once we have shown that $\mathcal{S}'$ is closed.

For this goal, notice that for each sequence $\mathbf{A} = (A_n) \in \mathcal{S}'$ with

$$A_n := P_{h(n)}PAPP_{h(n)} + P_{h(n)}KP_{h(n)} + W_{h(n)}LW_{h(n)} + C_{h(n)}$$

the sequences $(A_n P_{h(n)})$ and $(W_{h(n)}A_n W_{h(n)})$ converge *-strongly to $W(\mathbf{A}) := PAP + K$ and $\widetilde{W}(\mathbf{A}) := JQAQJ + L = PJAJP + L$, respectively. The first of these assertions is evident. The second one follows since, by (15),

$$\begin{aligned}
W_{h(n)}PAPW_{h(n)} &= JJW_{h(n)}PAPW_{h(n)}JJ \\
&= JQU_{-h(n)}PAPU_{h(n)}QJ \to JQAQJ
\end{aligned}$$

*-strongly. By the Banach-Steinhaus theorem, the linear mappings W and $\widetilde{W}$ are continuous. Thus, if $(\mathbf{A}_k)$ is a Cauchy sequence in $\mathcal{S}'$, then $(W(\mathbf{A}_k)) = (PA_k P + K_k)$ is a Cauchy sequence in $\mathcal{A}_{AP,h}(\mathbb{Z}^+)$. As in the proof of Lemma 3.5 one concludes that this sequence converges to an operator $PAP + K$ with $A \in \mathcal{A}_{AP,h}(\mathbb{Z})$ and with a compact operator K. Further, $(\widetilde{W}(\mathbf{A}_k)) = (PJA_k JP + L_k)$ is a Cauchy

sequence, too. Since $\|PJA_kJP - PJAJP\| \to 0$ as we have just seen, (L_k) is a Cauchy sequence which converges to a compact operator L. Moreover, standard arguments show that the set of all sequences in $\mathcal{F}_h$ which tend to zero in the norm is closed in $\mathcal{F}_h$. This finally shows that the sequence $(\mathbf{A}_k)$ converges in the norm of $\mathcal{F}_h$ to a sequence of the form

$$\mathbf{A} := (P_{h(n)}PAPP_{h(n)} + P_{h(n)}KP_{h(n)} + W_{h(n)}LW_{h(n)} + C_{h(n)})$$

with $\|C_{h(n)}\| \to 0$ which clearly belongs to $\mathcal{S}'$. Thus, $\mathcal{S}'$ is closed.

For the reverse implication $\mathcal{S}' \subseteq \mathcal{S}_{AP,h}(\mathbb{Z}^+)$ we have to show that

$$(P_{h(n)}KP_{h(n)} + W_{h(n)}LW_{h(n)} + C_{h(n)}) \in \mathcal{S}_{AP,h}(\mathbb{Z}^+)$$

for arbitrary compact operators K and L and arbitrary zero sequences $(C_{h(n)})$. But this is clear since all finite sections sequences for Toeplitz operators with continuous and complex-valued generating function belong to $\mathcal{S}_{AP,h}(\mathbb{Z}^+)$, hence, $\mathcal{S}_{\mathbb{C}}(\mathbb{Z}^+) \subseteq \mathcal{S}_{AP,h}(\mathbb{Z}^+)$, and since all sequences of the form $(P_nKP_n+W_nLW_n+C_n)$ with compact operators K, L and with a zero sequence (C_n) belong to $\mathcal{S}_{\mathbb{C}}(\mathbb{Z}^+)$ as mentioned above. $\qquad\square$

In the preceding proof, we have defined linear mappings W and $\widetilde{W}$ on $\mathcal{S}'$. Due to the coincidence of $\mathcal{S}'$ with $\mathcal{S}_{AP,h}(\mathbb{Z}^+)$ these mappings are defined on the algebra $\mathcal{S}_{AP,h}(\mathbb{Z}^+)$, and it is easy to see that they act as *-homomorphisms from this algebra into $\mathcal{A}_{AP,h}(\mathbb{Z}^+)$.

As in proof of Theorem 1.54 in [11], a twice application of the perturbation theorem gives the following stability result for sequences in $\mathcal{S}_{AP,h}(\mathbb{Z}^+)$.

Theorem 5.2. *A sequence* $\mathbf{A} = (A_n) \in \mathcal{S}_{AP,h}(\mathbb{Z}^+)$ *is stable if and only if the two operators* $W(\mathbf{A})$ *and* $\widetilde{W}(\mathbf{A})$ *are invertible.*

Corollary 5.3. *The algebra* $\mathcal{S}_{AP,h}(\mathbb{Z}^+)/\mathcal{G}$ *is* *-*isomorphic to the* C^*-*subalgebra of* $L(l^2(\mathbb{Z}^+)) \times L(l^2(\mathbb{Z}^+))$ *which consists of all pairs* $(W(\mathbf{A}), \widetilde{W}(\mathbf{A}))$ *with* $\mathbf{A}$ *belonging to* $\mathcal{S}_{AP,h}(\mathbb{Z}^+)$.

Indeed, since $W(\mathcal{G}) = 0$ for each sequence $\mathbf{G} \in \mathcal{G}$, the mapping

$$\mathcal{S}_{AP,h}(\mathbb{Z}^+)/\mathcal{G} \to L(l^2(\mathbb{Z}^+)) \times L(l^2(\mathbb{Z}^+)), \quad \mathbf{A} + \mathcal{G} \mapsto (W(\mathbf{A}), \widetilde{W}(\mathbf{A}))$$

is correctly defined. It turns out that this mapping is a *-homomorphism which, by Theorem 5.2, preserves spectra. Elementary C^*-arguments show that then this mapping is an isomorphism.

6. Spectral approximation

Another corollary to Theorem 5.2 states that the algebra $\mathcal{S}_{AP,h}(\mathbb{Z}^+)$ is fractal in the following sense. Let again $\mathcal{F}$ stand for the algebra of all matrix sequences with dimension function δ. For each strongly monotonically increasing sequence

$\eta : \mathbb{Z}^+ \to \mathbb{Z}^+$, let $\mathcal{F}_\eta$ refer to the algebra of all matrix sequences with dimension function $\delta \circ \eta$. There is a natural *-homomorphism $R_\eta : \mathcal{F} \to \mathcal{F}_\eta$ given by

$$R_\eta : (A_n) \mapsto (A_{\eta(n)});$$

thus, $A_{\eta(n)}$ is a $\delta(\eta(n)) \times \delta(\eta(n))$-matrix.

Definition 6.1. *A C^*-subalgebra $\mathcal{A}$ of $\mathcal{F}$ with $\mathcal{G} \subseteq \mathcal{A}$ is called* fractal *if, for every strongly monotonically increasing sequence $\eta : \mathbb{Z}^+ \to \mathbb{Z}^+$, there is a mapping $\pi_\eta : R_\eta \mathcal{A} \to \mathcal{F}/\mathcal{G}$ such that*

$$\pi_\eta(R_\eta \mathbf{A}) = \mathbf{A} + \mathcal{G} \qquad \text{for each sequence } \mathbf{A} \in \mathcal{A}.$$

Thus, the coset $\mathbf{A} + \mathcal{G} \in \mathcal{A}/\mathcal{G}$ can be reconstructed from each infinite subsequence of $\mathbf{A}$.

Theorem 6.2. *The subalgebra $\mathcal{S}_{AP,h}(\mathbb{Z}^+)$ of $\mathcal{F}$ is fractal.*

This follows immediately from Corollary 5.3 in combination with Theorem 1.69 in [11].

Fractal subalgebras of $\mathcal{F}$ are distinguished by their excellent convergence properties. For a general account on this topic, see the third chapter of [11]. Here we will mention only a few facts which arise immediately from Corollary 5.3 and from the general results presented in [11].

For each element A on a unital C^*-algebra, let $\sigma(A)$ refer to the spectrum of A and $\sigma_{\text{sing}}(A)$ to the set of all square roots of the points in $\sigma(A^* A)$. Thus, for an $n \times n$-matrix A, $\sigma_{\text{sing}}(A)$ is just the set of the singular values of that matrix.

Corollary 6.3. *Let $\mathbf{A} := (A_n) \in \mathcal{S}_{AP,h}(\mathbb{Z}^+)$ be a self-adjoint sequence. Then the spectra $\sigma(A_n)$ converge in the Hausdorff metric to the spectrum of the coset $\mathbf{A} + \mathcal{G}$ in $\mathcal{S}_{AP,h}(\mathbb{Z}^+)/\mathcal{G}$ which, on its hand, coincides with $\sigma(W(\mathbf{A})) \cup \sigma(\widetilde{W}(\mathbf{A}))$.*

Corollary 6.4. *Let $\mathbf{A} := (A_n) \in \mathcal{S}_{AP,h}(\mathbb{Z}^+)$. Then the sets of the singular values $\sigma_{\text{sing}}(A_n)$ converge in the Hausdorff metric to $\sigma_{\text{sing}}(\mathbf{A} + \mathcal{G})$ in $\mathcal{S}_{AP,h}(\mathbb{Z}^+)/\mathcal{G}$ which is equal to $\sigma_{\text{sing}}(W(\mathbf{A})) \cup \sigma_{\text{sing}}(\widetilde{W}(\mathbf{A}))$.*

Let $\varepsilon > 0$. The ε-*pseudospectrum* $\sigma^{(\varepsilon)}(A)$ of an element A of a C^*-algebra with identity element I is the set of all $\lambda \in \mathbb{C}$ for which either $A - \lambda I$ is not invertible or $\|(A - \lambda I)^{-1}\| \geq 1/\varepsilon$.

Corollary 6.5. *Let $\varepsilon > 0$ and $\mathbf{A} := (A_n) \in \mathcal{S}_{AP,h}(\mathbb{Z}^+)$. Then the ε-pseudospectra $\sigma^{(\varepsilon)}(A_n)$ converge in the Hausdorff metric to $\sigma^{(\varepsilon)}(\mathbf{A} + \mathcal{G})$ in $\mathcal{S}_{AP,h}(\mathbb{Z}^+)/\mathcal{G}$ which coincides with $\sigma^{(\varepsilon)}(W(\mathbf{A})) \cup \sigma^{(\varepsilon)}(\widetilde{W}(\mathbf{A}))$.*

Another consequence of Corollary 5.3 is related with Fredholm sequences and the splitting phenomenon of their singular values. Given an $n \times n$-matrix A, let $0 \leq \sigma_1(A) \leq \sigma_2(A) \leq \cdots \leq \sigma_n(A) = \|A\|$ refer to the singular values of A, counted with respect to their multiplicity. A sequence $\mathbf{A} = (A_n) \in \mathcal{F}$ is a *Fredholm sequence* if there is a non-negative integer k such that

$$\liminf_{n \to \infty} \sigma_{k+1}(A_n) > 0,$$

and the smallest number k with this property is the α-*number* of $\mathbf{A}$. We denote it by $\alpha(\mathbf{A})$.

Corollary 6.6. *A sequence* $\mathbf{A} := (A_n) \in \mathcal{S}_{AP,h}(\mathbb{Z}^+)$ *is Fredholm if and only if its strong limit* $W(\mathbf{A})$ *is a Fredholm operator. In this case,* $\widetilde{W}(\mathbf{A})$ *is a Fredholm operator, too,*

$$\alpha(\mathbf{A}) = \dim \ker W(\mathbf{A}) + \dim \ker \widetilde{W}(\mathbf{A}), \tag{18}$$

and, moreover, $\lim_{n\to\infty} \sigma_{\alpha(\mathbf{A})}(A_n) = 0$.

The first part of the assertion holds for general band-dominated operators; see Theorem 5.7 (b) in [18]. The identity (18) and the final assertion follow from Theorem 6.12 in [11].

Corollary 6.3 can be refined in the following way. Given a self-adjoint sequence $\mathbf{A} := (A_n) \in \mathcal{S}_{AP,h}(\mathbb{Z}^+)$ and an open interval $U \subseteq \mathbb{R}$, let $N_n(U)$ refer to the number of eigenvalues of A_n in U, counted with respect to their multiplicity. A point $\lambda \in \mathbb{R}$ is called *essential* for $\mathbf{A}$ if, for every open interval U containing λ,

$$\lim_{n\to\infty} N_n(U) = \infty,$$

and $\lambda \in \mathbb{R}$ is called a *transient* point for $\mathbf{A}$ if there is an open interval U containing λ such that

$$\sup_n N_n(U) < \infty.$$

For a self-adjoint sequence $\mathbf{A} \in \mathcal{S}_{AP,h}(\mathbb{Z}^+)$, Theorem 5.1 establishes the existence of a self-adjoint operator $A \in \mathcal{A}_{AP,h}(\mathbb{Z})$ as well as of compact operators K and L such that $W(\mathbf{A}) = PAP + K$ and $\widetilde{W}(\mathbf{A}) = PJAJP + L$. Then, by Corollary 5.3,

$$\sigma(\mathbf{A} + \mathcal{G}) = \sigma(W(\mathbf{A})) \cup \sigma(\widetilde{W}(\mathbf{A})) = \sigma(PAP + K) \cup \sigma(PJAJP + L).$$

Corollary 6.7. *Let* $\mathbf{A} := (A_n) \in \mathcal{S}_{AP,h}(\mathbb{Z}^+)$ *be self-adjoint. Then every point* $\lambda \in \sigma(A)$ *is essential, and every point* $\lambda \in \mathbb{R} \setminus \sigma(A)$ *is transient for* $\mathbf{A}$*. Moreover, for every point* $\lambda \in \mathbb{R} \setminus \sigma(A)$*, the sequence* $\mathbf{A} - \lambda\mathbf{P}$ *(with* $\mathbf{P} := (P_n) \in \mathcal{S}_{AP,h}(\mathbb{Z}^+)$*) is Fredholm, and there is an open interval* $U \subseteq \mathbb{R}$ *containing* λ *such that*

$$\sup_n N_n(U) = \alpha(\mathbf{A} - \lambda\mathbf{P}).$$

The proof follows basically from Theorem 7.12 in [11]. The first assertion of the corollary is mentioned explicitly there. Moreover, it is shown there that the set of all essential points coincides with the union of the essential spectra of $PAP + K$ and $PJAJP + L$. Since K and L are compact, this set coincides with the union of the essential spectra of PAP and $PJAJP$ which, by Lemma 3.6, coincides with the union of the spectra of A and JAJ. Clearly, the spectra of A and JAJ coincide for each operator A.

For the Fredholmness of the sequence $\mathbf{A} - \lambda\mathbf{P}$ in case $\lambda \in \mathbb{R} \setminus \sigma(A)$ notice that both sets $\sigma(PAP + K) \setminus \sigma(A)$ and $\sigma(PJAJP + L) \setminus \sigma(A)$ consist of isolated eigenvalues of finite multiplicity due to a theorem by H. Weyl. $\square$

The first assertion of Corollary 6.7 implies in particular that each real number is either essential or transient for $\mathbf{A}$. This property is usually referred to as the *Arveson dichotomy* of that sequence. Arveson studied this effect in a couple of papers [1, 2, 3]. He proved the dichotomy for the sequence of the finite sections of a self-adjoint band operator and also for a large class of self-adjoint band-dominated operators. The case of general self-adjoint band-dominated operators is dealt with in [18].

7. Test calculations

In this section we shall demonstrate how Corollaries 6.3 and 6.7 can be used to determine numerically the spectrum of the Almost Mathieu operator for some choices of the parameters α, λ and θ. The computations were carried out by Matlab with help of an ordinary PC. We thank V. Vassiliev for his kind assistance while performing these tests.

For each of the triples

$$\left(\tfrac{2}{5}, 2, 0\right), \quad \left(\tfrac{2}{5}, 2, \tfrac{1}{2}\right), \quad \left(\tfrac{2}{7}, 2, 0\right), \quad \left(\tfrac{2}{7}, 2, \tfrac{1}{2}\right), \quad \left(\tfrac{\sqrt{2}}{2}, 2, 0\right), \quad \left(\tfrac{\sqrt{5}-1}{2}, 2, \tfrac{1}{2}\right),$$

in place of $(\alpha, \lambda, \theta)$, we choose a distinguished sequence h_j of the corresponding Almost Mathieu operator which depends only on

$$\alpha_j \in \{2/5, 2/7, \sqrt{2}/2, (\sqrt{5} - 1)/2\},$$

namely

$$\alpha_1 = \tfrac{2}{5} \quad : \quad h_1(k) = 5k$$
$$\alpha_2 = \tfrac{2}{7} \quad : \quad h_2(k) = 7k$$
$$\alpha_3 = \tfrac{\sqrt{2}}{5} \quad : \quad h_3(k) = \tfrac{1}{2}((1 + \sqrt{2})^k + (1 - \sqrt{2})^k)$$
$$\alpha_4 = \tfrac{\sqrt{5}-1}{2} \quad : \quad h_4(k) = \tfrac{5+\sqrt{5}}{10}\left(\tfrac{1+\sqrt{5}}{2}\right)^k + \tfrac{5-\sqrt{5}}{10}\left(\tfrac{1-\sqrt{5}}{2}\right)^k$$

For irrational α_k, this choice has been done via continued fractions. Notice that the sequences h_3 and h_4 are rapidly growing. For instance, $h_3(13) = 47321$ and $h_4(23) = 46368$. The results are plotted in Figures 1–8. One clearly observes that the spectra of $P_{h_j(k)} H_{\alpha, \lambda, \theta} P_{h_j(k)}$ rapidly stabilize and that they approach a set of Cantor type in case of irrational α, whereas they fill out collections of small intervals for α being rational.

To illustrate the effects of distinguished sequences, Figure 9 presents the corresponding eigenvalues for the triple $(\tfrac{\sqrt{5}-1}{2}, 2, \tfrac{1}{2})$, but now with $h_5(k) = 2k$ in place of h_4, which is clearly not a distinguished sequence for the corresponding Almost Mathieu operator.

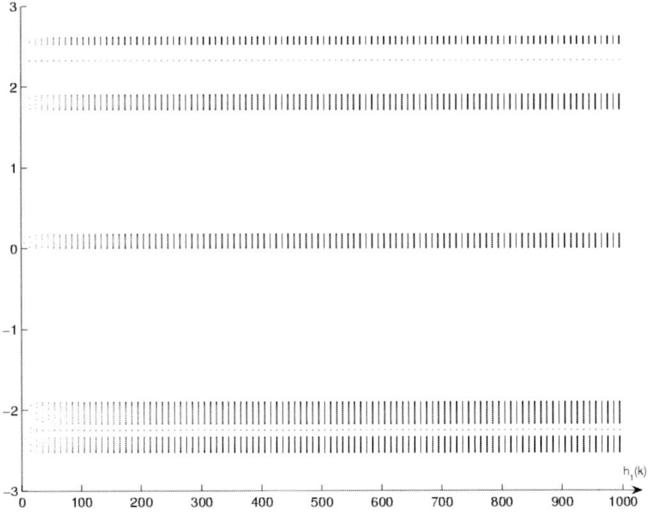

FIGURE 1. Eigenvalues of $P_{h_1(k)}H_{\alpha,\lambda,\theta}P_{h_1(k)}$ with $\alpha = 2/5$, $\lambda = 2$, $\theta = 0$.

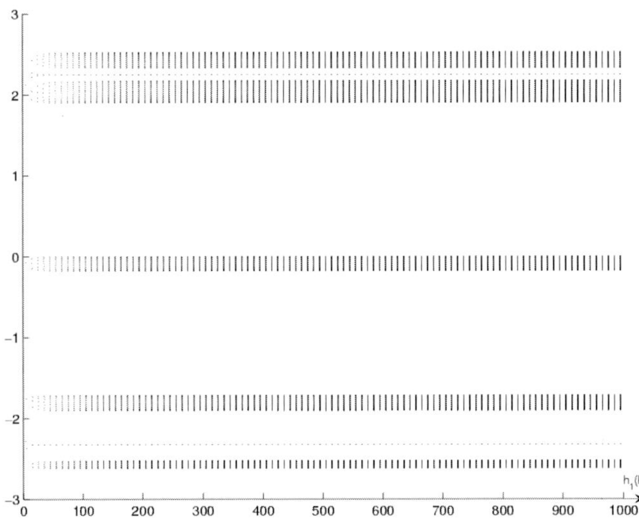

FIGURE 2. Eigenvalues of $P_{h_1(k)}H_{\alpha,\lambda,\theta}P_{h_1(k)}$ with $\alpha = 2/5$, $\lambda = 2$, $\theta = 1/2$.

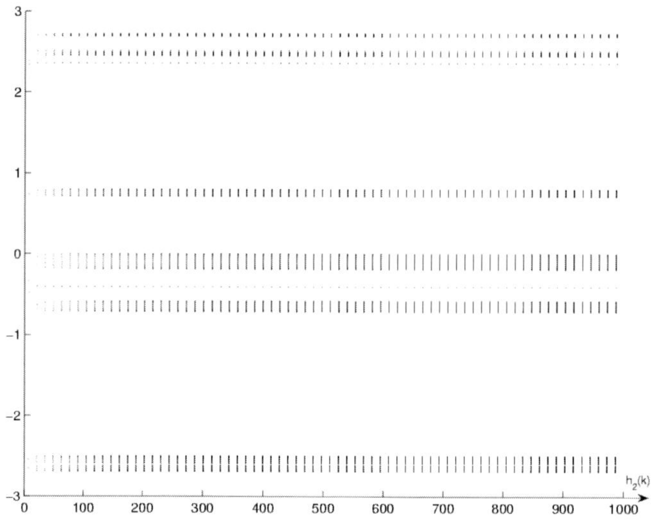

FIGURE 3. Eigenvalues of $P_{h_2(k)} H_{\alpha, \lambda, \theta} P_{h_2(k)}$ with $\alpha = 2/7$, $\lambda = 2$, $\theta = 0$.

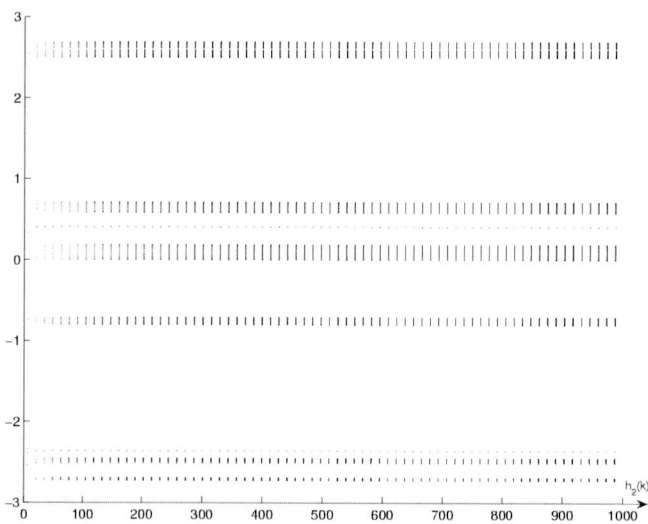

FIGURE 4. Eigenvalues of $P_{h_2(k)} H_{\alpha, \lambda, \theta} P_{h_2(k)}$ with $\alpha = 2/7$, $\lambda = 2$, $\theta = 1/2$.

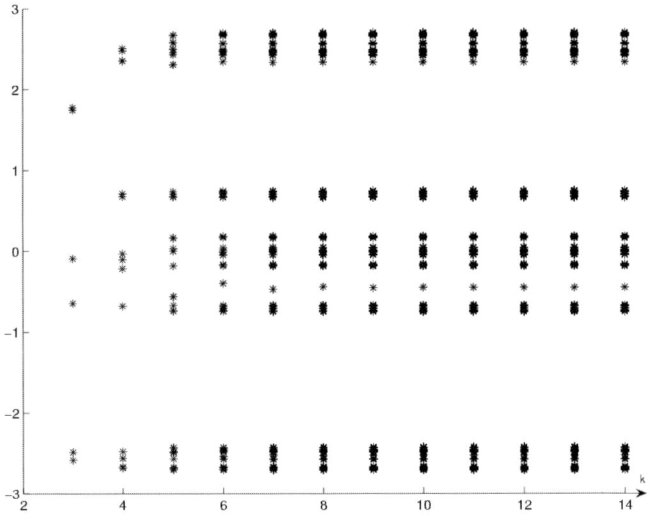

FIGURE 5. Eigenvalues of $P_{h_3(k)}H_{\alpha,\lambda,\theta}P_{h_3(k)}$ with $\alpha = \sqrt{2}/2$, $\lambda = 2$, $\theta = 0$.

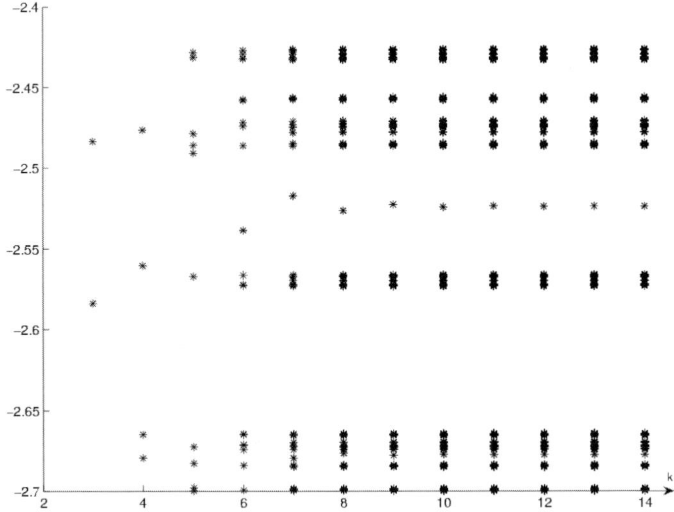

FIGURE 6. The eigenvalues of $P_{h_3(k)}H_{\alpha,\lambda,\theta}P_{h_3(k)}$ with $\alpha = \sqrt{2}/2$, $\lambda = 2$, $\theta = 0$ which lie in the interval $(-2.4, -2.8)$.

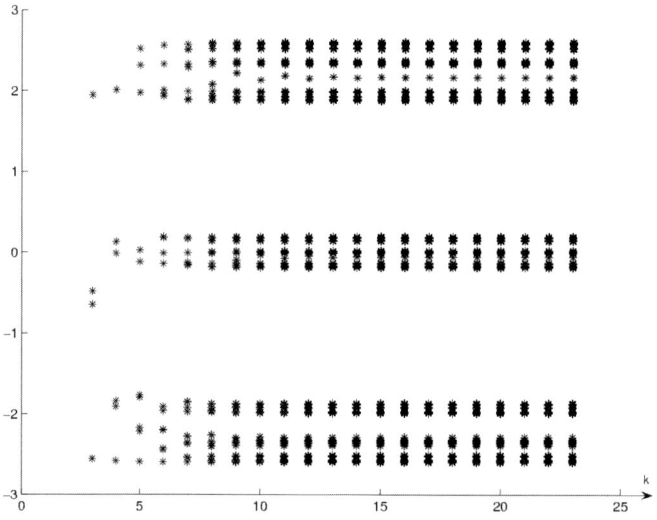

FIGURE 7. Eigenvalues of $P_{h_4(k)}H_{\alpha,\lambda,\theta}P_{h_4(k)}$ with $\alpha = (\sqrt{5}-1)/2$, $\lambda = 2$, $\theta = 0.5$.

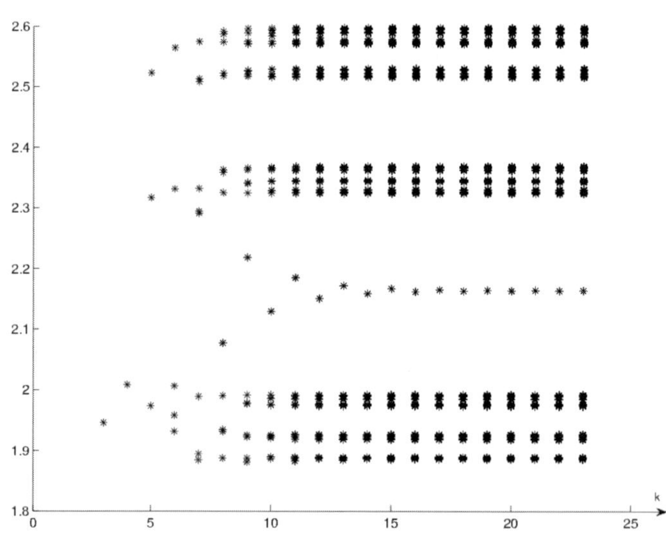

FIGURE 8. The eigenvalues of $P_{h_4(k)}H_{\alpha,\lambda,\theta}P_{h_4(k)}$ with $\alpha = (\sqrt{5}-1)/2$, $\lambda = 2$, $\theta = 0.5$, which lie in the interval $(1.8, 2.6)$.

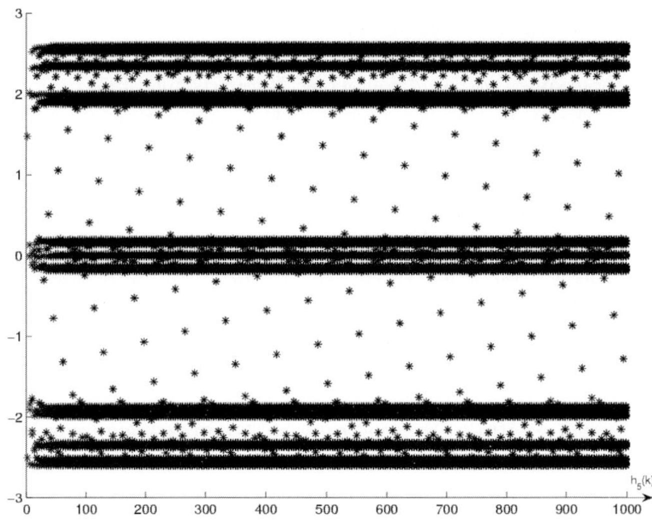

FIGURE 9. Eigenvalues of $P_{h_5(k)} H_{\alpha,\lambda,\theta} P_{h_5(k)}$ with $\alpha = (\sqrt{5}-1)/2$, $\lambda = 2$, $\theta = 0.5$.

References

[1] W. ARVESON, Improper filtrations for C^*-algebras: Spectra of unilateral tridiagonal operators. – Acta Sci. Math. (Szeged) **57**(1993), 11–24.

[2] W. ARVESON, C^*-algebras and numerical linear algebra. – J. Funct. Anal. **122**(1994), 333–360.

[3] W. ARVESON, The role of C^*-algebras in infinite dimensional numerical linear algebra. – Contemp. Math. **167**(1994), 115–129.

[4] F.P. BOCA, Rotation C^*-Algebras and Almost Mathieu Operators. – Theta Series in Advanced Mathematics **1**, The Theta Foundation, Bucharest 2001.

[5] A. BÖTTCHER, Infinite matrices and projection methods. – *In:* P. Lancaster (Ed.), Lectures on Operator Theory and its Applications, Fields Institute Monographs Vol. 3, Amer. Math. Soc., Providence, Rhode Island 1995, 1–72.

[6] A. BÖTTCHER, S.M. GRUDSKY, B. SILBERMANN, Norms of inverses, spectra, and pseudospectra of large truncated Wiener-Hopf operators and Toeplitz matrices. – New York J. Math. **3**(1997), 1–31.

[7] A. BÖTTCHER, B. SILBERMANN, Analysis of Toeplitz Operators. – Akademie-Verlag, Berlin 1989 and Springer-Verlag, Berlin, Heidelberg, New York 1990.

[8] A. BÖTTCHER, B. SILBERMANN, Introduction to Large Truncated Toeplitz Matrices. – Springer-Verlag, Berlin, Heidelberg 1999.

[9] N. BROWN, AF embeddings and the numerical computation of spectra in irrational rotation algebras. – Preprint 2004.

[10] I. GOHBERG, I. FELDMAN, Convolution Equations and Projection Methods for Their Solution. – Nauka, Moskva 1971 (Russian, Engl. transl.: Amer. Math. Soc. Transl. of Math. Monographs, Vol. 41, Providence, Rhode Island, 1974).

[11] R. HAGEN, S. ROCH, B. SILBERMANN, C^*-Algebras and Numerical Analysis. – Marcel Dekker, Inc., New York, Basel 2001.

[12] A.V. KOZAK, I.B. SIMONENKO, Projection methods for the solution of multidimensional convolution equations. – Sib. Mat. Zh. **21**(1980), 2, 119–127 (Russian).

[13] S. PRÖSSDORF, B. SILBERMANN, Numerical Analysis for Integral and Related Operator Equations. – Akademie-Verlag, Berlin, 1991, and Birkhäuser Verlag, Basel, Boston, Stuttgart 1991.

[14] V.S. RABINOVICH, S. ROCH, B. SILBERMANN, Fredholm theory and finite section method for band-dominated operators. – Integral Equations Oper. Theory **30**(1998), 4, 452–495.

[15] V.S. RABINOVICH, S. ROCH, B. SILBERMANN, Algebras of approximation sequences: Finite sections of band-dominated operators. – Acta Appl. Math. **65**(2001), 315–332.

[16] V.S. RABINOVICH, S. ROCH, B. SILBERMANN, Limit Operators and Their Applications in Operator Theory. – Operator Theory: Adv. and Appl. **150**, Birkhäuser Verlag, Basel, Boston, Berlin 2004.

[17] S. ROCH, Spectral approximation of Wiener-Hopf operators with almost periodic generating function. – Numer. Funct. Anal. Optimization **21**(2000), 1-2, 241–253.

[18] S. ROCH, Finite sections of band-dominated operators. – Preprint 2355 TU Darmstadt, July 2004, 98 p., submitted to Memoirs Amer. Math. Soc.

[19] A. ROGOZHIN, B. SILBERMANN, On the approximation numbers for the finite sections of block Toeplitz matrices. – Submitted to London Math. Soc.

Vladimir S. Rabinovich
Instituto Politecnico Nacional
ESIME Zacatenco
Avenida IPN
Mexico, D. F. 07738, Mexico
e-mail: `vladimir.rabinovich@gmail.com`

Steffen Roch
Department of Mathematics
Technical University of Darmstadt
Schlossgartenstrasse 7
D-64289 Darmstadt, Germany
e-mail: `roch@mathematik.tu-darmstadt.de`

Bernd Silbermann
Faculty of Mathematics
Technical University of Chemnitz
D-09107 Chemnitz, Germany
e-mail: `bernd.silbermann@mathematik.tu-chemnitz.de`

Operator Theory:
Advances and Applications, Vol. 170, 229–248

On the Toeplitz Operators with Piecewise Continuous Symbols on the Bergman Space

N. Vasilevski

To I.B. Simonenko in occasion of his 70th birthday.

Abstract. The paper is devoted to the study of Toeplitz operators with piecewise continuous symbols. We clarify the geometric regularities of the behavior of the essential spectrum of Toeplitz operators in dependence on their crucial data: the angles between jump curves of symbols at a boundary point of discontinuity and on the limit values reached by a symbol at that boundary point. We show then that the curves supporting the symbol discontinuities, as well as the number of such curves meeting at a boundary point of discontinuity, do not play any essential role for the Toeplitz operator algebra studied. Thus we exclude the curves of symbol discontinuity from the symbol class definition leaving only the set of boundary points (where symbols may have discontinuity) and the type of the expected discontinuity. Finally we describe the C^*-algebra generated by Toeplitz operators with such symbols.

Mathematics Subject Classification (2000). Primary 47B35; Secondary 47C15.

Keywords. Toeplitz operator, Bergman space, piece-wise continuous symbol, C^*-algebra.

1. Introduction

Let $\mathbb{D}$ be the unit disk in $\mathbb{C}$ and $\gamma = \partial\mathbb{D}$ be its boundary. Consider $L_2(\mathbb{D})$ with respect to the standard Lebesgue plane measure and its subspace, the Bergman space $\mathcal{A}^2(\mathbb{D})$, which consists of functions analytic in $\mathbb{D}$. Let $B_{\mathbb{D}}$ stand for the orthogonal Bergman projection of $L_2(\mathbb{D})$ onto $\mathcal{A}^2(\mathbb{D})$. Given a function $a(z) \in L_\infty(\mathbb{D})$, the Toeplitz operator T_a with symbol $a = a(z)$ is defined on $\mathcal{A}^2(\mathbb{D})$ as follows:

$$T_a : \varphi \in \mathcal{A}^2(\mathbb{D}) \longmapsto B_{\mathbb{D}}(a\varphi) \in \mathcal{A}^2(\mathbb{D}).$$

In the paper we study Toeplitz operators with piecewise continuous symbols. The first results in this direction date from the early 1980s (see [3, 4, 5, 6]) and

This work was partially supported by CONACYT Project 46936, México.

show that essentially the situation is the same as in the case of Toeplitz operators with piecewise continuous symbols on the Hardy space. The exact result is given in Theorem 2.2 below.

The next essential advance in this direction was made by M. Loaiza [2] after about 20 years of silence. She described the case of piecewise continuous symbols having more then two limit values at the boundary point of discontinuity. This result was made possible due to recent work [7] describing the commutative C^*-algebras of Toeplitz operators on the Bergman space.

We recall that for piecewise continuous symbols the product of two Toeplitz operators is not in general a compact perturbation of a Toeplitz operator. Thus the algebra generated by such operators has a quite complicated structure, coinciding with the uniform closure of the set of all elements of the form

$$\sum_{k=1}^{p}\prod_{j=1}^{q_k} T_{a_{j,k}}. \tag{1.1}$$

It is very interesting and important to understand the nature of the operators forming the algebra and, in particular, to know *whether this Toeplitz operator algebra contains any other Toeplitz operator, apart from its initial generators.* Note that this question has remained unanswered since the very first work on the subject.

In the paper we present some recent advances in the area. In Section 2 we recall the previous results, especially on algebras generated by Toeplitz operators with piecewise continuous symbols, which are relevant to the main content of the paper. In Section 3 we show how the results of [2] allow us to understand the geometric regularities of the behavior of the essential spectrum of Toeplitz operators in dependence on their crucial data: the angles between jump curves of symbols at a boundary point of discontinuity and on the limit values reached by a symbol at that boundary point. Section 4 is devoted to the local analysis of Toeplitz operators at a point of discontinuity. The results of [7] permit us to get a highly unexpected result, which partially answers the above question. We show that the closure of elements of the form (1.1) contains many Toeplitz operators, and the symbols of these Toeplitz operators belong to a much wider class of discontinuous functions, as compared with the symbols of the initial generators. In particular, it turns out that the algebra considered in [6] already contained all operators from the algebra considered in [2], though previously there were no means to realize this fact. The main conclusion of the section is that we can start from very different sets of symbols and obtain exactly the same operator algebra as a result. That is, the curves supporting the symbol discontinuities, as well as the number of such curves meeting at a boundary point of discontinuity, do not play in fact any essential role for the Toeplitz operator algebra studied. This observation motivates us to exclude the curves of symbol discontinuity from the very beginning and to leave in the symbol class definition only the set of boundary points (where symbols may have discontinuity) and the type of the expected discontinuity. We do this in the final Section 5 introducing the so-called boundary piecewise continuous symbols and describing the algebra generated by Toeplitz operators with such symbols.

2. Preliminaries

In this section we recall some well-known results relevant to the main content of the paper.

Given a linear space (or algebra) $\mathcal{A} \subset L_\infty(\mathbb{D})$, we denote by $\mathcal{T}(\mathcal{A})$ the C^*-algebra generated by all Toeplitz operators T_a with $a \in \mathcal{A}$, and we denote by $\operatorname{Sym} \mathcal{T}(\mathcal{A}) = \mathcal{T}(\mathcal{A})/\mathcal{K}$ its (Fredholm) symbol, or Calkin algebra. Here $\mathcal{K}$ is the ideal of all compact operators on $\mathcal{A}^2(\mathbb{D})$.

We start with the description of the algebra generated by Toeplitz operators with continuous symbols, which goes back to L. Coburn [1].

Theorem 2.1. *The algebra $\mathcal{T}_C = \mathcal{T}(C(\overline{\mathbb{D}}))$ is irreducible and contains the entire ideal $\mathcal{K}$ of compact operators on $\mathcal{A}^2(\mathbb{D})$. Each operator $T \in \mathcal{T}(C(\overline{\mathbb{D}}))$ is of the form*

$$T = T_a + K,$$

where $a \in C(\overline{\mathbb{D}})$ and K is a compact operator. The homomorphism

$$\operatorname{sym} : \mathcal{T}_C \longrightarrow \operatorname{Sym} \mathcal{T}_C = \mathcal{T}_C/\mathcal{K} \cong C(\gamma)$$

is given by

$$\operatorname{sym} : T = T_a + K \longmapsto a_{|_\gamma}.$$

The operator $T \in \mathcal{T}_C$ is Fredholm if and only if its symbol is invertible, i.e., the function $\operatorname{sym} T \neq 0$ on γ, and

$$\operatorname{Ind} T = -\frac{1}{2\pi}\{\operatorname{sym} T\}_\gamma.$$

The situation changes if we extend the symbol class from continuous to piecewise continuous functions. The corresponding results were obtained in [3, 4, 5, 6]. To introduce them we proceed as follows. Denote by ℓ a union of a finite number of piecewise smooth curves in $\overline{\mathbb{D}}$. We will assume that the intersection $\gamma \cap \ell$ consists of a finite number of endpoints of ℓ: $T = \gamma \cap \ell = \{t_1, \ldots, t_m\}$, and each $t_p \in T$ is the endpoint for only one curve from ℓ.

Denote by $PC(\overline{\mathbb{D}}, \ell)$ the algebra of all functions $a(z)$, continuous in $\overline{\mathbb{D}} \setminus \ell$, and having left and right limit values at all points of ℓ. In particular, at each point $t_p \in T$ any function $a \in PC(\overline{\mathbb{D}}, \ell)$ has two limit values: $a(t_p - 0)$ and $a(t_p + 0)$, following the positive orientation of γ.

Let $\widehat{\gamma}$ be the boundary γ, cut by points $t_p \in T$. The pair of points, which correspond to a point $t_p \in T$, $p = \overline{1, m}$, we denote by $t_p - 0$ and $t_p + 0$, following the positive orientation of γ. Let $\overline{X} = \sqcup_{p=1}^{m} \Delta_p$ be the disjoint union of segments $\Delta_p = [0, 1]$. Denote by Γ the union $\widehat{\gamma} \cup \overline{X}$ with the following point identification

$$t_p - 0 \equiv 0_p, \qquad t_p + 0 \equiv 1_p,$$

where $t_p \pm 0 \in \widehat{\gamma}$, 0_p and 1_p are the endpoints of Δ_p, $p = 1, 2, \ldots, m$.

Theorem 2.2. *The C^*-algebra $\mathcal{T}_{PC} = \mathcal{T}(PC(\overline{\mathbb{D}}, \ell))$ is irreducible and contains the ideal $\mathcal{K}$ of compact operators. The (Fredholm) symbol algebra $\operatorname{Sym} \mathcal{T}_{PC} = \mathcal{T}_{PC}/\mathcal{K}$ is isomorphic to the algebra $C(\Gamma)$. The homomorphism*

$$\operatorname{sym} : \mathcal{T}_{PC} \to \operatorname{Sym} \mathcal{T}_{PC} = \mathcal{T}_{PC}/\mathcal{K} \cong C(\Gamma)$$

is generated by the following mapping of generators of $\mathcal{T}_{PC}$

$$\operatorname{sym} : T_a \longmapsto \begin{cases} a(t), & t \in \widehat{\gamma} \\ a(t_p - 0)(1 - x) + a(t_p + 0)x, & x \in [0, 1] \end{cases},$$

where $t_p \in \ell \cap \gamma$, $p = 1, 2, \ldots, m$.

An operator $T \in \mathcal{T}_{PC}$ is Fredholm if and only if its symbol is invertible, i.e., the function $\operatorname{sym} T \neq 0$ on Γ, and

$$\operatorname{Ind} T = -\frac{1}{2\pi}\{\operatorname{sym} T\}_{\Gamma}.$$

We note that for piecewise continuous symbols the product of two Toeplitz operators is in general not longer a compact perturbation of a Toeplitz operator. The algebra $\mathcal{T}_{PC}$ does not coincide with the set of all operators of the form $T_a + K$ as in case of continuous symbols. It has a much more complicated structure, coinciding with the uniform closure of the set of all elements of the form

$$\sum_{k=1}^{p} \prod_{j=1}^{q_k} T_{a_{j,k}}, \tag{2.1}$$

where $a_{j,k} \in PC(\overline{D}, \ell)$, $p, q_k \in \mathbb{N}$.

At this stage an important question arises: *does the algebra $\mathcal{T}_{PC}$ contain any other Toeplitz operator, apart from its initial generators?*

An unexpected (partial) answer to this question will be provided in last two sections of the paper.

The key result permitting one to handle local situations for a wider class of discontinuous symbols was given in [7] and is as follows.

We start from $L_2(\Pi)$ over the upper half-plane Π with the usual Lebesgue plane measure and its Bergman subspace $\mathcal{A}^2(\Pi)$. Denote by $\mathcal{A}_\infty$ the C^*-algebra of bounded measurable homogeneous functions on Π of order zero, or functions depending only on the polar coordinate θ. Introduce the Toeplitz operator algebra $\mathcal{T}(\mathcal{A}_\infty)$, which is generated by all operators T_a with $a(\theta) \in \mathcal{A}_\infty$.

Theorem 2.3. *Let $a = a(\theta) \in \mathcal{A}_\infty$. Then the Toeplitz operator T_a acting on $\mathcal{A}^2(\Pi)$ is unitary equivalent to the multiplication operator $\gamma_a I = R T_a R^*$ acting on $L_2(\mathbb{R})$. The function $\gamma_a(\lambda)$ is given by*

$$\gamma_a(\lambda) = \frac{2\lambda}{1 - e^{-2\pi\lambda}} \int_0^\pi a(\theta) \, e^{-2\lambda\theta} \, d\theta, \qquad \lambda \in \mathbb{R}. \tag{2.2}$$

Analyzing formula (2.2) we note that for each $a(\theta) \in L_\infty(0, \pi)$ the function $\gamma_a(\lambda)$ is continuous at all finite points $\lambda \in \mathbb{R}$. For a "very large λ" ($\lambda \to +\infty$) the exponent $e^{-2\lambda\theta}$ has a very sharp maximum at the point $\theta = 0$, and thus the major contribution to the integral in (2.2) for these "very large λ" is determined by values of $a(\theta)$ in a neighborhood of the point 0. The major contribution for a "very large negative λ" ($\lambda \to -\infty$) is determined by values of $a(\theta)$ in a neighborhood of π, due to a very sharp maximum of $e^{-2\lambda\theta}$ at $\theta = \pi$ for these values of λ. In particular, if $a(\theta)$ has limits at the points 0 and π, then

$$\lim_{\lambda \to +\infty} \gamma_a(\lambda) = \lim_{\theta \to 0} a(\theta),$$

$$\lim_{\lambda \to -\infty} \gamma_a(\lambda) = \lim_{\theta \to \pi} a(\theta).$$

Corollary 2.4. *The algebra $\mathcal{T}(\mathcal{A}_\infty)$ is commutative. The isomorphic imbedding*

$$\tau_\infty : \mathcal{T}(\mathcal{A}_\infty) \longrightarrow C_b(\mathbb{R})$$

is generated by the following mapping of generators of the algebra $\mathcal{T}(\mathcal{A}_\infty)$

$$\tau_\infty : T_a \longmapsto \gamma_a(\lambda),$$

where $a = a(\theta) \in \mathcal{A}_\infty$.

The above result was the starting point for the study of the algebra generated by Toeplitz operators with piecewise continuous symbols having more than two limit values at the boundary points, done by M. Loaiza [2]. We list here the principal local situation and the final result in a form convenient for us.

Via a Möbius transformation the principal local situation in [2] is reduced to the following upper half-plane setting. Given a finite number of different points on $[0, \pi]$,

$$0 = \theta_0 < \theta_1 < \theta_2 < \cdots < \theta_{n-1} < \theta_n = \pi,$$

we denote by $\mathcal{A}(\Lambda)$ with $\Lambda = \{\theta_1, \theta_2, \ldots, \theta_{n-1}\}$ the algebra of piecewise constant functions on $[0, \pi]$ with jump points in Λ, and let $H(\mathcal{A}(\Lambda))$ be the algebra of homogeneous of zero order functions on Π whose restrictions onto the upper half-circle (parameterized by $\theta \in [0, \pi]$) belong to $\mathcal{A}(\Lambda)$. Note that each (piecewise constant) function $a \in H(\mathcal{A}(\Lambda))$ has n limit values at the origin.

Denote by V_k, $k = 1, 2, \ldots, n$, the cone on the upper half-plane Π, supported on $(\theta_{k-1}, \theta_k]$. Then the n-dimensional algebra $H(\mathcal{A}(\Lambda))$ consists of all functions having the form

$$a(z) = a_1 \chi_{V_1}(z) + a_2 \chi_{V_2}(z) + \cdots + a_n \chi_{V_n}(z),$$

where $(a_1, a_2, \ldots, a_n) \in \mathbb{C}^n$, and $\chi_k(z)$ are the characteristic functions of the cones V_k, $k = 1, 2, \ldots, n$.

The Toeplitz C^*-algebra $\mathcal{T}(H(\mathcal{A}(\Lambda)))$ is obviously generated by n *commuting* Toeplitz operators $T_{\chi_{V_k}}$, $k = 1, 2, \ldots, n$, and we have

$$\gamma_{\chi_{V_k}}(\lambda) = \frac{2\lambda}{1 - e^{-2\pi\lambda}} \int_{\theta_{k-1}}^{\theta_k} e^{-2\lambda\theta}\, d\theta = \frac{e^{-2\theta_k\lambda} - e^{-2\theta_{k-1}\lambda}}{e^{-2\pi\lambda} - 1}, \qquad \lambda \in \mathbb{R}. \qquad (2.3)$$

Each function $\gamma_{\chi_{V_k}}$ is continuous on $\overline{\mathbb{R}}$ and

$$\lim_{\lambda \to -\infty} \gamma_{\chi_{V_1}}(\lambda) = 0, \qquad \lim_{\lambda \to +\infty} \gamma_{\chi_{V_1}}(\lambda) = 1,$$

$$\lim_{\lambda \to -\infty} \gamma_{\chi_{V_k}}(\lambda) = 0, \qquad \lim_{\lambda \to +\infty} \gamma_{\chi_{V_k}}(\lambda) = 0, \qquad k = 2, 3, \ldots, n-1,$$

$$\lim_{\lambda \to -\infty} \gamma_{\chi_{V_n}}(\lambda) = 1, \qquad \lim_{\lambda \to +\infty} \gamma_{\chi_{V_n}}(\lambda) = 0.$$

Furthermore, each function $\gamma_{\chi_{V_k}}$ is non-negative and

$$\sum_{k=0}^{n} \gamma_{\chi_{V_k}}(\lambda) \equiv 1;$$

thus the set

$$\Delta(\Lambda) = \{t = (t_1, t_2, \ldots, t_n) : t_k = \gamma_{\chi_{V_k}}(\lambda), \quad \lambda \in \overline{\mathbb{R}}, \quad k = 1, \ldots, n\} \tag{2.4}$$

is a continuous curve lying on the standard $(n-1)$-dimensional simplex, and connecting the vertices $(1, 0, \ldots, 0)$ and $(0, \ldots, 0, 1)$.

In the following figure we present the behavior of the set $\Delta(\Lambda)$ for the case $n = 3$ in dependence of the angles (θ_1, θ_2).

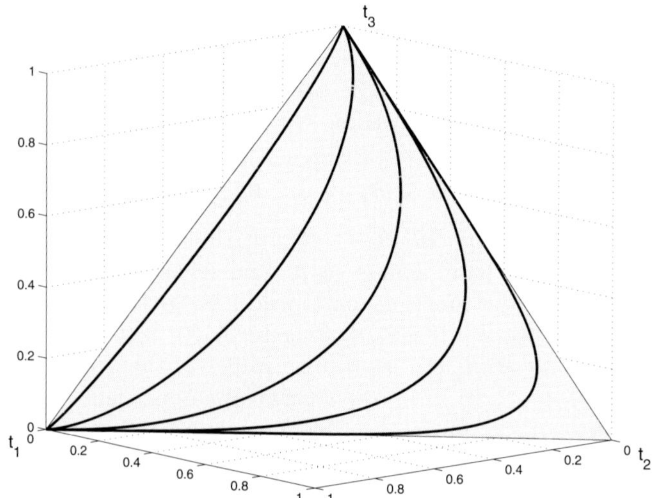

FIGURE 1. The angles (θ_1, θ_2) left to right: $(0.48\pi, 0.52\pi)$, $(0.4\pi, 0.6\pi)$, $(0.3\pi, 0.7\pi)$, $(0.2\pi, 0.8\pi)$, $(0.1\pi, 0.9\pi)$.

Theorem 2.5. *Given a set* $\Lambda = \{\theta_1, \theta_2, \ldots, \theta_{n-1}\}$, *the Toeplitz C^*-algebra* $\mathcal{T}(H(\mathcal{A}(\Lambda)))$ *is isomorphic and isometric to* $C(\Delta(\Lambda))$. *The isomorphism*

$$\tau : \mathcal{T}(H(\mathcal{A}(\Lambda))) \longrightarrow C(\Delta(\Lambda))$$

*is generated by the following mapping of generators of the algebra $T(H(\mathcal{A}(\Lambda)))$:
if $a(z) = a_1\chi_{V_1}(z) + a_2\chi_{V_2}(z) + \cdots + a_n\chi_{V_n}(z)$, then*

$$\tau : T_a \longmapsto a_1 t_1 + a_2 t_2 + \cdots + a_n t_n,$$

where $t = (t_1, t_2, \ldots, t_n) \in \Delta(\Lambda)$.

Proof. The C^*-algebra $T(H(\mathcal{A}(\Lambda)))$ is commutative, and is generated by n operators $T_{\chi_{V_k}}$, $k = 1, 2, \ldots, n$. Thus it is isomorphic and isometric to the algebra of all
continuous functions on the joint spectrum of the above operators, which coincides
obviously with $\Delta(\Lambda)$. $\square$

Consider now the general case of Toeplitz operators with piecewise continuous
symbols. Denote by ℓ a piecewise smooth curve in the closed unit disk $\overline{\mathbb{D}}$, satisfying
the following properties: there are a finite number of points (nodes), which divide
ℓ into simple oriented smooth curves ℓ_j, $j = \overline{1, k}$. We assume that the endpoints of
ℓ are among the nodes. We will refer to a node using symbols u_{q,r_q}, where r_q is the
number of curves meeting at this node, and q corresponds the node numbering.
Denote by T the set of all nodes from $\ell \cap \gamma$, and assume that T consists of m points.
For each node $t_{q,r_q-1} \in T$ there are $r_q - 1$ curves meeting at t_{q,r_q-1}, $q = 1, \ldots, m$.
We assume as well that locally near t_{q,r_q-1} these curves are hypercycles, that is,
there is a Möbius transformation of the unit disk to the upper half-plane under
which the node t_{q,r_q-1} goes to the origin and the curves meeting at t_{q,r_q-1} are
mapped to curves which near origin are straight line segments meeting at the
origin.

Let now $PC(\overline{\mathbb{D}}, \ell)$ be the algebra of all functions $a(z)$, continuous in $\overline{\mathbb{D}} \setminus \ell$,
and having left and right limit values at all points of ℓ_j: $a^+(z)$ and $a^-(z)$. On the
nodes of type $t_{q,r_q-1} \in T$ the functions from $PC(\overline{\mathbb{D}}, \ell)$ have r limit values. We
denote them by $a^{(1)}_{t_{q,r_q-1}}, \ldots, a^{(r)}_{t_{q,r_q-1}}$, counting counter-clockwise.

Let $T = T(PC(\overline{\mathbb{D}}, \ell))$ be the C^*-algebra generated by all Toeplitz operators
T_a with symbols $a \in PC(\overline{\mathbb{D}}, \ell)$.

For each node $t_{q,r_q-1} \in T$ introduce the ordered set

$$\Lambda_q = \{\theta_1, \theta_2, \ldots, \theta_{r_q-1}\}$$

of the angles which the $r_q - 1$ curves meeting at the node t_{q,r_q-1} form with the
boundary γ, counting them counter-clockwise. Introduce as well the corresponding
curve

$$\Delta(\Lambda_q) = \{(t_1, t_2, \ldots, t_{r_q}) : t_k = \gamma_{\chi_{V_k}}(\lambda), \ \lambda \in \mathbb{R}, \ k = 1, \ldots, r_q\}, \tag{2.5}$$

where each $t_k = \gamma_{\chi_{V_k}}(\lambda)$, $k = 1, \ldots, r_q$, is given by, see (2.3),

$$t_k = \gamma_{\chi_{V_k}}(\lambda) = \frac{2\lambda}{1 - e^{-2\pi\lambda}} \int_{\theta_{k-1}}^{\theta_k} e^{-2\lambda\theta} \, d\theta = \frac{e^{-2\theta_k\lambda} - e^{-2\theta_{k-1}\lambda}}{e^{-2\pi\lambda} - 1}, \qquad \lambda \in \mathbb{R},$$

Observe that the curve $\Delta(\Lambda_q)$ lies on the standard $(r_q - 1)$-dimensional simplex
and connects its vertices $(1, 0, \ldots, 0)$ and $(0, \ldots, 0, 1)$.

Denote by $\widehat{\gamma}$ the set γ, cut by points $t_{q,r_q-1} \in T = \ell \cap \gamma$. The pair of points which correspond to a point $t_{q,r_q-1} \in T$ we denote by $t_{q,r_q-1} - 0$ and $t_{q,r_q-1} + 0$, following the positive orientation of γ. Let $\overline{X} = \cup_q \Delta(\Lambda_q)$ be the disjoint union of the sets (2.5). Denote by Γ the union $\widehat{\gamma} \cup \overline{X}$ with the following point identification

$$t_{q,r_q-1} - 0 \equiv (1, 0, \ldots, 0) \qquad t_{q,r_q-1} + 0 \equiv (0, \ldots, 0, 1),$$

where $t_{q,r_q-1} \pm 0 \in \widehat{\gamma}$, and $(1, 0, \ldots, 0)$ and $(0, \ldots, 0, 1)$ are the vertices of $\Delta(\Lambda_q)$.

Now the final result reads as follows.

Theorem 2.6. *The C^*-algebra $\mathcal{T} = \mathcal{T}(PC(\overline{\mathbb{D}}, \ell))$ is irreducible and contains the ideal $\mathcal{K}$ of compact operators. The symbol algebra $\operatorname{Sym} \mathcal{T} = \mathcal{T}/\mathcal{K}$ is isomorphic to the algebra $C(\Gamma)$. Identifying them, the symbol homomorphism*

$$\operatorname{sym} : \mathcal{T} \to \operatorname{Sym} \mathcal{T} = C(\Gamma)$$

is generated by the following mapping of generators of $\mathcal{T}$

$$\operatorname{sym} : T_a \longmapsto \begin{cases} a(t), & t \in \widehat{\gamma} \\ a^{(1)}_{t_{q,r_q-1}} t_1 + a^{(2)}_{t_{q,r_q-1}} t_2 + \cdots + a^{(r_q)}_{t_{q,r_q-1}} t_{r_q}, \\ \qquad\qquad t = (t_1, t_2, \ldots, t_{r_q}) \in \Delta(\Lambda_q) \end{cases},$$

where $t_{q,r_q-1} \in T$.

An operator $T \in \mathcal{T}$ is Fredholm if and only if its symbol is invertible, i.e., the function $\operatorname{sym} T \neq 0$ on Γ, and

$$\operatorname{Ind} T = -\frac{1}{2\pi} \{\operatorname{sym} T\}_\Gamma.$$

3. Essential and local spectra

The results given by Theorem 2.6 permit us, in particular, to describe easily the essential spectrum of T_a and to understand the geometric regularities of its behavior.

Indeed, given a symbol $a \in PC(\overline{\mathbb{D}}, \ell)$, the essential spectrum $\operatorname{ess} - \operatorname{sp} T_a$ of the operator T_a, which is obviously equal to $\operatorname{Im} \operatorname{sym} T_a$, consists of two parts. Its regular part is the image of the symbol restricted on the boundary points of continuity, i.e., $\operatorname{sym} T_a|_{\widehat{\gamma}} = a|_{\widehat{\gamma}}$. The complementary part is a finite number of additional arcs, each one of which is the restriction of $\operatorname{sym} T_a$ onto the curve $\Delta(\Lambda_q)$, corresponding to the boundary point of discontinuity t_{q,r_q-1}.

We note that each such curve $\operatorname{sym} T_a|_{\Delta(\Lambda_q)}$ describes as well the spectrum of the local representative at the point t_{q,r_q-1} of the initial operator T_a.

Let us assume that t_0 is a boundary point of discontinuity for functions from $PC(\overline{\mathbb{D}}, \ell)$ an which n curves from ℓ intersect. As previously, introduce the ordered set

$$\Lambda = \{\theta_1, \theta_2, \ldots, \theta_{n-1}\}$$

of the angles which the above n curves form with the boundary γ, counting them counter-clockwise. As above, we add $\theta_0 = 0$ and $\theta_n = \pi$. Given a symbol $a \in PC(\overline{\mathbb{D}}, \ell)$, introduce the ordered set

$$A = \{a_1, a_2, \ldots, a_n\},$$

where each a_k, $k = 1, 2, \ldots, n$, is the limit value of a at the point t_0 reached from the region between the $(k-1)$th and kth curves.

The local representative at the point t_0 of the operator T_a can be taken as the Toeplitz operator $T_{A,\Lambda}$ with piecewise constant symbol

$$a_{A,\Lambda}(\theta) = a_1 \chi_{V_1}(\theta) + \cdots + a_n \chi_{V_n}(\theta) \in H(\mathcal{A}(\Lambda)),$$

where each χ_{V_k} is the characteristic function of the cone V_k supported on $(\theta_{k-1}, \theta_k]$.

That is the spectrum of $T_{a_{A,\Lambda}}$, which the same as the corresponding portion of the essential spectrum of T_a, is governed by the sets A and Λ and is given by the formula

$$\operatorname{sp} T_{a_{A,\Lambda}} = \{a_1 t_1 + \cdots + a_n t_n : t = (t_1, t_2, \ldots, t_n) \in \Delta(\Lambda)\}. \tag{3.1}$$

It is instructive to understand the geometric regularities of its behavior.

We start with the simplest case of just two limit values. Let $A = (a_1, a_2)$ and $\Lambda = \{\theta_1\}$. In this case the spectrum $\operatorname{sp} T_{a_{A,\Lambda}}$ *does not depend* on Λ, is *uniquely* determined by A, and is the straight line segment connecting the points a_1 and a_2. This is an effect of low dimension: *each curve connecting the vertices of a one-dimensional simplex is the simplex itself, and is the straight line segment connecting the vertices.*

Passing to $n > 2$ we consider first the most transparent case $n = 3$. In this case the curve $\Delta(\Lambda)$ lies on a two-dimensional simplex, which has the same dimension as the complex plane where the spectrum lies.

As we already know (see Figure 1), the continuous curve $\Delta(\Lambda)$ connecting the vertices $v_1 = (1, 0, 0)$ and $v_3 = (0, 0, 1)$ *does depend* essentially on Λ. Then by (3.1), the spectrum $\operatorname{sp} T_{a_{A,\Lambda}}$, geometrically, is the image of the curve $\Delta(\Lambda)$ under the projection of the two-dimensional simplex to the complex plane such that each its vertex v_k is projected to a_k, $k = 1, 2, 3$, and $a_k \in A$. That is, the set A determines the triangle to which the simplex is projected, while the set Λ determines the shape of the curve $\Delta(\Lambda)$, whose projection into the already defined triangle gives the spectrum.

In the next two pictures we illustrate this for three different sets Λ, being the first, third, and fifth set of angles of Figure 1. That is, we consider the following sets of angles $(0.48\pi, 0.52\pi)$, $(0.3\pi, 0.7\pi)$, and $(0.1\pi, 0.9\pi)$, ordered as generated from less to more curved lines. For the first picture the set A is given by $(0.1 + 0.1i, 0.9i, 0.9 + 0.5i)$, while $A = (0.1 + 0.1i, 1 + 0.2i, 0.9 + 0.5i)$, for the second picture. For both sets we leave the same values of a_1 and a_3, making the pictures "one-parametric" in dependence on a_2.

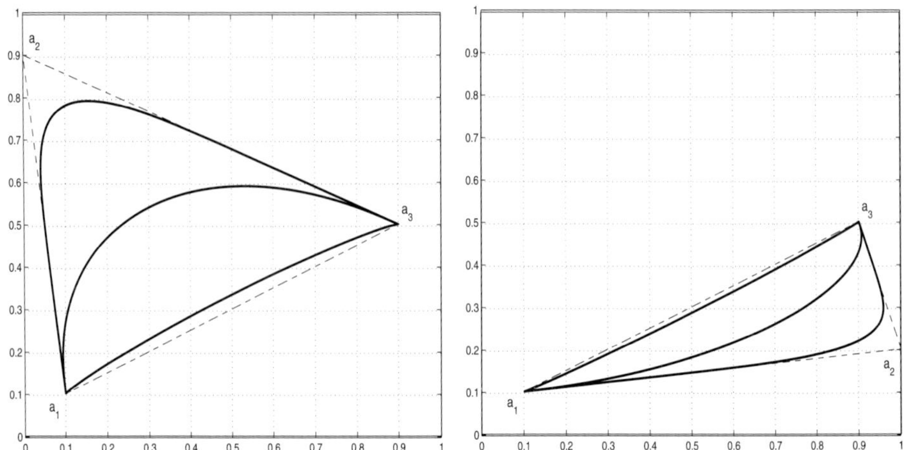

FIGURE 2. Spectra of $T_{a_{A,\Lambda}}$ for three limit values symbols.

The case $n > 3$ maintains in principle the same features. The spectrum $\operatorname{sp} T_{a_{A,\Lambda}}$ is the image of the curve $\Delta(\Lambda)$ under the projection of, now, the $(n-1)$-dimensional simplex onto a certain convex polygon in the complex plane such that each vertex v_k is projected to a_k, $k = 1, 2, , \ldots, n$, and $a_k \in A$. The curve $\Delta(\Lambda)$ connecting the vertices $v_1 = (1, 0, \ldots, 0)$ and $v_n = (0, \ldots, 0, 1)$ again *does depend* essentially on Λ. The set A determines the polygon to which the simplex is projected, while the set Λ determines the shape of the curve $\Delta(\Lambda)$, whose projection into the already defined polygon gives the spectrum. The only difference is that now this convex polygon has n or *less* vertices, depending on the way, prescribed by A, in which the $(n-1)$-dimensional simplex is projected onto the two-dimensional polygon. That is, the projections of some vertices may (or may not) be in the interior of the polygon.

In the next two pictures we present the cases of five limit values symbols for which the 4-dimensional simplex is projected onto a pentagon and a triangle, respectively. We consider the following sets A

$$(0.2 + 0.1i, 0.4 + 0.9i, 0.8 + 0.1i, 0.1 + 0.7i, 0.9 + 0.8i)$$

and

$$(0.2 + 0.1i, 0.5 + 0.6i, 0.1 + 0.9i, 0.3 + 0.4i, 0.9 + 0.8i),$$

maintaining the same values of a_1 and a_5 for both cases. Both pictures represent three spectra for the following sets Λ

$$(0.46\pi, 0.48\pi, 0.52\pi, 0.54\pi), \ (0.2\pi, 0.2\pi, 0.7\pi, 0.8\pi),$$
$$(0.0002\pi, 0.01\pi, 0.99\pi, 0.9998\pi),$$

and which again correspond to lines ordered from less to more curved.

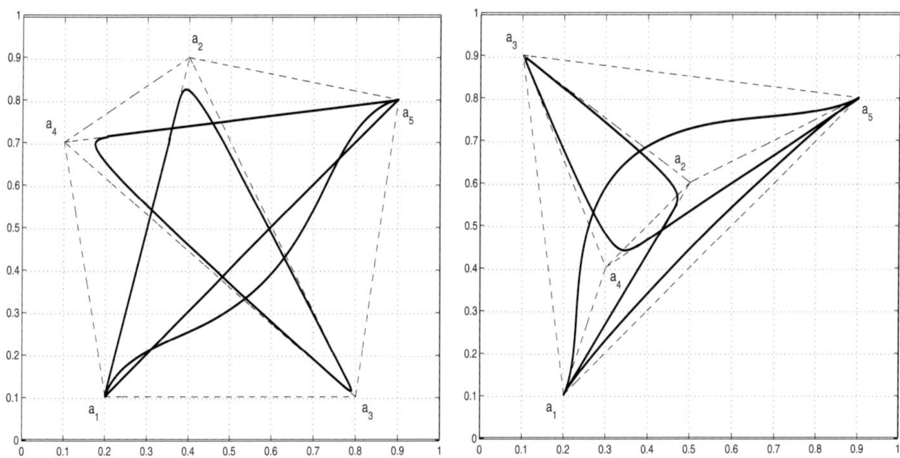

FIGURE 3. Spectra of $T_{a_{A,\Lambda}}$ for five limit values symbols (pentagon and triangle).

We note that the spectrum $\operatorname{sp} T_{a_{A,\Lambda}}$ becomes more rectilinear and more stable under the perturbations of $a_k \in A$, $k = 2, \ldots, n-1$, for bigger values of the angles θ_1 and $\pi - \theta_{n-1}$. In this case the spectrum approaches the straight line segment connecting the images of the vertices $(1, 0, \ldots, 0)$ and $0, \ldots, 0, 1)$ when the sum of these angles tends to π. The opposite, in a sense, tendency appears when the angles between the curves intersecting at t_0 and the boundary of the domain tend to 0. In that case the spectrum approaches the union of straight line segments passing in sequence through the images of the vertices $(1, 0, \ldots, 0)$, $(0, 1, 0, \ldots, 0)$, $\ldots (0, \ldots, 0, 1)$.

4. Local analysis at a point of discontinuity

Although the description given by Theorem 2.6 proves to be useful, it hides, at the same time, some essential properties of the above Toeplitz operator algebras. In particular, it turns out that each Toeplitz operator algebra $\mathcal{T}(PC(\overline{\mathbb{D}}, \ell))$, besides the initial generators T_a with symbols $a \in PC(\overline{\mathbb{D}}, \ell)$, contains many another Toeplitz operators with much more general symbols.

We show this here for the model situation at a point of discontinuity. We introduce first a number of symbol sets. Denote by $L_\infty^{\{0,\pi\}}(0, \pi)$ the C^*-subalgebra of $L_\infty(0, \pi)$ which consists of all functions having limits at the points 0 and π. Let $C[0, \pi]$ be, as usual, the algebra of all continuous functions on $[0, \pi]$; denote by $PC([0, \pi], \Lambda)$, where $\Lambda = \{\theta_1, \theta_2, \ldots, \theta_{n-1}\}$, the algebra of all piece-wise continuous functions on $[0, \pi]$, continuous in $[0, \pi] \setminus \Lambda$ and having one-sided limit values at the points of Λ. Let $PCo([0, \pi]), \Lambda)$ be the subalgebra of $PC([0, \pi], \Lambda)$ consisting of all piece-wise constant functions. Given a function $a_0(\theta)$, denote by $L(1, a_0)$ the linear two-dimensional space, generated by 1 and the function a_0.

Note that $PCo([0,\pi],\{\theta_1\}) = L(1,\chi_{[0,\theta_1]})$, where $\chi_{[0,\theta_1]}(\theta)$ is the characteristic function of $[0,\theta_1]$.

For a continuous function a_0, a set Λ, and an arbitrary point $\theta_k \in \Lambda$, we have the following chain of proper inclusions

$$\begin{matrix} L(1,a_0) \subset C[0,\pi] \\ PCo([0,\pi],\{\theta_k\}) \subset PCo([0,\pi]),\Lambda \end{matrix} \subset PC([0,\pi],\Lambda) \subset L_\infty^{\{0,\pi\}}(0,\pi). \qquad (4.1)$$

Given a linear set $\mathcal{A}$, the subset of $L_\infty(0,\pi)$, denote by $H(\mathcal{A})$ the subset of $\mathcal{A}_\infty$ which consists of all homogeneous functions of zero order on the upper half-plane whose restrictions onto the upper half of the unit circle (parameterized by $\theta \in [0,\pi]$) belong to $\mathcal{A}$. Further let $\mathcal{T}(H(\mathcal{A}))$ be the the C^*-algebra generated by all Toeplitz operators T_a with symbols $a \in H(\mathcal{A})$.

Note that for any real nonconstant function a_0, the algebra $\mathcal{T}(H(L(1,a_0)))$ is a C^*-algebra with identity generated by a *single* self-adjoint element, the Toeplitz operator T_{a_0}.

Let $\mathcal{A}$ be any of the sets in (4.1), consider the C^*-algebra $\mathcal{T}(H(\mathcal{A}))$.

For the largest set (algebra) $L_\infty^{\{0,\pi\}}(0,\pi)$ we have

Theorem 4.1. *The C^*-algebra $\mathcal{T}(H(L_\infty^{\{0,\pi\}}(0,\pi)))$ is isomorphic and isometric to $C(\overline{\mathbb{R}})$, where $\overline{\mathbb{R}} = \mathbb{R} \cup \{-\infty\} \cup \{+\infty\}$ is the two-point compactification of $\mathbb{R}$. The isomorphic isomorphism*

$$\tau_\infty : \mathcal{T}(H(L_\infty^{\{0,\pi\}}(0,\pi))) \longrightarrow C(\overline{\mathbb{R}})$$

is generated by the mapping of generators of the algebra $\mathcal{T}(H(L_\infty^{\{0,\pi\}}(0,\pi)))$

$$\tau_\infty : T_a \longmapsto \gamma_a(\lambda), \qquad (4.2)$$

where $a = a(\theta) \in H(L_\infty^{\{0,\pi\}}(0,\pi))$.

Proof. We need to show only that the mapping (4.2) is onto. The inclusion

$$\tau_\infty(\mathcal{T}(H(L_\infty^{\{0,\pi\}}(0,\pi)))) \subset C(\overline{\mathbb{R}})$$

is trivial. The inverse inclusion will follow from the next theorem. $\qquad \square$

Passing to another extreme, the smallest possible set, we have

Theorem 4.2. *Let $a_0(\theta) \in L_\infty^{\{0,\pi\}}(0,\pi)$ be a real-valued function such that the function $\gamma_{a_0}(\lambda)$ separates the points of $\overline{\mathbb{R}}$. Then the C^*-algebra $\mathcal{T}(H(L(1,a_0)))$ is isomorphic and isometric to $C(\overline{\mathbb{R}})$. The isomorphic isomorphism*

$$\tau_\infty : \mathcal{T}(H(L(1,a_0))) \longrightarrow C(\overline{\mathbb{R}})$$

is generated by the same mapping of generators of the algebra $\mathcal{T}(H(L(1,a_0)))$

$$\tau_\infty : T_a \longmapsto \gamma_a(\lambda),$$

Proof. Follows directly from the Stone-Weierstrass theorem. $\qquad \square$

Corollary 4.3. *Given a point $\theta_0 \in (0,\pi)$, the C^*-algebra $\mathcal{T}(H(PCo([0,\pi],\{\theta_0\})))$ is isomorphic and isometric to $C(\overline{\mathbb{R}})$.*

Proof. As it was already mentioned $PCo([0,\pi],\{\theta_0\}) = L(1,\chi_{[0,\theta_0]})$. All we need to prove is that the real-valued function

$$\gamma_{\chi_{[0,\theta_0]}}(\lambda) = \frac{2\lambda}{1 - e^{-2\pi\lambda}} \int_0^{\theta_0} e^{-2\lambda\theta}\,d\theta = \frac{e^{-2\theta_0\lambda} - 1}{e^{-2\pi\lambda} - 1}$$

separates the points of $\overline{\mathbb{R}}$. We show that the function $\gamma_{\chi_{[0,\theta_0]}}$ is strictly increasing by a simple but somewhat lengthy procedure.

After the scaling $t = 2\pi\lambda$, $\theta_0 = \alpha\pi$, with $\alpha \in (0,1)$, we have

$$\gamma(t) = \frac{e^{-\alpha t} - 1}{e^{-t} - 1}, \quad t \in \mathbb{R}.$$

First let $t > 0$, and calculate

$$\gamma'(t) = \frac{\alpha e^{-\alpha t}(1 - e^{-t}) - e^{-t}(1 - e^{-\alpha t})}{(1 - e^{-t})^2}.$$

To show that $\gamma'(t) > 0$, it is equivalent to show that

$$\alpha e^{-\alpha t}\frac{e^t - 1}{e^t} - e^{-t}\frac{e^t - 1}{e^{\alpha t}} > 0,$$

or that

$$\alpha(e^t - 1) - (e^{\alpha t} - 1) > 0$$

or

$$\sum_{k=1}^{\infty}(\alpha - \alpha^k)\frac{t^k}{k!} > 0.$$

The last inequality is evident because $\alpha \in (0,1)$.

Pass now to $t < 0$. Substituting $x = -t$, $x \in \mathbb{R}_+$, we have

$$\gamma(t(x)) = \frac{e^{\alpha x} - 1}{e^x - 1}$$

and

$$\gamma'(t(x)) = \frac{\alpha e^{\alpha x}(e^x - 1) - e^x(e^{\alpha x} - 1)}{(e^x - 1)^2}.$$

Now we need to show that the function $\gamma(t(x))$ is strictly decreasing, or that $\gamma'(t(x)) < 0$. This is equivalent to

$$e^x(e^{\alpha x} - 1) - \alpha e^{\alpha x}(e^x - 1) > 0$$

or to

$$(1 - \alpha)(e^x - 1) - (e^{(1-\alpha)x} - 1) > 0,$$

or to

$$\sum_{k=1}^{\infty}[(1 - \alpha) - (1 - \alpha)^k]\frac{x^k}{k!} > 0.$$

Again the last inequality is evident because $\alpha \in (0,1)$. $\qquad\square$

Let now $a_0(\theta) = \frac{\theta}{\pi}$, this function $a_0(\theta)$ is obviously real-valued and continuous on $[0,\pi]$.

Corollary 4.4. *The C^*-algebra $T(H(L(1, a_0)))$ is isomorphic and isometric to $C(\overline{\mathbb{R}})$.*

Proof. We have

$$
\begin{aligned}
\gamma_{a_0}(\lambda) &= \frac{2\lambda}{\pi(1 - e^{-2\pi\lambda})} \int_0^\pi \theta e^{-2\lambda\theta}\, d\theta \\
&= \frac{1}{\pi(1 - e^{-2\pi\lambda})} \left[-\pi e^{-2\pi\lambda} - \frac{1}{2\lambda}(e^{-2\pi\lambda} - 1) \right] = \frac{1}{2\pi\lambda} - \frac{1}{e^{2\pi\lambda} - 1}.
\end{aligned}
$$

The function $\gamma_{a_0}(\lambda)$ is continuous on $\overline{\mathbb{R}}$, and

$$
\begin{aligned}
\lim_{\lambda \to 0} \gamma_{a_0}(\lambda) &= \gamma_{a_0}(0) = \frac{1}{2} \\
\lim_{\lambda \to -\infty} \gamma_{a_0}(\lambda) &= \gamma_{a_0}(-\infty) = 1 \\
\lim_{\lambda \to +\infty} \gamma_{a_0}(\lambda) &= \gamma_{a_0}(+\infty) = 0.
\end{aligned}
$$

To finish the proof we need to show that the function $\gamma_{a_0}(\lambda)$ separates the points of $\overline{\mathbb{R}}$. To do this we show that the function

$$
\gamma(t) = \frac{1}{t} - \frac{1}{e^t - 1}, \qquad t \in \mathbb{R}
$$

is strictly decreasing, or that $\gamma'(t) < 0$ for all $t \neq 0$.

The function

$$
\gamma'(t) = -\frac{1}{t^2} + \frac{e^t}{(e^t - 1)^2}
$$

is even. Thus it is sufficient to prove that

$$
\frac{e^t}{(e^t - 1)^2} < \frac{1}{t^2},
$$

or that

$$
t^2 e^t < (e^t - 1)^2,
$$

for each $t > 0$. The last inequality is easy to check, comparing coefficients of the power series

$$
t^2 e^t = \sum_{n=2}^\infty \frac{1}{(n-2)!} t^n,
$$

$$
(e^t - 1)^2 = \sum_{n=2}^\infty \frac{2(2^{n-1} - 1)}{n!} t^n. \qquad \square
$$

Remark 4.5. The above statements show that in spite of the fact that the generating sets of symbols in (4.1) are quite different, the resulting Toeplitz C^*-algebras are the same. Moreover, this (common) C^*-algebra with identity can be generated by a *single* Toeplitz operator with either *continuous*, or *piece-wise constant* symbol. Further, although the algebraic operations with Toeplitz operators do not give a Toeplitz operator, in general, the resulting (single-generated) algebra is extremely rich in Toeplitz operators: *each Toeplitz operator with symbol from $H(L_\infty^{\{0,\pi\}}(0,\pi))$ belongs to this algebra.*

We give now a number of illustrating examples. Consider $\mathcal{A}^2(\Pi)$ and the Toeplitz operator T_+ with symbol $a_+(z) = \chi_+(\operatorname{Re} z) = \chi_+(x)$, where χ_+ is the characteristic function of the positive half-line. We have as well that $a_+(z) = a_+(re^{i\theta}) = \chi_{[0,\pi/2]}(\theta)$, and thus $a_+ \in H(PCo([0,\pi],\{\pi/2\}))$.

The Toeplitz operator $T_+ \in \mathcal{T}(H(PCo([0,\pi],\{\pi/2\})))$ is unitary equivalent to the multiplication operator $\gamma_{a_+} I$, where, by (2.2),

$$\gamma_{a_+}(\lambda) = \frac{2\lambda}{1 - e^{-2\pi\lambda}} \int_0^\pi \chi_{[0,\pi/2]}(\theta)\, e^{-2\lambda\theta}\, d\theta = \frac{e^{-\pi\lambda} - 1}{e^{-2\pi\lambda} - 1}, \qquad \lambda \in \mathbb{R}.$$

The operator T_+ is obviously self-adjoint and $\operatorname{sp} T_+ = [0,1]$. Thus for any function f continuous on $[0,1]$ the operator $f(T_+)$ is well defined by the standard functional calculus in C^*-algebras, furthermore the operator $f(T_+)$ belongs to the same algebra $\mathcal{T}(H(PCo([0,\pi],\{\pi/2\})))$.

Example. Consider the family of functions f_α parameterized by $\alpha \in [0,1]$ and given as follows:

$$f_\alpha(x) = x^{2(1-\alpha)} \frac{(1-x)^{2\alpha} - x^{2\alpha}}{(1-x) - x}, \qquad x \in [0,1]. \tag{4.3}$$

Each function f_α is continuous on $[0,1]$, and $f_\alpha(0) = 0$, $f_\alpha(1) = 1$. Let us mention as well some particular cases

$$f_0(x) \equiv 0, \qquad f_{\frac{1}{2}}(x) = x, \qquad f_1(x) \equiv 1.$$

Then

$$f_\alpha(T_+) = T_{\chi_{[0,\alpha\pi]}} \in \mathcal{T}(H(PCo([0,\pi],\{\pi/2\}))),$$

where the symbol $\chi_{[0,\alpha\pi]}$ of the operator $T_{\chi_{[0,\alpha\pi]}}$ belongs to $H(PCo([0,\pi],\{\alpha\pi\}))$.

Proof. We will exploit the isomorphism between the Toeplitz operator algebra and the functional algebra given in Corollary 2.4. Introduce

$$x = \gamma_{a_+}(\lambda) = \frac{e^{-\pi\lambda} - 1}{e^{-2\pi\lambda} - 1} = \frac{1}{e^{-\pi\lambda} + 1} \in [0,1],$$

which is equivalent to

$$\lambda = \lambda(x) = -\frac{1}{\pi} \ln \frac{1-x}{x}.$$

Then for the operator $T_{\chi_{[0,\alpha\pi]}}$ the corresponding function $\gamma_{\chi_{[0,\alpha\pi]}}$ is given by

$$\gamma_{\chi_{[0,\alpha\pi]}}(\lambda) = \frac{2\lambda}{1 - e^{-2\pi\lambda}} \int_0^\pi \chi_{[0,\lambda\pi]}(\theta)\, e^{-2\lambda\theta}\, d\theta = \frac{e^{-2\alpha\pi\lambda} - 1}{e^{-2\pi\lambda} - 1}, \qquad \lambda \in \mathbb{R}.$$

Substituting $\lambda = \lambda(x)$ we have

$$\gamma_{\chi_{[0,\alpha\pi]}}(\lambda(x)) = \frac{e^{2\alpha\pi\frac{1}{\pi}\ln\frac{1-x}{x}} - 1}{e^{2\pi\frac{1}{\pi}\ln\frac{1-x}{x}} - 1}$$

$$= \frac{\left(\frac{1-x}{x}\right)^{2\alpha} - 1}{\left(\frac{1-x}{x}\right)^{2} - 1} = x^{2(1-\alpha)}\frac{(1-x)^{2\alpha} - x^{2\alpha}}{(1-x) - x}. \qquad \square$$

Note that the above-mentioned particular cases of f_α lead to the equalities

$$f_0(T_+) = 0, \qquad f_{\frac{1}{2}}(T_+) = T_+, \qquad f_1(T_+) = I,$$

as it should be.

In the next example we present a connection between Toeplitz operators with piece-wise constant symbols having just two and more than two limit values at the single point of discontinuity.

Example. Given a finite ordered set of numbers $0 < \alpha_1 < \alpha_2 < \cdots < \alpha_{n-1} < 1$, introduce

$$\Lambda = \{\alpha_1\pi, \alpha_2\pi, \ldots, \alpha_{n-1}\pi\};$$

for convenience we add $\alpha_0 = 0$ and $\alpha_n = 1$. Let further $A = \{a_1, a_2, \ldots, a_n\}$ be an ordered set of complex numbers.

Given both A and Λ, we define the piece-wise constant symbol

$$a_{A,\Lambda}(\theta) = \sum_{k=1}^{n} a_k \chi_{(\alpha_{k-1}\pi, \alpha_k\pi]} \in PCo([0, \pi], \Lambda)$$

and the function $f_{A,\Lambda} = f_{A,\Lambda}(x)$ continuous on $[0, 1]$

$$f_{A,\Lambda}(x) = \sum_{k=1}^{n} a_k \frac{(1 - x)^{2\alpha_k}x^{2(1-\alpha_k)} - (1 - x)^{2\alpha_{k-1}}x^{2(1-\alpha_{k-1})}}{(1 - x) - x}.$$

Then

$$f_{A,\Lambda}(T_+) = T_{a_{A,\Lambda}} \in \mathcal{T}(H(PCo([0, \pi], \{\pi/2\}))).$$

Proof. Consider the Toeplitz operator $T_{a_{A,\Lambda}}$. Using (2.3) we have

$$\gamma_{a_{A,\Lambda}}(\lambda) = \frac{2\lambda}{1 - e^{-2\pi\lambda}}\int_0^\pi a_{A,\Lambda}(\theta)\, e^{-2\lambda\theta}\, d\theta$$

$$= \sum_{k=1}^{n} a_k \frac{e^{-2\alpha_k\pi\lambda} - e^{-2\alpha_{k-1}\pi\lambda}}{e^{-2\pi\lambda} - 1}, \qquad \lambda \in \mathbb{R}.$$

Substitute $\lambda = \lambda(x)$ as in the previous example. Then after a simple calculation we have

$$\gamma_{a_{A,\Lambda}}(\lambda(x)) = \sum_{k=1}^{n} a_k \frac{(1 - x)^{2\alpha_k}x^{2(1-\alpha_k)} - (1 - x)^{2\alpha_{k-1}}x^{2(1-\alpha_{k-1})}}{(1 - x) - x}. \qquad \square$$

Theorem 4.2 and Corollary 4.3 imply, in particular, that each Toeplitz operator with $H(L_\infty^{\{0,\pi\}}(0,\pi))$-symbol can be obtained in a similar way. The exact formula for the corresponding continuous function $f(x)$, though forcedly rather implicit, is given in the next example.

Example. Given a function $a = a(\theta) \in H(L_\infty^{\{0,\pi\}}(0,\pi))$, let

$$f_a(x) = \frac{2x^2}{\pi} \frac{\ln(1-x) - \ln x}{(1-x) - x} \int_0^\pi a(\theta) \left(\frac{1-x}{x} \right)^{\frac{2\theta}{\pi}} d\theta.$$

Then

$$f_a(T_+) = T_a.$$

Remark 4.6. In the above examples we have considered the Toeplitz operator T_+ as the starting operator by a very simple reason: in this specific case the generically transcendental equation $x = \gamma_a(\lambda)$ admits an explicit solution.

We can start as well from any Toeplitz operator T_α having the symbol $\chi_{[0,\alpha\pi]}$, where $\alpha \in (0,\pi)$. Indeed, as follows from the proof of Corollary 4.3, the function $\gamma_{\chi_{[0,\alpha\pi]}}(\lambda)$ is strictly increasing. This implies that the function $f_\alpha(x)$ (see (4.3)), which maps $[0,1]$ onto $[0,1]$, is strictly increasing as well. Thus the function $f_\alpha^{-1}(x)$ is well defined and continuous on $[0,1]$.

Finally, given $\alpha, \beta \in (0,\pi)$, A, Λ, and $a = a(\theta) \in H(L_\infty^{\{0,\pi\}}(0,\pi))$, we have

$$T_\alpha \in \mathcal{T}(H(PCo([0,\pi],\{\alpha\pi\})))$$

and

$$
\begin{aligned}
(f_\beta \circ f_\alpha^{-1})(T_\alpha) &= T_\beta, \\
(f_{A,\Lambda} \circ f_\alpha^{-1})(T_\alpha) &= T_{a_{A,\Lambda}}, \\
(f_a \circ f_\alpha^{-1})(T_\alpha) &= T_a,
\end{aligned}
$$

where all Toeplitz operators from the right-hand side of the above equalities belong to $\mathcal{T}(H(PCo([0,\pi],\{\alpha\pi\})))$.

5. Boundary piecewise continuous functions

The above examples show that studying the algebra generated by Toeplitz operators, whose symbols admit discontinuities at a finite number of boundary points, we can start from any symbol algebra selected from a wide variety of symbol classes. Moreover, the curve ℓ, entering in the definition of the symbol algebra $PC(\overline{\mathbb{D}},\ell)$, does not play in fact any significant role. In all such cases the resulting C^*-algebra will contain all Toeplitz operators whose symbols admit a "homogeneous type discontinuity" in each boundary point of discontinuity, locally described by the algebra $H(L_\infty^{\{0,\pi\}}(0,\pi))$.

Thus it seems reasonable to include the Toeplitz operators with such symbols among the generators of the algebra from the very beginning. In this case the definition proceeds as follows.

Let $T = \{t_1, t_2, \ldots, t_m\}$ be a finite set of distinct points on the unit circle $\gamma = \partial \mathbb{D}$. Introduce the linear space $BPC(\mathbb{D}, T)$ (BPC stands for *Boundary Piecewise Continuous*) which consists of all functions $a(z)$ obeying the following properties:

  (i) $a(z) \in L_\infty(\mathbb{D})$;

 (ii) $a(z)$ has limit values at all boundary point $t \in \gamma \setminus T$, and the function $a(t)$ constructed by these limit values is continuous in $\gamma \setminus T$;

(iii) at each point $t_0 \in T$ the function $a(z)$ has a "homogeneous type discontinuity", which means that there exist a Möbius transformation $z = z_{t_0}(w)$ of the upper half-plane Π to the unit disk $\mathbb{D}$ with $t_0 = z_{t_0}(0)$ and a homogeneous function of order zero $a_{t_0}(w) \in H(L_\infty^{\{0,\pi\}}(0,\pi))$ such that

$$\lim_{w \to 0} [a(z_{t_0}(w)) - a_{t_0}(w)] = 0.$$

Let us make several comments on this definition. The set $BPC(\mathbb{D}, T)$ in fact is a C^*-algebra, although only the linear space structure is important for our purposes. The function $a(t)$, as a function of the boundary points, belongs to $PC(\gamma, T)$; that is, for each point $t_0 \in T$ the following limits

$$\lim_{t \to t_0,\, t \prec t_0} a(t) = a(t_0 - 0) \qquad \text{and} \qquad \lim_{t \to t_0,\, t_0 \prec t} a(t) = a(t_0 + 0)$$

are well defined. Property (iii) of the above definition can be alternatively done in geometric terms of $\mathbb{D}$ as follows. For each point $t_0 \in T$ there are a hyperbolic pencil $\mathcal{P}_{t_0}$ of geodesics in $\mathbb{D}$, such that t_0 is the endpoint of its axis, and a function $\widetilde{a}_{t_0}(z)$ which is constant on cycles of $\mathcal{P}_{t_0}$ and whose values on (each) geodesic are given by an L_∞-function having limit values at the endpoints of the geodesic on γ (points at infinity in hyperbolic geometry), such that

$$\lim_{z \to t_0} [a(z) - \widetilde{a}_{t_0}(z)] = 0.$$

Consider now the C^*-algebra $\mathcal{T}_{BPC} = \mathcal{T}(BPC(\mathbb{D}, T))$ generated by all Toeplitz operators T_a with symbols $a \in BPC(\mathbb{D}, T)$.

Let, as above, $\widehat{\gamma}$ be the set γ, cut by points $t_p \in T$. The pair of points which correspond to a point $t_p \in T$, $p = \overline{1, m}$, we denote by $t_p - 0$ and $t_p + 0$, following the positive orientation of γ. Let $\overline{X} = \sqcup_{p=1}^m \Delta_p$ be the disjoint union of segments $\Delta_p = [0, 1]$. Denote by Γ the union $\widehat{\gamma} \cup \overline{X}$ with the following point identification

$$t_p - 0 \equiv 1_p, \qquad t_p + 0 \equiv 0_p,$$

where $t_p \pm 0 \in \widehat{\gamma}$, 0_p and 1_p are the endpoints of Δ_p, $p = 1, 2, \ldots, m$.

Than we have obviously

Theorem 5.1. *The C^*-algebra $\mathcal{T}_{BPC} = \mathcal{T}(BPC(\mathbb{D}, T))$ is irreducible and contains the ideal $\mathcal{K}$ of compact operators. The symbol algebra $\mathrm{Sym}\,\mathcal{T}_{BPC} = \mathcal{T}_{BPC}/\mathcal{K}$ is isomorphic to the algebra $C(\Gamma)$. Identifying them, the symbol homomorphism*

$$\mathrm{sym} : \mathcal{T}_{BPC} \to \mathrm{Sym}\,\mathcal{T}_{BPC} = C(\Gamma)$$

is generated by the following mapping of generators of $\mathcal{T}_{BPC}$

$$\mathrm{sym} : T_a \longmapsto \begin{cases} a(t), & t \in \widehat{\gamma} \\ \gamma_{a_{t_p}}\left(\dfrac{1-2x}{\sqrt{1-(1-2x)^2}}\right), & x \in [0,1] \end{cases},$$

where a_{t_p} is the function defined by the above property (iii) *for $a(z)$ at the point $t_p \in \Lambda$, $p = 1, 2, \ldots, m$, and*

$$\gamma_{a_{t_p}}(\lambda) = \frac{2\lambda}{1 - e^{-2\pi\lambda}} \int_0^\pi a_{t_p}(\theta)\, e^{-2\lambda\theta}\, d\theta, \qquad \lambda \in \mathbb{R}.$$

An operator $T \in \mathcal{T}_{BPC}$ is Fredholm if and only if its symbol is invertible, i.e., the function $\mathrm{sym}\, T \neq 0$ on Γ, and

$$\mathrm{Ind}\, T = -\frac{1}{2\pi}\{\mathrm{sym}\, T\}_\Gamma.$$

Proof. Easily follows from the standard local principle, Theorem 4.1 and Theorem 2.6. $\qquad\square$

We mention that the algebras described by Theorems 2.2, 2.6, and 5.1 consist of the same operators, in spite of the fact that their initial generators are quite different. That is, as it turned out, the first algebra generated by Toeplitz operators with discontinuous symbols, which was described by Theorem 2.2, already contained all the operators with $BPC(\mathbb{D}, T)$-symbols. For about twenty years there was no way to see this. At the same time Theorem 5.1 gives a transparent description for all Toeplitz operators for all $BPC(\mathbb{D}, T)$-symbols.

We end the paper formulating two open problems.

PROBLEM 1. Extend the description of Toeplitz operator algebra from $BPC(\mathbb{D}, T)$-symbols to a rotation invariant symbol set containing $BPC(\mathbb{D}, T)$.

PROBLEM 2. Extend the description of Toeplitz operator algebra from $BPC(\mathbb{D}, T)$-symbols to a Möbius invariant symbol set containing $BPC(\mathbb{D}, T)$. This class of symbols can be naturally called $BPC(\mathbb{D})$.

References

[1] L.A. Coburn. Singular integral operators and Toeplitz operators on odd spheres. *Indiana Univ. Math. J.*, 23(5):433–439, 1973.

[2] M. Loaiza. On an algebra of Toeplitz operators with piecewise continuous symbols. *Integr. Equat. Oper. Th.*, 51(1):141–153, 2005.

[3] G. McDonald. Toeplitz operators on the ball with piecewise continuous symbol. *Illinois J. Math.*, 23(2):286–294, 1979.

[4] N.L. Vasilevski. Banach algebras that are generated by certain two-dimensional integral operators. II. (Russian). *Math. Nachr.*, 99:135–144, 1980.

[5] N.L. Vasilevski. Banach algebras generated by two-dimensional integral operators with a Bergman kernel and piecewise continuous coefficients. I. *Soviet Math. (Izv. VUZ)*, 30(3):14–24, 1986.

[6] N.L. Vasilevski. Banach algebras generated by two-dimensional integral operators with a Bergman kernel and piecewise continuous coefficients. II. *Soviet Math. (Izv. VUZ)*, 30(3):44–50, 1986.

[7] N.L. Vasilevski. Bergman space structure, commutative algebras of Toeplitz operators and hyperbolic geometry. *Integr. Equat. Oper. Th.*, 46:235–251, 2003.

N. Vasilevski
Departamento de Matemáticas
CINVESTAV del I.P.N.
México, D.F., México
e-mail: `nvasilev@math.cinvestav.mx`

Operator Theory:
Advances and Applications, Vol. 170, 249–256

Asymptotics of a Class of Operator Determinants

Harold Widom

For I.B. Simonenko on the occasion of his 70th birthday

Abstract. In previous work of C.A. Tracy and the author asymptotic formulas were derived for certain operator determinants whose interest lay in the fact that quotients of them gave solutions to the cylindrical Toda equations. In the present paper we consider a more general class of operators which retain some of the properties of those cited and we find analogous asymptotics for the determinants.

Mathematics Subject Classification (2000). Primary 47G10; Secondary 47B35.

Keywords. Operator determinant, Wiener-Hopf, Kac-Ahieser theorem.

1. Introduction and statement of the result

There are innumerable instances in the mathematics and mathematical physics literature where the problem arises of determining the asymptotics of operator determinants of the form $\det\,(I+K_\alpha)$ where K_α is a trace class operator depending on a parameter. The first general result is probably the continuous analogue of the strong Szegö limit theorem due to M. Kac [3], and independently by N.I. Ahieser [1], where K_α is an integral operator of convolution type on an interval of length α, and $\alpha \to \infty$. (Thus the result is generally called the Kac-Ahieser theorem.) There have been many generalizations of this result, but by no means all such problems fall into this category and their results cannot be used directly. This paper is concerned with some of these other cases.

The impetus for the present work was [6], where the authors considered operator kernels of the form

$$K(u, v) = \int \frac{e^{-t[(1-\omega)u+(1-\omega^{-1})u^{-1}]}}{-\omega u + v}\, d\rho(\omega)$$

The author was supported by National Science Foundation grant DMS-0243982.

acting on $L^2(\mathbf{R}^+)$, where ρ can be any finite complex measure supported on a compact subset of $\{\omega \in \mathbf{C} : \Re\omega < 1,\ \Re\omega^{-1} < 1\}$. The interest in these kernels was due to the fact that logarithms of certain ratios of determinants give solutions to the cylindrical Toda equations. The main interest is in the limit $t \to 0+$. In the cited paper asymptotics of the form bt^a were determined, with the constants a and b having integral representations. Thus, they were explicitly determined. This was done, naturally, only under certain assumptions on ρ.

If one makes the substitutions $u \to e^x$, $v \to e^y$ then operators on $L^2(\mathbf{R}^+)$ become operators on $L^2(\mathbf{R})$. In the present paper we consider a more general class of operators on $L^2(\mathbf{R})$ which retain some of the properties of those cited and we find analogous asymptotics for the determinants. The approach we use is different from that in [6] in that we deal with the determinants directly rather than through the resolvent. Because of the generality of the setting we do not obtain an explicit integral representation for the constant factor in the asymptotics; rather it itself is given in terms of operator determinants. At the end we shall indicate what special property of the particular operators of [6] allows the evaluation of these determinants.

Our setting is a family of trace class operators K_α on $L^2(\mathbf{R})$ which converge strongly as $\alpha \to \infty$ to an operator K with kernel $k(x-y)$, where $k \in L^1(\mathbf{R})$. Thus each $\det(I + K_\alpha)$ is defined but $\det(I + K)$ is not. The problem is to find the asymptotics of $\det(I + K_\alpha)$.

A basic requirement is that if $\hat{k}$ is the Fourier transform of k,

$$\hat{k}(\xi) = \int_{-\infty}^{\infty} e^{ix\xi}\, k(x)\, dx,$$

and if we set $\sigma(\xi) = 1 + \hat{k}(\xi)$ then

$$\sigma(\xi) \neq 0, \quad \arg\sigma(\xi)|_{-\infty}^{\infty} = 0. \tag{1}$$

This assures that the Wiener-Hopf operators $I + \chi^{\pm} K \chi^{\pm}$ are invertible.[1] We shall also assume that $|x|^{1/2}\, k(x) \in L^2(\mathbf{R})$ which, together with (1), will allow us to use the Kac-Ahieser theorem.[2]

The first connection between K_α and K we require is that

$$\chi^{-}(K_\alpha - K)\chi^{+} = o_1(1),$$
$$\chi^{+}(K_\alpha - K)\chi^{-} = o_1(1).$$

Here $\chi^{\pm}$ denotes multiplication by $\chi_{\mathbf{R}^{\pm}}$ and $o_1(1)$ denotes any family of operators whose trace norms are $o(1)$. (In particular, $\chi^{-} K \chi^{+}$ and $\chi^{-} K \chi^{+}$ are trace class.)

To state what characterizes our family K_α we introduce the translation operator T_a defined by $T_a f(x) = f(x - a)$. Our main assumption is that there are

---

[1] See, for example, [2, §I.8].

[2] This says that if $W_\alpha(\sigma) = I + \chi_{(-\alpha, \alpha)} K \chi_{(-\alpha, \alpha)}$ acting on $L^2(-\alpha, \alpha)$ then $\det W_\alpha(\sigma) \sim G(\sigma)^{2\alpha} E(\sigma)$, where $G(\sigma) = \exp(s(0))$, $E(\sigma) = \exp\{\int_0^\infty x\, s(x)\, s(-x)\, dx\}$, and $s(x)$ is the inverse Fourier transform of $\log\sigma(\xi)$. There is also the alternative expression $E(\sigma) = \det(W(\sigma)\, W(\sigma^{-1}))$.

operators $K_\pm$ such that

$$\chi^+ (K_\alpha - T_\alpha K_+ T_{-\alpha}) \chi^+ = o_1(1),$$
$$\chi^- (K_\alpha - T_{-\alpha} K_- T_\alpha) \chi^- = o_1(1),$$

(2)

and the operators

$$K_{11} := K_- - \chi^+ K \chi^+$$

and

$$K_{22} := K_+ - \chi^- K \chi^-$$

are trace class.[3]

Our result is that

$$\det(I + K_\alpha) \sim G(\sigma)^{2\alpha} E(\sigma)$$

$$\times \det \left(I + (I + \chi^+ K\chi^+)^{-1} K_{11}\right) \det \left(I + (I + \chi^- K\chi^-)^{-1} K_{22}\right),$$

(3)

where $G(\sigma)$ and $E(\sigma)$ are the constants in the Kac-Ahieser theorem.

In the classical example where $K_\alpha = \chi_{(-\alpha,\alpha)} K \chi_{(-\alpha,\alpha)}$ we take $K_+ = \chi^- K \chi^-$ and $K_- = \chi^+ K \chi^+$ (the operators in (2) are then identically zero) so $K_{11} = K_{22} = 0$.

In the example which arises from the simplest case in [6] (after symmetrization and variable change) K_α is the integral operator with kernel

$$K_\alpha(x,y) = \lambda \, \frac{e^{-(e^{x-\alpha}+e^{y-\alpha}+e^{-x-\alpha}+e^{-y-\alpha})}}{\cosh\left[(x-y)/2\right]},$$

and the other kernels are

$$K(x,y) = \lambda \, \frac{1}{\cosh\left[(x-y)/2\right]},$$
$$K_+(x,y) = \lambda \, \frac{e^{-(e^x+e^y)}}{\cosh\left[(x-y)/2\right]},$$
$$K_-(x,y) = \lambda \, \frac{e^{-(e^{-x}+e^{-y})}}{\cosh\left[(x-y)/2\right]}.$$

The determinants in (3) can be evaluated in this case.

Remark. The cases where the conditions (1) are satisfied are the easier ones. There is great interest in the integrable systems community in operators for which $\sigma(\xi)$ may have zeros. (In the above example this occurs when $\lambda \in (-\infty, -\pi^{-1}]$.) Asymptotic formulas have been derived in some of these cases [5, 4], but as far as we know none have been proved rigorously. A hope is that the approach presented here may be applicable, at least to some extent, to these.

---

[3] The reason for the subscripts is that K_{11} and K_{22} will appear in the diagonal entries of a 2×2 matrix kernel.

2. Derivation of the result

We think of $L^2(\mathbf{R})$ as $L^2(\mathbf{R}^-) \oplus \mathbf{L}^2(\mathbf{R}^+)$ and the corresponding matrix representation. Condition (2) tells us that with error $o_1(1)$ the matrix representation of K_α is

$$\begin{pmatrix} \chi^- T_{-\alpha} K_- T_\alpha \chi^- & \chi^- K \chi^+ \\ \chi^+ K \chi^- & \chi^+ T_\alpha K_+ T_{-\alpha} \chi^+ \end{pmatrix}.$$

The upper-left corner of this matrix may be written

$$\chi^- \left(T_{-\alpha} \chi^+ K \chi^+ T_\alpha + T_{-\alpha} K_{11} T_\alpha \right) \chi^-. \tag{4}$$

We shall use, here and below, the fact that K commutes with translations and the relation

$$\chi_J T_a = T_a \chi_{J-a} \tag{5}$$

for any a and any set J. Thus (4) may be written

$$\chi_{(-\alpha,0)} K \chi_{(-\alpha,0)} + \chi^- T_{-\alpha} K_{11} T_\alpha \chi^-.$$

Similarly for the lower-right corner, and so the matrix representation of K_α is $o_1(1)$ plus

$$\begin{pmatrix} \chi_{(-\alpha,0)} K \chi_{(-\alpha,0)} + \chi^- T_{-\alpha} K_{11} T_\alpha \chi^- & \chi^- K \chi^+ \\ \chi^+ K \chi^- & \chi_{(0,\alpha)} K \chi_{(0,\alpha)} + \chi^+ T_\alpha K_{22} T_{-\alpha} \chi^+ \end{pmatrix}.$$

The upper-right corner is trace class, so if we multiply it on either side by $\chi_{(-\infty,-\alpha)}$ or $\chi_{(\alpha,\infty)}$ the result is $o_1(1)$. Similarly for the lower-left corner. Hence with this error the above equals

$$\begin{pmatrix} \chi_{(-\alpha,0)} K \chi_{(-\alpha,0)} & \chi_{(-\alpha,0)} K \chi_{(0,\alpha)} \\ \chi_{(0,\alpha)} K \chi_{(-\alpha,0)} & \chi_{(0,\alpha)} K \chi_{(0,\alpha)} \end{pmatrix}$$
$$+ \begin{pmatrix} \chi^- T_{-\alpha} K_{11} T_\alpha \chi^- & 0 \\ 0 & \chi^+ T_\alpha K_{22} T_{-\alpha} \chi^+ \end{pmatrix}.$$

The operator

$$I + \begin{pmatrix} \chi_{(-\alpha,0)} K \chi_{(-\alpha,0)} & \chi_{(-\alpha,0)} K \chi_{(0,\alpha)} \\ \chi_{(0,\alpha)} K \chi_{(-\alpha,0)} & \chi_{(0,\alpha)} K \chi_{(0,\alpha)} \end{pmatrix}$$

is just $W_\alpha(\sigma)$ in its matrix representation and so its determinant is asymptotically equal to $G(\sigma)^{2\alpha} E(\sigma)$, by the Kac-Ahieser theorem.[4]

---

[4]Strictly speaking $W_\alpha(\sigma)$ acts on $L^2(-\alpha, \alpha)$ while the above operator acts on $L^2(\mathbf{R})$. But the determinants are the same.

The next step is to factor out this operator from $I + K_\alpha$ and determine the asymptotics of the determinant of the result. For convenience we write the above operator as $D + E$, where

$$D = \begin{pmatrix} I + \chi_{(-\alpha,0)} \, K \, \chi_{(-\alpha,0)} & 0 \\ 0 & I + \chi_{(0,\alpha)} \, K \, \chi_{(0,\alpha)} \end{pmatrix}$$

and

$$E = \begin{pmatrix} 0 & \chi_{(-\alpha,0)} \, K \, \chi_{(0,\alpha)} \\ \chi_{(0,\alpha)} \, K \, \chi_{(-\alpha,0)} & 0 \end{pmatrix},$$

so the operator whose determinant we now want is $o_1(1)$ plus[5]

$$I + (D+E)^{-1} \begin{pmatrix} \chi^- \, T_{-\alpha} \, K_{11} \, T_\alpha \, \chi^- & 0 \\ 0 & \chi^+ \, T_\alpha \, K_{22} \, T_{-\alpha} \, \chi^+ \end{pmatrix}. \tag{6}$$

In our notation D stands for "diagonal" and E stands for "error" because, as we shall now show, the contribution of E will be $o_1(1)$.

The difference between the above operator and the one with E replaced by 0 equals

$$(D+E)^{-1} \, E \, D^{-1} \begin{pmatrix} \chi^- \, T_{-\alpha} \, K_{11} \, T_\alpha \, \chi^- & 0 \\ 0 & \chi^+ \, T_\alpha \, K_{22} \, T_{-\alpha} \, \chi^+ \end{pmatrix}.$$

The left-most operator $(D+E)^{-1}$ is $W_\alpha(\sigma)^{-1}$, which has uniformly bounded norm. We shall show that the product of the remaining ones is $o_1(1)$.

We use the notations

$$W_\alpha^+ = I + \chi_{(0,\alpha)} \, K \, \chi_{(0,\alpha)}, \quad W_\alpha^- = I + \chi_{(-\alpha,0)} \, K \, \chi_{(-\alpha,0)}$$

and

$$W^+ = I + \chi^+ \, K \, \chi^+, \quad W^- = I + \chi^- \, K \, \chi^-$$

because these operators arise so often. Observe that by (5)

$$W_\alpha^+ \, T_\alpha = T_\alpha \, W_\alpha^-, \quad (W_\alpha^+)^{-1} \, T_\alpha = T_\alpha \, (W_\alpha^-)^{-1}. \tag{7}$$

The upper-right corner of the product in question is

$$\chi_{(-\alpha,0)} \, K \, \chi_{(0,\alpha)} \, (W_\alpha^+)^{-1} \, \chi^+ \, T_\alpha \, K_{22} \, T_{-\alpha} \, \chi^+.$$

By (5) again and the fact that K commutes with T_α we see that this equals

$$\chi_{(-\alpha,0)} \, T_\alpha \, K \, \chi_{(-\alpha,0)} \, (W_\alpha^-)^{-1} \, \chi_{(-\alpha,\infty)} \, K_{22} \, \chi_{(-\alpha,\infty)}.$$

The operator $\chi_{(-\alpha,\infty)} \, K_{22} \, \chi_{(-\alpha,\infty)}$ on the right converges in trace norm to K_{22}, the operator $K \, \chi_{(-\alpha,0)} \, (W_\alpha^-)^{-1}$ in the middle converges strongly to $K \, \chi^- \, (W^-)^{-1}$,[6]

---

[5]The reason the error remains $o_1(1)$ after multiplying by $W_\alpha(\sigma)^{-1}$ is that these operators have uniformly bounded norms. See [2, §III.1].

[6]See [2, §III.1] also for the fact that $(W_\alpha^-)^{-1}$ converges strongly to $(W^-)^{-1}$.

and the operator $\chi_{(-\alpha,0)} \, T_\alpha$ on the left converges strongly to zero. Hence the entire product converges in trace norm to zero. A similar argument applies to the lower-left corner of the product.

We have shown that with error $o_1(1)$ we may replace E by zero in (6), which then becomes

$$I + \begin{pmatrix} \chi^-\,(W_\alpha^-)^{-1}\,T_{-\alpha}\,K_{11}\,T_\alpha\,\chi^- & 0 \\[2ex] 0 & \chi^+\,(W_\alpha^+)^{-1}\,T_\alpha\,K_{22}\,T_{-\alpha}\,\chi^+ \end{pmatrix}.$$

This operator, which acts on $L^2(\mathbf{R}^-) \oplus \mathbf{L^2}(\mathbf{R}^+)$, can be extended in an obvious way, without change of notation or determinant, to one acting on $L^2(\mathbf{R}) \oplus \mathbf{L^2}(\mathbf{R})$. By (7) and (5) we may now rewrite it as

$$I + \begin{pmatrix} T_{-\alpha}\,\chi_{(-\infty,\alpha)}\,(W_\alpha^+)^{-1}\,K_{11}\,\chi_{(-\infty,\alpha)}\,T_\alpha & 0 \\[2ex] 0 & T_\alpha\,\chi_{(-\alpha,\infty)}\,(W_\alpha^-)^{-1}\,K_{22}\,\chi_{(-\alpha,\infty)}\,T_{-\alpha} \end{pmatrix}.$$

If we left-multiply by the unitary operator $\begin{pmatrix} T_\alpha & 0 \\ 0 & T_{-\alpha} \end{pmatrix}$ and right-multiply by its inverse $\begin{pmatrix} T_{-\alpha} & 0 \\ 0 & T_\alpha \end{pmatrix}$ the operator becomes

$$I + \begin{pmatrix} \chi_{(-\infty,\alpha)}\,(W_\alpha^+)^{-1}\,K_{11}\,\chi_{(-\infty,\alpha)} & 0 \\[2ex] 0 & \chi_{(-\alpha,\infty)}\,(W_\alpha^-)^{-1}\,K_{22}\,\chi_{(-\infty,\alpha)} \end{pmatrix}.$$

The determinant is unchanged, and the error term remains $o_1(1)$. By an argument already used, this operator (plus the error term $o_1(1)$) converges in trace norm to

$$I + \begin{pmatrix} (W^+)^{-1}\,K_{11} & 0 \\[2ex] 0 & (W^-)^{-1}\,K_{22} \end{pmatrix},$$

and so its determinant converges to the determinant of this one. The determinant of this one equals the product of determinants on the right side of (3), which is therefore now established.

3. Final remarks

Remark 1. There is the possibility that one of the operators on the right side of (3) is not invertible, in which case the product of determinants is zero. Since

$$I + (I + \chi^+\,K\chi^+)^{-1}\,K_{11} = (I + \chi^+\,K\,\chi^+)^{-1}\,(I + K_-),$$

the invertibility of this operator is equivalent to that of $I + K_-$. Similarly the invertibility of the other operator is equivalent to that of $I + K_+$. These are separate issues.

Remark 2. Here is a rough explanation of why the determinants on the right side of (3) are sometimes evaluable. The first one, for example, is

$$\det\left[I + (I + \chi^+ K \chi^+)^{-1} K_{11}\right].$$

If we introduce a parameter λ then

$$\frac{d}{d\lambda} \log \det\left[I + (I + \lambda\chi^+ K \chi^+)^{-1}\lambda K_{11}\right]$$

is equal to the trace of

$$(I + \lambda K_-)^{-1} K_- - (I + \lambda\chi^+ K \chi^+)^{-1} \chi^+ K \chi^+.$$

Now $(I + \lambda\chi^+ K \chi^+)^{-1}$ is the direct sum of I acting on $L^2(\mathbf{R}^-)$ and $W(1+\lambda\hat{k})^{-1}$ acting on $L^2(\mathbf{R}^+)$, so it is known – it is expressible in terms of the Wiener-Hopf factors of $1 + \lambda\hat{k}$. The pleasant fact is that the first operator, while not of this form, can be brought to this form in the cases considered in [6].

In the particular one mentioned in the introduction the kernel of K_- equals (when λ there is taken to be 1)

$$2\,e^{-(x+y)/2}\,\frac{e^{-(e^{-x}+e^{-y})}}{e^x + e^y}.$$

This has the integral representation

$$2\,e^{-(x+y)/2}\int_0^\infty e^{-(e^{u-x}+e^{u-y})}\,e^u\,du.$$

Hence if M is the integral operator from $L^2(\mathbf{R}^+)$ to $L^2(\mathbf{R})$ with kernel

$$M(x,u) = \sqrt{2}\,e^{-x/2}\,e^{-e^{u-x}}\,e^{u/2}$$

and N is the integral operator from $L^2(\mathbf{R})$ to $L^2(\mathbf{R}^+)$ with kernel

$$N(v,y) = \sqrt{2}\,e^{-y/2}\,e^{-e^{v-y}}\,e^{v/2}$$

then $K_- = MN$. The kernel of NM, an operator on $L^2(\mathbf{R}^+)$, is

$$\int_{-\infty}^\infty N(u,x)\,M(x,v)\,dx = 2\,\frac{e^{(u+v)/2}}{e^u + e^v} = \frac{1}{\cosh[(u-v)/2]}.$$

This is itself a Wiener-Hopf operator so the inverse of $I + \lambda NM$ may be written down, and we have

$$(I + \lambda K_-)^{-1}\lambda K_- = I - (I + \lambda MN)^{-1} = \lambda M\,(I + \lambda NM)^{-1} N.$$

This enables one, at least in principle, to compute the logarithmic derivative of the determinant and hence the determinant itself.

References

[1] N.I. Ahieser, *The continuous analogue of some theorems on Toeplitz matrices* (Russian), Ukrain. Mat. Zh. **16**:4 (1964) 445–462.

[2] I.C. Gohberg and I.F. Feldman, Convolution Equations and Projection Methods for their Solution, Transl. Math. Monog. **41**, Amer. Math. Soc., 1974.

[3] M. Kac, *Toeplitz matrices, translation kernels and a related problem in probability theory*, Duke Math. J. **21** (1954) 501–510.

[4] A.V. Kitaev: *Method of isometric deformation for "degenerate" third Painlevé equation.* J. Soviet Math. **46** (1989) 2077–2082.

[5] B.M. McCoy, C.A. Tracy, and T.T. Wu, *Painlevé functions of the third kind.* J. Math. Phys. **18** (1977) 1058–1092.

[6] C.A. Tracy and H. Widom, *Asymptotics of a class of solutions to the cylindrical Toda equations*, Commun. Math. Phys. **190** (1998) 697–721.

Harold Widom
Department of Mathematics
University of California
Santa Cruz, CA 95064, USA
e-mail: widom@math.ucsc.edu